Co-Branding Strategies for Smart Luxury Products

Theodore Tarnanidis
International Hellenic University, Greece

Bruno Sousa
Polytechnic Institute of Cavado and Ave, Portugal

Maro Vlachopoulou
University of Macedonia, Greece

Evridiki Papachristou
International Hellenic University, Greece

Vijaya Kittu Manda
PBMEIT, India

IGI Global
Scientific Publishing
Publishing Tomorrow's Research Today

Vice President of Editorial	Melissa Wagner
Director of Acquisitions	Mikaela Felty
Director of Book Development	Jocelynn Hessler
Production Manager	Mike Brehm
Cover Design	Jose Rosado

Published in the United States of America by
IGI Global Scientific Publishing
701 East Chocolate Avenue
Hershey, PA, 17033, USA
Tel: 717-533-8845
Fax: 717-533-7115
Website: https://www.igi-global.com E-mail: cust@igi-global.com

Library of Congress Cataloging-in-Publication Data

Names: Tarnanidis, Theodore, 1977- editor | Sousa, Bruno Barbosa editor | Vlachopoulou, Maro, 1956- editor | Papachristou, Eyridiki, 1975- editor | Manda, Vijaya Kittu, 1980- editor
Title: Co-branding strategies for smart luxury products / edited by: Theodore Tarnanidis, Bruno Sousa, Maro Vlachopoulou, Evridiki Papachristou, and
Vijaya Kittu Manda.
Description: Hershey, PA : IGI Global Scientific Publishing, [2026] | Includes bibliographical references and index. | Summary: "The book is intended to analyze the impact of co-branding marketing strategies for luxury and smart consumer products in general, and its implications for businesses of all sizes across diverse contexts and markets"-- Provided by publisher.
Identifiers: LCCN 2025011178 (print) | LCCN 2025011179 (ebook) | ISBN 9798337330105 hardcover | ISBN 9798337330112 paperback | ISBN 9798337332611 ebook
Subjects: LCSH: Luxuries--Marketing | Selling--Luxuries | Luxury goods industry--Technological innovations | Embedded Internet devices--Marketing
Classification: LCC HD9999.L852 C63 2025 (print) | LCC HD9999.L852 (ebook) | DDC 658.8--dc23/eng/20250903
LC record available at https://lccn.loc.gov/2025011178
LC ebook record available at https://lccn.loc.gov/2025011179

British Cataloguing in Publication Data
A Cataloguing in Publication record for this book is available from the British Library.

Editorial Advisory Board

Table of Contents

Preface xi

Chapter 1
Co-Branding: Strategies for Smart Luxury Products 1
Rismawati Rismawati, Universitas Muhammadiyah Palopo, Indonesia
Suaedi Suaedi, Universitas Cokroaminoto Palopo, Indonesia
Harman Preet, Chandigarh Group of Colleges Jhanjeri, Mohali, India
Andi Subhan Amir, Hasanuddin University, Indonesia

Chapter 2
Challenges and Risks in Co-Branding Smart Luxury Products 35
Venkatesh Ganapathy, ICFAI Business School, ICFAI Foundation for Higher Education, Bangalore, India

Chapter 3
Co-Creation, Co-Marketing, and Co-Branding Strategies for Smart Luxury Digital Ecosystems 65
Anuradha Yadav, REVA University, India
Vijaya Kittu Manda, PBMEIT, India
Theodore Tarnanidis, International Hellenic University, Greece
Ahmad Salman Farid, STAIN Mandailing Natal, Indonesia

Chapter 4
The Psychology of Smart Consumer Products: Behavioral Insights for Co-Branding Success 99
Jomon Jose M., Sahrdaya Institute of Management Studies, Kodakara, India
Jyothis R. Mathews, Ahalia School of Management, Palakkad, India

Chapter 5
Ethical and Sustainable Co-Branding in the Luxury Industry: Strategies for a New Era of Symbolic Value and Corporate Responsibility 167
Noelia Araújo Vila, University of Vigo, Spain
Lucía Rubio-Escuderos, University of Valencia, Spain

Chapter 6
Advanced Co-Branding Strategies for Smart Luxury Products: A Multi-Criteria Decision-Making Approach Using Neutrosophic Sets 203
Ajoy Kanti Das, Tripura University, India
Nandini Gupta, Bir Bikram Memorial College, India
Takaaki Fujita, Independent Researcher, Japan
Suman Das, National Institute of Technology, Agartala, India
Carlos Granados, Universidad de Antioquia, Colombia
Bijoy Krishna Debnath, Tezpur University, India
Rakhal Das, ICFAI University, Tripura, India

Chapter 7
Connected Couture: Co-Branding in the Era of Smart Luxury Fashion............ 259
Sushila Sahani, Lovely Professional University, India

Chapter 8
The Fusion of Co-Branding and Digital Fashion: Transforming Smart Luxury Products in the Digital Era 291
Sunanda Vincent Jaiwant, Christ University, India

Chapter 9
Exploring Consumer-Brand Relationships and Organizational Behavior in the Automotive Sector: An Embryonic Study 315
Abílio Bragança Milheiro, Polytechnic University of Cávado and Ave, Portugal
Bruno Barbosa Sousa, Polytechnic University of Cávado and Ave, Portugal
Estela Vilhena, Polytechnic Institute of Cávado and Ave, Portugal
Theodore Tarnanidis, International Hellenic University, Greece

Compilation of References 333

About the Contributors 395

Index 403

Detailed Table of Contents

Preface.. xi

Chapter 1
Co-Branding: Strategies for Smart Luxury Products.. 1
Rismawati Rismawati, Universitas Muhammadiyah Palopo, Indonesia
Suaedi Suaedi, Universitas Cokroaminoto Palopo, Indonesia
Harman Preet, Chandigarh Group of Colleges Jhanjeri, Mohali, India
Andi Subhan Amir, Hasanuddin University, Indonesia

Co-branding in the luxury sector has markedly progressed through the incorporation of smart technology, allowing businesses to augment exclusivity while fulfilling the demands of contemporary consumers. This research examines the strategic execution of co-branding in smart luxury products, emphasizing critical elements such as brand compatibility, market positioning, and consumer perception. The research examines successful and unsuccessful collaborations to elucidate the influence of AI, blockchain, and the metaverse on the future of luxury branding. Furthermore, it analyzes marketing and distribution tactics, encompassing omnichannel experiences, scarcity marketing, and experiential retail, to maintain brand equity and consumer engagement. The results indicate that an effectively implemented co-branding approach can produce a harmonious fusion of innovation and prestige, guaranteeing sustained success in the competitive luxury sector.

Chapter 2
Challenges and Risks in Co-Branding Smart Luxury Products 35
Venkatesh Ganapathy, ICFAI Business School, ICFAI Foundation for Higher Education, Bangalore, India

This chapter examines the intricate dynamics of co-branding within the luxury sector, particularly in relation to smart products, which pose distinctive challenges and risks. As the market for smart luxury items expands, brands are compelled to navigate the complexities of harmonizing advanced technology with their premium image. A primary concern is the potential for brand dilution; luxury brands, renowned for their exclusivity and tradition, may face threats to their identity when collaborating with tech companies—especially if the technological offerings fail to meet the high expectations of discerning consumers. Additionally, reputational risks are paramount; any ethical, technical, or public relations issues faced by a partner brand can adversely impact the luxury brand's reputation. The operational intricacies of co-branding ventures further complicate matters. Maintaining a delicate balance between innovation and tradition is essential, as brands strive to attract younger consumers without alienating their established clientele.

Chapter 3
Co-Creation, Co-Marketing, and Co-Branding Strategies for Smart Luxury Digital Ecosystems 65
Anuradha Yadav, REVA University, India
Vijaya Kittu Manda, PBMEIT, India
Theodore Tarnanidis, International Hellenic University, Greece
Ahmad Salman Farid, STAIN Mandailing Natal, Indonesia

The convergence of luxury and technology has redefined consumer expectations. This necessitated innovative, collaborative strategies in the smart luxury industry. This chapter examines how co-creation, co-marketing, and co-branding strategies drive value creation, consumer engagement, and competitive differentiation in this sector. The role of digital platforms, AI-driven personalization, and cross-industry partnerships in shaping modern luxury branding is analyzed. A synthesis of theoretical frameworks, industry case studies, and empirical research is done. Key findings highlight the importance of strategic alignment, brand compatibility, and technology adoption for successful collaborations. The chapter also explores crisis management and future trends, such as sustainability and metaverse integration. By bridging theory and practice, this work offers actionable insights for academics and luxury brand managers navigating the evolving digital landscape. The chapter demonstrates how smart luxury brands influence collaborative strategies, innovation, and sustainability initiatives.

Chapter 4
The Psychology of Smart Consumer Products: Behavioral Insights for Co-Branding Success 99
Jomon Jose M., Sahrdaya Institute of Management Studies, Kodakara, India
Jyothis R. Mathews, Ahalia School of Management, Palakkad, India

Co-branding in the luxury space is a travelled and lofty highway. Luxury brands have historically guarded their brand and market zealously, enduring only through exclusivity, legacy, and craftsmanship. Even the stodgiest of luxury players have been forced to adopt strategic partnerships in the push for personalization and digitisation of consumer lives, which means even faster. Imagine co-branding once in a high-end round table, with heritage extended to monogamy with innovation — the strategic marriage is for real. Smart tech in luxury commodities has added functionality, data driven design, and consumer engagement. The chapter will probe the biography and obituary of co-brands in luxury territories, dig beneath somekey drivers and innovative trends we can expect playing a pivotal role for brand binding alliances, and how this is shapingup as a future (sensory) landscape.

Chapter 5
Ethical and Sustainable Co-Branding in the Luxury Industry: Strategies for a New Era of Symbolic Value and Corporate Responsibility 167
Noelia Araújo Vila, University of Vigo, Spain
Lucía Rubio-Escuderos, University of Valencia, Spain

The luxury industry's transformation in recent decades reflects growing integration of ethical and sustainable values due to societal pressure and evolving consumer expectations. This chapter examines ethical and sustainable co-branding as an emerging strategy in luxury, addressing demands for transparency, responsibility, and environmental commitment. Using a qualitative, multiple case study approach, it analyzes four collaborations: Burberry x The RealReal, Chantecaille x Space for Giants, Stella McCartney x Parley for the Oceans, and IWC x Laureus Sport for Good. Findings indicate that co-branding can effectively integrate sustainable practices while preserving luxury's symbolic value, expanding audiences, reinforcing brand authenticity, and supporting the Sustainable Development Goals (SDGs). Challenges include the risk of greenwashing, value alignment between partners, and equity in collaboration.

Chapter 6
Advanced Co-Branding Strategies for Smart Luxury Products: A Multi-Criteria Decision-Making Approach Using Neutrosophic Sets 203
Ajoy Kanti Das, Tripura University, India
Nandini Gupta, Bir Bikram Memorial College, India
Takaaki Fujita, Independent Researcher, Japan
Suman Das, National Institute of Technology, Agartala, India
Carlos Granados, Universidad de Antioquia, Colombia
Bijoy Krishna Debnath, Tezpur University, India
Rakhal Das, ICFAI University, Tripura, India

The convergence of luxury branding and smart technology has led to a new paradigm in strategic co-branding. This chapter introduces a Multi-Criteria Decision-Making (MCDM) framework that integrates Neutrosophic Sets to address uncertainty, hesitation, and conflicting expert opinions in evaluating smart luxury co-branding partnerships. Traditional decision-making models often fail to capture the complexities of consumer perception, technological integration, and market acceptance, which involve varying degrees of truth, indeterminacy, and falsity. Neutrosophic MCDM enhances decision robustness by incorporating Neutrosophic Weighted Aggregation Operators and Similarity Measures to quantify brand synergy and compatibility. Case studies, including collaborations between luxury brands and technology firms, illustrate the framework's application in selecting optimal co-branding partners. The findings demonstrate that the Neutrosophic approach provides a more nuanced and data-driven methodology for evaluating strategic alliances in the smart luxury sector, ensuring informed MCDM.

Chapter 7
Connected Couture: Co-Branding in the Era of Smart Luxury Fashion............ 259
Sushila Sahani, Lovely Professional University, India

The luxury fashion industry is undergoing a major transformation driven by digital innovation and tech partnerships. This chapter explores how collaborations between fashion brands and tech companies are reshaping the meaning of luxury for modern consumers. It highlights the growing demand for personalization, immersive digital experiences, and sustainability as key influences on brand strategies. It discusses the rise of smart luxury products from sensor-embedded clothing to AI-powered personalization. It also delves into how technologies such as artificial intelligence, IoT, and data analytics are driving new fashion experiences. Digital fashion's role is explored through examples like Balenciaga's Fortnite collaboration and Gucci's presence on Roblox, showcasing new forms of digital ownership and exclusivity. The chapter also examines how technology is helping brands meet ethical and sustainability goals, while noting concerns like privacy and tech's environmental footprint.

Chapter 8
The Fusion of Co-Branding and Digital Fashion: Transforming Smart Luxury Products in the Digital Era.. 291
Sunanda Vincent Jaiwant, Christ University, India

In the evolving landscape of luxury products, co-branding strategies and digital fashion integration have become essential tools for enhancing brand value and consumer engagement. Smart luxury products, which incorporate advanced technology, require a unique blend of traditional craftsmanship and digital innovation to maintain exclusivity while catering to tech-savvy consumers. This chapter explores the strategic collaborations between luxury brands and technology firms, analysing their impact on brand equity, consumer perception, and market performance. It examines case studies of successful co-branding initiatives and evaluates the role of digital fashion in enhancing personalization and sustainability in luxury consumption.

Chapter 9
Exploring Consumer-Brand Relationships and Organizational Behavior in the Automotive Sector: An Embryonic Study 315
Abílio Bragança Milheiro, Polytechnic University of Cávado and Ave, Portugal
Bruno Barbosa Sousa, Polytechnic University of Cávado and Ave, Portugal
Estela Vilhena, Polytechnic Institute of Cávado and Ave, Portugal
Theodore Tarnanidis, International Hellenic University, Greece

The automotive sector is highly characterized by innovation and the development of competitive strategies, but also by specific consumer behavior and strong relationships of trust with brands. From an exploratory perspective, this chapter aims to present some discussion around the importance that brand attachment may represent in the behavior of luxury car consumers. A brand that has significant commercial value and brand awareness facilitates, in many cases, strong feelings of affection and loyalty in its consumers (and their respective repurchase and recommendation). Emotional variables play a critical role in the decision-making process of luxury car consumers. As a preliminary perspective, the study will present some initial insights for marketing in the (luxury) automotive sector and will present lines of research for future studies.

Compilation of References 333

About the Contributors 395

Index 403

Preface

Co-branding is a marketing strategy that has the power to bring together two brands and harness their strengths to create a product with greater value. The importance of innovation, exclusivity, and brand reputation is especially significant in industries such as luxury smart consumer products. In multicultural markets, it refers to a strategic alliance between a luxury brand and a technology brand, designed to appeal to a range of cultural preferences, values, and consumption patterns. Companies can now deliver innovation while maintaining exclusivity thanks to the fusion of luxury and technology in their branding strategies. By using this approach, the product can resonate with global audiences while still maintaining the exclusivity and prestige associated with luxury. Therefore, this book examines the use of co-branding strategies for the advancement of smart consumer products, and the convergence of luxury and technology in branding.

Below is an overview of each chapter:

CHAPTER OVERVIEW

Chapter 1: Co-Branding: Strategies for Smart Luxury Products

This chapter introduces the foundational principles of co-branding in the luxury domain, focusing on brand compatibility, strategic alignment, and the impact of emerging technologies such as AI and blockchain. Through an analysis of both successful and unsuccessful collaborations, it highlights how co-branding can preserve brand equity while promoting innovation and consumer engagement.

Chapter 2: Challenges and Risks in Co-Branding Smart Luxury Products

The second chapter examines the potential pitfalls of co-branding in smart luxury, including brand dilution, reputational risk, and the challenge of balancing innovation with heritage. It underscores the importance of aligning values and expectations between partners to maintain credibility and market positioning.

Chapter 3: Co-Creation, Co-Marketing, and Co-Branding Strategies for Smart Luxury Digital Ecosystems

Focusing on the broader collaborative strategies beyond branding alone, this chapter explores how co-creation and co-marketing enhance consumer involvement and brand differentiation. It investigates digital ecosystems and the role of AI, sustainability, and immersive technologies in reshaping luxury branding.

Chapter 4: The Psychology of Smart Consumer Products: Behavioral Insights for Co-Branding Success

This chapter delves into the psychological drivers behind luxury consumer behavior. It explores how personalization, sensory experience, and data-driven design inform co-branding strategies and how emotional and cognitive factors affect the success of these collaborations.

Chapter 5: Ethical and Sustainable Co-Branding in the Luxury Industry

With increasing consumer demand for ethical responsibility, this chapter examines how co-branding can integrate sustainability while maintaining luxury's symbolic value. Through real-world case studies, it highlights how transparency, authenticity, and aligned values are critical to success in this evolving landscape.

Chapter 6: Advanced Co-Branding Strategies for Smart Luxury Products

Introducing a novel decision-making model using Neutrosophic Sets, this chapter presents a data-driven framework for evaluating potential co-branding partnerships. It addresses complexity and uncertainty in the decision-making process, offering a scientific approach to assess synergy, compatibility, and market fit.

Chapter 7: Connected Couture: Co-Branding in the Era of Smart Luxury Fashion

This chapter explores the fashion industry's embrace of smart technology through co-branding. It discusses how personalization, digital experiences, and sustainability are shaping the future of luxury fashion. Examples like digital fashion collaborations and AI-integrated garments illustrate how brands are redefining exclusivity.

Chapter 8: The Fusion of Co-Branding and Digital Fashion

Building on the previous chapter, this contribution further examines how digital fashion and smart luxury intersect. It focuses on strategic alliances that drive brand equity and enhance consumer engagement. The chapter emphasizes personalization, market performance, and the role of digital experiences in brand storytelling.

Chapter 9: Exploring Consumer-Brand Relationships and Organizational Behavior in the Automotive Sector

Closing the volume, this exploratory chapter shifts focus to the luxury automotive industry. It discusses how brand attachment and emotional loyalty influence consumer behavior in this sector. Preliminary insights are offered into how co-branding might evolve within automotive luxury markets, providing avenues for future research.

This publication serves as a valuable reference for academics, researchers, undergraduate students, postgraduate students, consultants, technology developers, and policymakers who want to study the evolution of co-branding challenges for luxury and smart products for businesses of all sizes and markets. We are confident that readers will find this book to be essential for improving their current work and education. We express our final gratitude to the contributors of the accepted chapters.

Chapter 1

Co-Branding: Strategies for Smart Luxury Products

Rismawati Rismawati
https://orcid.org/0000-0001-8046-6764
Universitas Muhammadiyah Palopo, Indonesia

Suaedi Suaedi
https://orcid.org/0009-0004-2196-2142
Universitas Cokroaminoto Palopo, Indonesia

Harman Preet
https://orcid.org/0009-0004-0649-1856
Chandigarh Group of Colleges Jhanjeri, Mohali, India

Andi Subhan Amir
https://orcid.org/0000-0002-7680-5556
Hasanuddin University, Indonesia

ABSTRACT

Co-branding in the luxury sector has markedly progressed through the incorporation of smart technology, allowing businesses to augment exclusivity while fulfilling the demands of contemporary consumers. This research examines the strategic execution of co-branding in smart luxury products, emphasizing critical elements such as brand compatibility, market positioning, and consumer perception. The research examines successful and unsuccessful collaborations to elucidate the influence of AI, blockchain, and the metaverse on the future of luxury branding. Furthermore, it analyzes marketing and distribution tactics, encompassing omnichannel experiences, scarcity marketing, and experiential retail, to maintain brand equity and consumer engagement. The results indicate that an effectively implemented co-branding ap-

DOI: 10.4018/979-8-3373-3010-5.ch001

proach can produce a harmonious fusion of innovation and prestige, guaranteeing sustained success in the competitive luxury sector.

1. THE TRANSFORMATION OF LUXURY IN THE DIGITAL ERA

1.1 The Notion of Luxury and Intelligent Products

Luxury has always been linked to exclusivity, craftsmanship, and legacy (Cabigiosu, 2020; Heyes, 2021; Wei & Li, 2020; Wirtz et al., 2020). Historically, luxury products were characterized by their scarcity, exceptional quality, and careful craftsmanship, primarily serving affluent consumers. The origins of luxury brands date back centuries, with European maisons like Louis Vuitton, Chanel, and Rolex establishing industry standards in fashion, accessories, and horology. Traditionally, luxury denoted more than mere material ownership; it symbolized status, prestige, and an aspirational lifestyle.

In recent decades, the luxury landscape has seen substantial transformation due to the advent of digitalization and technological innovation. The advent of smart technology in luxury goods has altered consumer expectations, resulting in a new category termed smart luxury products (Heyes, 2021; Javornik et al., 2021). These goods incorporate sophisticated technical capabilities while preserving the exclusivity and workmanship characteristic of luxury companies. Smartwatches, linked vehicles, sophisticated fashion items, and advanced home automation systems are now integral components of the developing luxury ecosystem.

The amalgamation of technology and luxury is propelled by evolving customer preferences and a need for improved functionality without sacrificing beauty. Innovative partnerships, such by the Apple-Hermès smartwatch, have illustrated how luxury companies may effectively incorporate technology while preserving their fundamental identity (Javornik et al., 2021). Prestigious vehicle brands like as Tesla and Bentley integrate AI-driven interfaces and IoT functionalities to enhance customer experience. Likewise, intelligent fashion innovations—such as Louis Vuitton's LED-integrated purses or Prada's collaboration with wearable technology companies—illustrate the amalgamation of luxury and digitalization.

Key characteristics of smart luxury products distinguish them from conventional luxury and mass-market smart technology. These characteristics include:

1. Innovative Technology: Smart luxury goods leverage cutting-edge innovations such as artificial intelligence, blockchain authentication, and biometric security while ensuring seamless usability.

2. Exclusivity & Customization: Personalization is a hallmark of smart luxury, offering bespoke experiences such as custom-designed interfaces, limited-edition releases, and high-end concierge services.
3. Aesthetic & Craftsmanship: Unlike standard smart products, luxury tech items emphasize premium materials, artisanal craftsmanship, and timeless design.
4. Sustainability & Ethical Sourcing: Modern luxury consumers demand eco-conscious and responsibly sourced products, leading brands to integrate sustainable materials and ethical production processes into smart innovations.
5. Enhanced User Experience: Beyond aesthetics, smart luxury products focus on intuitive, user-friendly experiences that align with the prestige and sophistication expected by high-net-worth individuals.

The evolution of luxury in the digital age is an ongoing transformation, pushing brands to innovate while preserving their essence. As co-branding becomes a strategic tool for luxury-tech collaborations, understanding this shift is crucial for brands aiming to thrive in a technologically advanced, high-end market (Serdari, 2020).

1.2 The Evolving Consumer Perspective in Luxury Markets

Luxury consumption has experienced a significant metamorphosis in recent years, influenced by technological breakthroughs, demographic changes, and shifting societal ideals. Historically, luxury was seen as a status symbol, designated for an exclusive class that esteemed exclusivity, lineage, and workmanship. Modern luxury customers, especially Millennials and Gen Z, have redefined the idea by prioritizing experiences, innovation, and social responsibility. This transition has necessitated luxury firms to reevaluate their strategies, adopting digital transformation and ethical issues while preserving their esteemed allure (Holmqvist et al., 2020).

A major shift in luxury consumer expectations is the requirement for hyper-personalization and digital integration (Cabigiosu, 2020). Contemporary affluent consumers desire things that exhibit exceptional workmanship and integrate seamlessly with their lifestyles via smart technology and AI-driven personalization. The emergence of intelligent luxury items, such Hermès-Apple smartwatches, Tesla's AI-enhanced automobiles, and Prada's connected fashion accessories, underscores how firms are employing co-branding methods to appeal to technologically adept premium consumers (Javornik et al., 2021). Furthermore, experiential luxury—encompassing virtual reality (VR) fashion shows, tailored concierge services, and exclusive digital memberships—has gained heightened appeal, transitioning emphasis from simple product ownership to customized brand experiences.

The digital transformation of luxury brand interaction has further expedited this shift. In contrast to the conventional luxury retail strategy, which preserved exclusivity via physical boutiques and tailored in-store experiences, contemporary luxury firms are adopting multichannel tactics. Luxury fashion brands and watch manufacturers have used AI-driven chatbots, virtual consultations, and blockchain verification systems to improve consumer engagement (Ashfaq et al., 2020; Guercini et al., 2020). Social media, influencer marketing, and luxury e-commerce sites like Farfetch and Net-a-Porter have transformed the manner in which firms engage with their audience. The amalgamation of NFTs (Non-Fungible Tokens), the metaverse, and Web3 techniques shows the transformative impact of digital innovation on luxury spending.

In addition to technology, sustainability and ethical issues have emerged as crucial factors affecting consumer decisions. Contemporary luxury consumers are progressively aware of environmental consequences, equitable labor methods, and ethical sourcing (Cabigiosu, 2020; Serdari, 2020). Brands such as Gucci, Stella McCartney, and Louis Vuitton are at the forefront by integrating sustainable materials, circular economy practices, and carbon-neutral strategies. Transparency and accountability in supply chains, enabled by blockchain technology, are emerging as conventional expectations, ensuring that luxury continues to represent both grandeur and responsibility.

As consumer attitudes shift, co-branding between luxury and technology companies will be crucial in shaping the future of premium markets. Luxury businesses may maintain relevance and preserve their legacy during rapid digital transition by balancing innovation, customisation, and sustainability.

1.3 The Function of Co-Branding in the Expansion of Luxury Brands

In the dynamic realm of luxury, co-branding has surfaced as a potent technique for brand promotion and market distinction. Co-branding denotes a strategic alliance between two brands that capitalize on each other's strengths to develop a distinctive product or service, hence augmenting brand equity and accessing new consumer demographics. Co-branding is particularly vital in the luxury sector, enabling high-end businesses to incorporate technical advancements, sustainability efforts, and digital experiences while preserving exclusivity and distinction (Javornik et al., 2021).

The importance of co-branding in luxury markets is its capacity to combine craftsmanship with innovation, producing items that appeal to contemporary, tech-savvy, and experience-oriented consumers (Rathi et al., 2022). In contrast to mass-market collaborations, luxury co-branding necessitates meticulous brand alignment to maintain exclusivity and brand distinction. Successful collaborations generally entail historical luxury companies teaming with technological innovators to deliver premium experiences, merging luxury aesthetics with advanced technology.

A number of successful premium co-branding ventures have established benchmarks for this method. A prominent example is the Apple x Hermès smartwatch, which merges Apple's cutting-edge technology with Hermès' renowned leather workmanship, targeting affluent consumers desiring both digital functionality and luxurious style (Holmqvist et al., 2020). The Bentley x Breitling collaboration exemplifies the fusion of the automotive and watch industries, resulting in limited-edition timepieces inspired by Bentley's high-performance automobiles. Likewise, Louis Vuitton's partnership with Master & Dynamic produced opulent wireless earphones, merging fashion, artistry, and exceptional sound technology. These collaborations illustrate how luxury brands may augment their market allure, broaden their audience, and sustain brand desirability through intelligent co-branding initiatives.

Co-branding offers substantial benefits, although it also has inherent hazards. A significant difficulty is brand dilution, wherein an exclusive luxury brand may forfeit its premium allure if linked with a partner that does not correspond with its brand values. Another risk is customer distrust, particularly if the partnership seems inauthentic or too commercialized (Reinikainen et al., 2021). Moreover, there exists a risk of misalignment in quality perception, wherein the reputation of one brand may be adversely affected by inconsistencies in product durability, design philosophy, or technological reliability.

Notwithstanding these concerns, strategic co-branding continues to be a vital growth strategy for luxury firms seeking innovation while preserving their exclusivity (Bian & Haque, 2020). Through the strategic selection of complementary partners, luxury brands may enhance their influence, access new consumer demographics, and solidify their heritage in a progressively digital and technology-oriented luxury market.

2. COMPREHENDING CO-BRANDING STRATEGIES IN LUXURY MARKETS

2.1 Categories of Co-Branding Collaborations in the Luxury Sector

Co-branding in the luxury sector manifests in diverse forms, each providing distinct value and strategic benefits. These partnerships allow brands to broaden their market presence, augment their technology proficiencies, and strengthen exclusivity (Javornik et al., 2021). Luxury co-branding relationships often categorize into three primary types: ingredient branding, complementary branding, and innovation-driven collaborations.

Ingredient Branding: The Convergence of Luxury and Advanced Technology. Ingredient branding entails incorporating a premium element from one brand into the luxury product of another, resulting in a distinguished and enhanced offering. This method is prevalent in the automotive and technology industries, as luxury automobile manufacturers partner with advanced technology companies to improve vehicle performance and opulence.

For example, the collaboration between Bang & Olufsen and Audi illustrates how premium audio systems enhance the driving experience. Bulgari's partnership with Intel in smartwatches integrates Intel's sophisticated CPUs into premium timepieces, merging artisanal workmanship with digital innovation. These alliances allow luxury businesses to preserve their exclusivity while gaining from advanced advancements that augment the product's appeal.

Complementary Branding: The Convergence of Fashion and Technology

Complementary branding transpires when two brands from distinct yet linked sectors collaborate their expertise to develop synergistic products (Guercini et al., 2020). This strategy is commonly observed in the fashion and technology sectors, where sophisticated luxury wearables and designer tech accessories attract contemporary consumers.

An exemplary case is the collaboration between Prada and LG, which produced a sophisticated, premium smartphone that integrated Prada's minimalist design with LG's mobile technology. A significant instance is Louis Vuitton's collaboration with Master & Dynamic, resulting in luxury wireless earphones that attract audiophiles who prioritize both aesthetics and exceptional sound quality. These agreements enable premium brands to preserve their exclusivity while diversifying into new product categories, appealing to younger, technologically adept audiences.

Collaborations Fueled by Innovation: The Smartwatch Revolution

The convergence of luxury and smart technology has led to innovation-focused partnerships, wherein brands jointly create advanced luxury items (Tang, 2021). This trend is especially pronounced in the wristwatch sector, where fashion brands partner with technology companies to produce elegant yet practical devices.

The Apple x Hermès wristwatch exemplifies the fusion of Apple's advanced technology and Hermès' artisanal craftsmanship, appealing to both luxury consumers and technology aficionados. Likewise, Tag Heuer's Connected Watch incorporates

Google's Wear OS, demonstrating that innovation-focused collaborations may enhance a brand's status while preserving its exclusivity.

Every form of co-branding in premium markets fulfills a strategic objective, allowing businesses to evolve while maintaining their identity. Success hinges on meticulous partner selection, alignment of values, and the preservation of exclusivity, so ensuring that the partnership elevates both companies' reputations and fortifies their market standing.

2.2 Strategic Alignment Between Luxury and Smart Technology Brands

The efficacy of co-branding alliances in the luxury sector relies on a robust strategic alignment between the partnering brands. When luxury brands collaborate with technology firms, the partnership must be founded on synergistic brand positioning, alignment of brand equity, and a meticulously developed customer perception strategy (Guercini et al., 2020). Inability to attain this congruence may result in brand dilution, credibility challenges, or market rejection.

Recognizing Synergy in Brand Positioning

For co-branding collaborations to succeed, the positioning of both companies must be mutually reinforcing. Luxury companies distinguish themselves by rarity, craftsmanship, legacy, and prestige, whereas smart technology brands are recognized for innovation, utility, and digital intelligence (Serdari, 2020). The primary difficulty is to ensure that these distinct traits integrate seamlessly instead of generating paradoxes.

The collaboration between Apple and Hermès exhibits ideal brand synergy—Apple's standing in high-tech innovation enhances Hermès' tradition of handmade production, resulting in a smartwatch that is both technologically sophisticated and aesthetically opulent. Tag Heuer's partnership with Google and Intel produced a smartwatch that preserved the classic sophistication of Swiss horology while incorporating advanced digital functionalities.

A discrepancy in posture can, however, estrange consumers. A luxury brand collaborating with a mass-market technology firm risks diminishing its exclusivity and compromising its brand perception (Cabigiosu, 2020; Wei & Li, 2020). Consequently, meticulous partner selection predicated on congruent values, target demographics, and brand reputation is imperative.

The Influence of Brand Equity Alignment

Brand equity, defined as the perceived worth and repute of a brand, is essential for the success of luxury co-branding. If two brands have comparable or complimentary levels of prestige, their partnership is likely to be positively regarded. Nonetheless, if a luxury brand collaborates with a technology brand that does not possess equivalent desirability, the luxury brand's reputation may be compromised.

Bentley's collaboration with Breitling in producing luxury timepieces is effective due to their mutual tradition of precision, exclusivity, and exceptional craftsmanship. A luxury brand collaborating with an economical tech company may convey ambiguous messages, resulting in consumer doubt or disapproval.

Consumer Perception of Hybrid Branding Approaches

The perception of co-branded luxury products by consumers considerably influences their success. Contemporary luxury consumers, particularly Millennials and Gen Z, are attracted to exclusivity, customization, and technology advancement. They value seamless brand integration, wherein intelligent technology elevates luxury without undermining its exclusivity.

Nonetheless, if a co-branded product is regarded as an artificial amalgamation rather than a seamless progression, it may struggle to achieve market approval. Brands must convey a persuasive narrative that elucidates the rationale for the collaboration and its enhancement of the customer experience (Xie et al., 2023). Storytelling, influencer collaborations, and experiential marketing are essential in creating consumer perception.

By establishing a strategic alignment between luxury and smart technology brands, firms may create co-branded goods that augment brand value, market relevance, and enduring consumer devotion while preserving the prestige of luxury.

2.3 Determinants of a Successful Co-Branding Partnership

Co-branding in the luxury sector necessitates a careful strategy to guarantee alignment, brand integrity, and sustained market success. Partnerships between luxury brands and technology firms present significant growth potential, contingent upon three essential factors: brand compatibility and shared values, the equilibrium between exclusivity and accessibility, and a long-term strategic vision alongside market positioning (Cabigiosu, 2020; Jain et al., 2021).

Brand Congruence and Common Principles

A successful luxury co-branding collaboration commences with brand compatibility. The collaborating brands must be congruent in heritage, values, and market positioning to facilitate a cohesive partnership. Luxury companies prioritize craftsmanship, exclusivity, and prestige, whereas technology brands concentrate on innovation, functionality, and connectivity. The essential factor for success is to ensure that these traits enhance one another instead of generating disputes.

The relationship between Hermès and Apple succeeded due to their mutual dedication to design superiority, user experience, and luxury branding. Apple's innovation-centric philosophy harmoniously integrated with Hermès' handcrafted expertise, resulting in a smartwatch that embodied both modernity and opulence. A collaboration between a luxury fashion firm and an inexpensive technology brand may lead to brand dilution, perplexing consumers and compromising the distinctiveness of the luxury brand.

Reconciling Exclusivity with Accessibility

Luxury brands flourish on exclusivity; however, the incorporation of smart technology frequently necessitates broader accessibility of products. Achieving an optimal equilibrium between these two elements is essential in co-branding efforts. When a luxury product becomes excessively accessible, it jeopardizes its aspirational allure. Conversely, if the product is excessively exclusive, it may estrange an expanding demographic of tech-savvy luxury clients who desire innovation.

Effective collaborations employ techniques including exclusive releases, high-end customization alternatives, and personalized service models. The Tag Heuer Connected Watch, created in partnership with Google and Intel, maintained its exclusivity through the use of premium materials, Swiss workmanship, and luxury branding, even while using widely available smart technologies (Jeddi & Bohr, 2020; Siepmann & Kowalczuk, 2021). Bentley's partnership with Naim Audio guaranteed that its premium audio systems remained a distinctive attribute of Bentley's luxury automobiles, rather than being a widely available product.

Long-Term Strategic Vision and Market Positioning

Luxury co-branding should be a long-term strategic endeavor rather than a transient marketing tactic. Brands must meticulously delineate their market positioning, product roadmap, and innovation trajectory to sustain relevance and consumer trust (Wei & Li, 2020). An effectively strategized collaboration takes into account prospective trends, shifting customer preferences, and sustainability considerations.

The collaboration between Lamborghini and MIT on electric vehicle technology exemplifies a progressive co-branding initiative, ensuring that performance-oriented premium automobiles maintain relevance in a future characterized by sustainability and electric mobility. Gucci's collaboration with The North Face merged fashion and outdoor technology, establishing a hybrid luxury sector with enduring cultural significance.

By emphasizing compatibility, exclusivity, and a long-term vision, luxury businesses may implement co-branding strategies that foster innovation, market expansion, and enduring brand prestige while adapting to the future of smart luxury.

3. CONSUMER BEHAVIOR AND BRAND PERCEPTION IN SMART LUXURY CO-BRANDING

3.1 The Psychological Aspects of Luxury Branding

Luxury brands elicit strong emotions and desires, transcending mere objects to symbolize prestige, exclusivity, and personal identity. Comprehending the psychological motivations for luxury consumption is crucial for developing effective co-branding strategies in smart luxury markets (Oliveira & Fernandes, 2020; Rathi et al., 2022). The success of a relationship between a luxury brand and a digital business is largely contingent upon consumer perceptions of the emotional, aspirational, and experiential value of the co-branded product.

The Emotional and Aspirational Significance of Luxury Goods. Luxury brands foster robust emotional ties with consumers, frequently targeting self-identity, social status, and personal gratification mechanisms (Bian & Haque, 2020). Possessing a premium product denotes achievement, refinement, and rarity, elevating it beyond mere functionality.

The emotional allure must be preserved when smart technology is incorporated into luxury products. The Apple x Hermès smartwatch preserves the prestigious Hermès insignia and artisanal leather straps, allowing buyers to enjoy the exclusivity of Hermès alongside Apple's advanced technology. The emotional bond with the premium brand persists, enhancing the aspiration to possess a distinctive and prestigious item.

Confidence and Reputation in Co-Branded Luxury Goods

Luxury companies are founded on legacy, craftsmanship, and a steadfast dedication to excellence, which cultivate profound consumer trust. Any co-branding initiative must maintain and augment this trust, ensuring that the use of smart technology does not compromise the brand's reputation.

Consumers frequently exhibit mistrust about co-branded products, questioning whether the collaboration elevates or undermines the status and authenticity of the luxury brand (Bian & Haque, 2020). If the technology partner lacks alignment in brand equity and perceived quality, the partnership may encounter pushback. Tag Heuer's collaboration with Google and Intel was successful due to their collective reputation for innovation, performance, and superior craftsmanship. A luxury brand partnering with a mass-market technology company may encounter skepticism and brand dilution.

Sensory & Experiential Marketing in Premium Luxury

Luxury consumption is intrinsically linked to sensory allure and immersive brand experiences. The tactile quality of artisanal leather and the aesthetic sophistication of high-end packaging both enhance the appeal of a luxury product (Correia & Kozak, 2021).

Intelligent luxury co-branding must transcend mere practicality and utilize multisensory interaction to preserve uniqueness. The relationship between Bentley and Naim Audio incorporates high-fidelity sound systems into luxury automobiles, augmenting the driving experience and brand image. Likewise, Louis Vuitton's Horizon Wireless Earphones integrate high-end aesthetics, sophisticated sound technology, and distinctive packaging to provide a product that exudes luxury in all dimensions.

Leveraging emotional ambitions, trust, and sensory experiences, co-branded luxury items can provide innovation while maintaining prestige, thereby sustaining consumer demand and brand loyalty in a progressively technology-oriented market.

3.2 Consumer Segmentation in the Smart Luxury Sector

The luxury market is no longer characterized by a singular consumer demographic. The emergence of smart luxury items has transformed the conventional luxury customer demographic, giving rise to a new cohort of technologically adept, experience-oriented purchasers (Jain & Rathi, 2023). To effectively execute co-branding strategies, luxury and technology firms must comprehend and address certain customer demographics, including high-net-worth individuals (HNWIs), Millennials and Gen Z, as well as digital influencers who affect market perceptions.

Affluent Individuals and Technologically Proficient Elites

High-net-worth individuals constitute the primary customers of luxury brands. They pursue exclusivity, legacy, and exceptional workmanship, while simultaneously appreciating innovation and advanced technology (Jain & Rathi, 2023; Rathi et al., 2022). This group is exceptionally discerning and anticipates a flawless amalgamation of luxury and usefulness in smart luxury co-branded products.

The relationship between Rolls-Royce and KEF in luxury automobile audio systems serves high-net-worth individuals seeking superior sound quality and customized automotive experiences. The Bentley x Breitling collaboration provides affluent clients with meticulously crafted timepieces that correspond to their opulent lifestyle and social standing. These consumers exhibit reduced price sensitivity yet possess a strong awareness of brand identity, rendering authenticity and exclusivity crucial in their purchase choices.

Millennials and Generation Z: The Emerging Luxury Consumers

Emerging consumers are shaping the future of smart luxury markets, as Millennials and Gen Z prioritize innovation, sustainability, and digital experiences over conventional luxury status symbols. In contrast to earlier generations, they emphasize customization, technology-driven experiences, and ethical manufacturing.

This transformation is seen in premium technology partnerships such as Apple x Hermès, where exquisite workmanship is integrated with contemporary technology and customization. Likewise, Louis Vuitton's Horizon Wireless Earphones attract younger consumers by merging stylish design with advanced audio technology. To engage digital-native consumers, marketers must prioritize storytelling, interactive digital campaigns, and sustainability initiatives.

The Influence of Social Media and Influencers on Perception. Social media has revolutionized luxury marketing, with influencers significantly impacting brand perception and consumer choices. In contrast to conventional luxury advertising that depended on exclusivity, contemporary luxury firms actively include consumers with interactive content, behind-the-scenes collaborations, and influencer endorsements (Zniva et al., 2023).

For instance, TAG Heuer's Connected Watch achieved prominence via celebrity sponsorships and technology influencers, establishing the smartwatch as an essential luxury tech accessory. Tesla's partnerships with prominent individuals have preserved its luxurious allure, notwithstanding its extensive availability. In an age where social media influences desirability, luxury businesses must deliberately utilize influencers and digital platforms to uphold aspirational allure, exclusivity, and technological sophistication in smart luxury co-branding efforts.

3.3 Administration of Brand Identity and Reputation in Co-Branding

The maintenance of brand identity and reputation is fundamental to luxury, rendering it crucial in any co-branding plan. Collaborations between luxury and technology firms can broaden market access and foster innovation, but they also entail risks. The difficulty resides in preserving exclusivity and reputation while using intelligent technologies (Li, 2021). Effective management of brand perception necessitates meticulous positioning, strategic communication, and engaging storytelling to guarantee that co-branded items resonate with consumers while preserving the integrity of the luxury brand.

Perils of Brand Dilution and Alterations in Perception

A principal risk of co-branding in luxury markets is brand dilution, wherein a luxury brand forfeits its exclusivity owing to excessive exposure or incompatible relationships. A co-branded product that is overly mass-market or too closely linked to a non-luxury brand may undermine the brand's perceived value (Huang, 2023; Xie et al., 2023).

Luxury fashion firms who collaborate with fast-fashion retailers frequently encounter backlash from conventional high-end buyers who perceive a compromise in the brand's exclusivity. Likewise, if a luxury watch manufacturer collaborates with a technological company recognized for mass production, it may undermine the perception of workmanship and enduring value that characterizes luxury timepieces. Meticulous brand alignment and regulated distribution are crucial to mitigate these dangers.

A significant problem is the alteration of perceptions, since consumers find it difficult to recognize the significance of a partnership (Sahne & Daronkola, 2025). A collaboration that seems contrived or only motivated by profit may engender suspicion instead of enthusiasm. Successful collaborations, exemplified by Apple x Hermès, circumvent this by fostering an organic balance between utility and luxurious aesthetics.

Communication Tactics for Positioning Co-Branded Products

An effective communication strategy is essential for establishing a co-branded product as desired and aspirational. Luxury brands should emphasize exclusivity, workmanship, and innovation in the marketing of a smart luxury product.

Principal strategies encompass:

1. Restricted releases and VIP access: Preserving the uniqueness of co-branded products via invitation-only acquisitions or premium-tier options.
2. Prominent endorsements: Utilizing brand ambassadors and influencers that represent both luxury and innovation to establish credibility.
3. Targeted digital storytelling: Employing luxury-oriented content platforms, immersive augmented reality/virtual reality experiences, and social media interaction to enhance exclusivity.

The Significance of Storytelling in the Success of Luxury Co-Branding. Storytelling serves as an effective instrument for maintaining brand identity in co-branding collaborations (Yu & Ko, 2021). Luxury brands must develop a story that intertwines heritage with innovation, illustrating how the cooperation enriches the brand's legacy rather than undermining it.

For example, Tag Heuer's Connected Watch is promoted not merely as a smartwatch but as an extension of Swiss horological brilliance in the digital era. By incorporating a captivating narrative that emphasizes craftsmanship, innovation, and exclusivity, firms may guarantee that co-branding enhances rather than undermines their market positioning. Effective brand management in luxury co-branding necessitates a strategic equilibrium between innovation and exclusivity, ensuring that technology augments rather than undermines the prestige and allure of the luxury brand.

4. ANALYSES OF CO-BRANDING ACHIEVEMENTS AND FAILURES

4.1 Notable Co-Branding Collaborations in Smart Luxury

The convergence of luxury and intelligent technology has led to numerous innovative co-branding partnerships. These collaborations illustrate how luxury businesses may effectively incorporate innovation while preserving exclusivity and reputation (Rathi et al., 2022). This document presents three exemplary case studies of successful co-branding that illustrate the potential of strategic luxury collaborations.

Apple and Hermès: Transforming Smartwatches into Luxury Fashion. The cooperation between Apple and Hermès exemplifies a successful luxury-tech collaboration, merging advanced technology with artisanal craftsmanship to produce the Apple Watch Hermès.

Apple, recognized for its ingenuity, elegant design, and digital environment, aimed to position its smartwatch inside the luxury market (Siepmann & Kowalczuk, 2021). Hermès, a brand celebrated for its artisanal leather products and enduring sophistication, offered the ideal complement. The partnership produced a unique series of smartwatches with hand-stitched Hermès leather bands, distinctive watch faces, and high-quality packaging.

The efficacy of this collaboration is rooted in the harmonious amalgamation of practicality and luxury, providing consumers with a product that attracts both technology aficionados and affluent individuals. The Apple Watch Hermès preserves the elegance and artistry of a luxury watch while offering smart connectivity and health-monitoring capabilities, establishing it as a status symbol in both the luxury and technology sectors.

Lamborghini and MIT: Leading the Advancement of Intelligent Electric Vehicles

Lamborghini's partnership with the Massachusetts Institute of Technology (MIT) exemplifies an innovative strategy for intelligent luxury transportation. This collaboration emphasizes the advancement of next-generation electric sports vehicles, enhancing performance, sustainability, and innovation (Du & Li, 2021).

In 2017, Lamborghini and MIT unveiled the Lamborghini Terzo Millennio, a visionary concept vehicle with self-healing carbon fiber, supercapacitor energy storage, and AI-augmented driving capabilities. This partnership merges Lamborghini's legacy of velocity, aesthetics, and exclusivity with MIT's proficiency in advanced materials and electric propulsion technology. The efficacy of this co-branding initiative is attributed to its progressive innovation, establishing Lamborghini as a frontrunner in sustainable luxury performance automobiles and bolstering its reputation for technological superiority.

Montblanc with Samsung: Integrating Luxury with Intelligent Accessories

Montblanc, a prestigious brand recognized for its exquisite writing utensils and luxury leather products, collaborated with Samsung to develop an elegant collection of smart accessories. This partnership produced high-quality leather smartwatch bands, styluses, and linked wearables, intended to enhance Samsung's Galaxy ecosystem (Siepmann & Kowalczuk, 2021).

The collaboration between Montblanc and Samsung underscores the potential of exquisite craftsmanship to elevate smart technology, catering to consumers who seek usefulness while maintaining elegance. This alliance effectively established smart accessories as premium lifestyle items by merging Montblanc's esteemed design aesthetics with Samsung's technological proficiency.

Each collaboration exemplifies the efficacy of effective co-branding within the smart luxury sector. The collaboration between Apple and Hermès boosted smartwatches to the premium sector, Lamborghini and MIT advanced electric performance, and Montblanc and Samsung enhanced the luxury tech accessory market. These examples demonstrate that when brand values are congruent, astute luxury co-branding can produce enduring, inventive, and exceptionally attractive items.

4.2 Insights from Co-Branding Failures

Although co-branding can foster brand innovation, market growth, and consumer involvement, not every partnership achieves success. Certain luxury co-branding collaborations fail because of discrepancies in brand values, audience incompatibility, and excessive commercialization, resulting in customer backlash, brand dilution, or financial detriment (Jain et al., 2021; Reinikainen et al., 2021). Comprehending these mistakes offers critical insights into what luxury firms ought to eschew when incorporating smart technology into their products.

Reasons for the Failure of Certain Luxury Co-Branding Initiatives

Luxury brands flourish due to exclusivity, legacy, and prestige, whereas technology businesses typically prioritize innovation, accessibility, and widespread adoption. When these foundational values conflict, co-branded products have difficulties in achieving traction.

A prominent instance is the collaboration between Vertu and Android, which sought to produce ultra-luxury smartphones by incorporating premium materials such as titanium and sapphire glass with the Android operating system. Notwithstanding its superior craftsmanship, the device faltered due to obsolete software, insufficient innovation, and an elevated price point. Consumers were reluctant to pay extra for luxury aesthetics devoid of advanced technology, underscoring the significance of functional superiority in conjunction with exclusivity (Correia & Kozak, 2021).

Discrepancy in Brand Values and Incongruence with Target Audience

A primary cause of co-branding failure is the incompatibility of ideals between luxury and technology brands, resulting in an ambiguous product identity and audience disorientation. If a relationship does not genuinely integrate brand identities, consumers may view the collaboration as contrived or insincere.

An illustration of inadequate brand alignment is the cooperation between Hummer and Victoria's Secret, which sought to promote a limited-edition luxury SUV influenced by a lingerie brand. The incongruity between robust automobile branding and high-fashion exclusivity perplexed both customer demographics, resulting in an unsuccessful marketing campaign (Rathi et al., 2022). Likewise, Dolce & Gabbana's collaboration with Motorola to produce a gold-plated Razr phone failed to appeal to luxury consumers or technology aficionados, since the device was deficient in both innovative features and enduring appeal.

Excessive Commercialization and Erosion of Exclusivity

Luxury brands depend on scarcity and distinction to sustain their attractiveness. When a co-branded product becomes overly accessible or unduly commercialized, it jeopardizes its aspirational value.

Burberry's excessive licensing in the early 2000s resulted in brand dilution, as the business applied its logo to mass-market accessories, fragrances, and technology products, diminishing its exclusivity and allure among affluent buyers. Conversely, successful collaborations—such as Apple x Hermès or Bentley x Breitling—preserve exclusivity via limited releases, high-end customization, and focused marketing techniques.

Essential Insights

1. The ineffectiveness of luxury co-branding initiatives frequently arises from inadequate strategic alignment, particularly when the exclusivity of luxury conflicts with the mass-market accessibility of technology.
2. Erroneously assessing consumer preferences, leading to items that do not resonate with either demographic.
3. Over-commercialization—diminishing exclusivity and undermining the perceived value of the luxury brand.

By analyzing these errors, firms may develop co-branding strategies that maintain exclusivity, guarantee product-market alignment, and establish cohesive synergies between luxury workmanship and intelligent innovation.

4.3 The Prospects of Co-Branding in Luxury Technology

The future of co-branding in smart luxury markets will be influenced by developing innovations, changing customer expectations, and an increased emphasis on sustainability as technology transforms the luxury business (Rathi et al., 2022; Vanhamme et al., 2021). Brands that effectively incorporate advanced technologies like AI, blockchain, and the metaverse, while upholding exclusivity and ethical standards, will secure a competitive edge.

Forecasts for Future Smart Luxury Partnerships

The forthcoming phase of co-branding collaborations in luxury technology will emphasize hyper-personalization, intelligent automation, and immersive experiences (Cheng et al., 2021). Numerous trends are poised to influence the future of smart luxury collaborations:

Luxury automotive brands will enhance AI-driven mobility solutions. In the wake of Lamborghini and MIT's success, more luxury automotive brands may partner with AI companies to create self-learning, voice-activated, and energy-efficient vehicle systems.

High-end fashion businesses will allocate resources to smart textiles. Brands like Chanel, Gucci, and Louis Vuitton may collaborate with biotech and wearable technology companies to develop temperature-regulating, biometric-tracking, and self-cleaning garments. Luxury watchmakers will incorporate quantum computers and biosensors. Future partnerships building on Tag Heuer's smartwatches are anticipated to incorporate health analytics, blockchain authentication, and solar-powered charging.

The Significance of AI, Blockchain, and Metaverse in Luxury Branding Technology has transitioned from a secondary consideration to a fundamental catalyst for value, exclusivity, and engagement in luxury branding (Indrakumari et al., 2021). Numerous significant breakthroughs will transform co-branding strategies: AI-driven customisation will improve client experiences through tailored recommendations, virtual fashion fittings, and anticipatory luxury trends. Artificial intelligence will enable brands to develop highly personalized smart luxury products that evolve according to consumer preferences over time.

Blockchain authentication will safeguard luxury assets by issuing digital certifications of authenticity for co-branded smart luxury products. This technology will mitigate counterfeiting threats and improve brand transparency. Metaverse luxury experiences will transform customer engagement with brands. Luxury businesses will collaborate with virtual reality (VR) and gaming companies to develop exclusive digital fashion collections, NFT-supported luxury assets, and interactive virtual showrooms.

5. APPROACHES TO SUSTAINABLE AND ETHICAL CO-BRANDING

With sustainability emerging as a paramount issue, the future of luxury co-branding will prioritize eco-friendly technologies (Milanesi et al., 2022). Brands will investigate sustainable materials via partnerships with biodegradable technology and circular fashion enterprises. Intelligent luxury wearables will be built with recyclable materials and reduced ecological footprint.

Carbon-neutral supply chains facilitated by AI and blockchain, guaranteeing ethical procurement of raw materials and minimizing waste in luxury production.

Regenerative luxury involves brands investing in eco-restoration initiatives and carbon offset schemes, thereby confirming their dedication to sustainability.

Principal Insights: The future of co-branding in luxury technology is characterized by the harmonious integration of innovation, exclusivity, and sustainability. As artificial intelligence, blockchain technology, and the metaverse transform luxury experiences, brands that emphasize personalization, security, and ethical innovation will become industry frontrunners. By forming strategic and progressive partnerships, premium brands may preserve their distinction while adapting to the digital revolution.

5.1 Essential Steps in Formulating a Co-Branding Strategy

An effective co-branding approach in the smart luxury sector necessitates a meticulous equilibrium of exclusivity, innovation, and brand integrity (Huang, 2023). Successful collaborations can enhance brand equity and broaden market reach, but unsuccessful partnerships may lead to brand dilution and customer distrust. To achieve success, luxury firms must adopt a strategic methodology in partner selection, market research, and the creation of premium product experiences.

Determining the Appropriate Co-Branding Partner

Choosing the optimal co-branding partner is essential for a successful relationship (Javornik et al., 2021). The selected partner must conform to the fundamental principles, brand reputation, and market positioning of the luxury brand. This alignment guarantees that the co-branded product elevates the appeal and perception of both brands instead of undermining their value.

Essential factors in partner selection encompass:

- Brand equity and reputation – Confirming that both brands possess comparable prestige and consumer trust.
- Complementary skills — A robust co-branding alliance capitalizes on the capabilities of each brand (e.g., a luxury fashion business partnering with a technological innovation).
- Target audience alignment - The partnership must appeal to a common or intersecting clientele, guaranteeing consumer attention and demand.

Notable instances encompass Apple x Hermès, wherein Apple's technological supremacy with Hermès' artisanal luxury produced a highly sought-after smartwatch. In contrast, inadequate partnerships—where a premium brand collaborates with a mass-market entity—can result in ambiguous brand positioning and diminished exclusivity.

Market Research and Analysis of Consumer Demand

Thorough market research is crucial to guarantee that a co-branded luxury product fulfills consumer expectations and conforms to emerging industry trends. This entails examining the present market demand for smart luxury items, such as the increasing interest in wearable technology, AI-enhanced fashion, or sustainable luxury. This involves analyzing:

- This entails examining the present market demand for smart luxury items, such as the increasing interest in wearable technology, AI-enhanced fashion, or sustainable luxury
- Consumer behavior and segmentation, such as the expectations of high-net-worth individuals compared to those of tech-savvy Millennials and Generation Z.
- Competitor landscape — Analyzing successful and unsuccessful co-branding initiatives to discern opportunities and potential hazards.

Brands ought to utilize data analytics, AI-generated consumer insights, and luxury trend forecasts to develop a co-branded product that is both novel and appealing. Creating a Distinctive and High-End Product Experience. Luxury brands must guarantee that the product experience is exclusive, immersive, and of superior quality. This can be accomplished via:

- Customized personalization – Providing tailored design alternatives, including engraving, material selections, or AI-enhanced personalization capabilities.
- Limited-edition releases — Preserving exclusivity via scarcity marketing and exclusive product launches for VIPs.
- Seamless integration of luxury and technology, ensuring that the smart elements augment rather than undermine the history of the luxury brand.

Tag Heuer's Connected Watch exemplifies the fusion of Swiss horological grandeur with sophisticated digital capabilities, rendering it an aspirational item for both luxury and technology enthusiasts. By adhering to these measures, luxury businesses may develop an effective co-branding strategy that enhances exclusivity, improves brand prestige, and aligns with the changing consumer expectations in the smart luxury market.

5.2 Marketing and Distribution Strategies for Smart Luxury Co-Branding

An effective co-branding strategy in the smart luxury sector necessitates not only the development of a novel product but also strategic marketing and distribution that enhance exclusivity, desirability, and digital engagement (Peng, 2021). Affluent consumers anticipate a cohesive omnichannel experience, exclusive product launches, and engaging brand narratives (Javornik et al., 2021). To optimize the effectiveness of a smart luxury cooperation, brands must implement a multifaceted strategy that integrates digital innovation with personalized luxury experiences.

Opulent Omnichannel Marketing and Digital Presence

Contemporary luxury buyers anticipate a meticulously tailored and digitally cohesive purchasing experience (Li, 2021; Milanesi et al., 2022). Luxury firms must adopt an advanced omnichannel marketing strategy that integrates online, offline, and mobile touchpoints to facilitate a cohesive shopping experience.

Essential methods comprise:

- Exclusive e-commerce platforms — Luxury companies are progressively establishing direct-to-consumer (DTC) online boutiques to exert control over branding and consumer experience (e.g., Louis Vuitton's and Rolex's limited online sales strategy).
- AI-driven personalization — Employing AI-based recommendations, chatbots, and concierge services to provide customized luxury shopping experiences.
- Augmented reality (AR) and virtual showrooms — Luxury businesses utilize AR to improve online shopping, enabling consumers to virtually experience products before to purchase (e.g., Gucci's AR-enabled sneaker fittings).

Limited Editions, Exclusivity, and Scarcity Marketing

Exclusivity remains the **core of luxury branding**. Overexposure can diminish the **perceived value of a luxury brand**, making **limited editions and scarcity marketing essential** (Jain et al., 2021).

Techniques used in luxury scarcity marketing include:

- Limited-edition co-branded releases – Premium luxury items ought to be launched in restricted quantities, guaranteeing elevated demand and resale value (e.g., Apple Watch Hermès limited stock strategy).
- VIP and exclusive access – Brands ought to provide preferential access to high-net-worth individuals (HNWIs) and loyal clientele prior to public release.
- Temporal product releases – Capitalizing on excitement via exclusive online flash sales, accessible only for a limited duration (e.g., Nike's SNKRS releases for premium sneakers).

Utilizing Experiential Retail and Immersive Technology

Affluent consumers prioritize experiences over mere things. To augment appeal, luxury co-branding initiatives should prioritize experiential retail, integrating physical and digital interactions (Javornik et al., 2021; Milanesi et al., 2022).

- Instances of successful experiential luxury marketing comprise:

Luxury firms ought to develop interactive flagship stores that provide technology-enhanced personalization, exclusive shopping experiences, and VIP access to limited edition products.

- Integration of Metaverse and NFTs – Brands can utilize Web3 technology to provide NFT-backed digital luxury items and virtual showrooms (e.g., Balenciaga's NFT fashion and Burberry's digital collectibles).
- Luxury technology pop-ups — Exclusive pop-up stores displaying sophisticated luxury advancements can generate excitement and exclusivity (e.g., Tag Heuer's smartwatch launch events).

Effective luxury co-branding marketing strategies should prioritize:

- Integrated omnichannel interaction, utilizing AI, AR, and VR for enhanced customisation.
- Utilization of exclusivity and scarcity marketing to enhance desirability via limited editions.
- Engaging experience retail, utilizing the metaverse, interactive showrooms, and exclusive VIP events.

By implementing these techniques, luxury brands may preserve their reputation in smart luxury collaborations while attracting digital-first consumers.

5.3 Assessing the Effectiveness and Outcomes of Co-Branding Strategies

An effective co-branding approach in the smart luxury sector necessitates ongoing evaluation and enhancement to guarantee that the partnership fulfills business goals while maintaining brand prestige (Li, 2021; Sahne & Daronkola, 2025). Assessing the efficacy of a luxury-tech collaboration necessitates monitoring key performance indicators (KPIs), scrutinizing consumer interaction, and appraising long-term brand equity and market positioning (Nurfitriansyah & Dirgantari, 2021).

Key Performance Indicators and Metrics for Assessing Co-Branding Efficacy

Brands must define measurable KPIs that match with their strategic objectives to assess the success of a co-branded luxury product. These encompass:

- Sales Performance & Revenue Growth – Monitoring initial sales volume, revenue augmentation, and profit margins to evaluate market response. Limited-edition co-branded luxury items ought to rapidly sell out while sustaining elevated price levels.
- Brand Awareness & Media Coverage - Assessing the quantity of media mentions, social media reach, and digital impressions to gauge public interest and exposure.
- Online Engagement Metrics — Evaluating website traffic, click-through rates (CTR), duration on product pages, and conversion rates to gauge consumer interest.

Resale Value and Scarcity Effect - Assessing demand in secondary luxury markets (e.g., The RealReal, StockX) to evaluate if the product retains exclusivity and appeal over time.

Consumer Feedback, Engagement, and Brand Fidelity

Success in luxury co-branding transcends financial criteria; firms must assess consumer perception and engagement (Jain et al., 2021; Thani & Sharma, 2021) to measure long-term loyalty.

- Customer Satisfaction & Reviews — Gathering post-purchase surveys, influencer endorsements, and internet evaluations to assess perceived value and product response.
- Luxury Community Engagement — Evaluating interactions within exclusive VIP clubs, loyalty programs, and members-only platforms to determine the perceptions of high-net-worth individuals (HNWIs) and brand loyalists on the co-branded product.
- Influencer and Celebrity Endorsements — Evaluating the effect of prominent partnerships on customer perception and appeal.

Prolonged Effects on Brand Equity and Market Positioning

A well-executed co-branding strategy should enhance the long-term standing of both brands in the smart luxury market, beyond merely achieving short-term success (Sahne & Daronkola, 2025; Yang-yu et al., 2021).

- Brand Equity Augmentation — Monitoring changes in brand appeal, perceived exclusivity, and readiness to pay a premium following collaboration.

- Competitive Market Position - Evaluating the impact of the collaboration on market standings relative to competitors in the luxury and technology industries.
- Recurring Collaborations & Innovation Pipeline — Assessing whether the co-branding agreement results in sustained collaborations or novel product developments, hence strengthening long-term strategic alignment.

Luxury businesses must utilize data-driven key performance indicators to assess co-branding effectiveness by monitoring revenue, brand visibility, and product demand.

Gather customer insights to enhance engagement strategies and loyalty initiatives.

Assess long-term brand positioning to ensure the partnership augments distinctiveness and commercial significance. By emphasizing performance indicators, consumer sentiment, and strategic brand alignment, luxury businesses may guarantee that intelligent luxury co-branding collaborations (Peng, 2021; Shang, 2022) foster innovation while preserving prestige and desirability.

6. THE DEVELOPMENT OF INTELLIGENT LUXURY CO-BRANDING

6.1 Comprehending Co-Branding in the Luxury Sector

Co-branding has emerged as a strategic instrument in the luxury sector, allowing firms to augment their market allure, stimulate innovation, and address shifting customer preferences. Historically, luxury branding has depended on legacy, craftsmanship, and exclusivity, establishing itself as aspirational and enduring. The emergence of smart technology and digital transformation has altered the luxury sector, leading to the development of smart luxury co-branding, wherein prestigious brands partner with technology firms to provide innovative, high-quality experiences (Al-Issa and Thanasi, 2024; Singh and Kaunert, 2024; Singh & Kaunert, 2024; Santos et al., 2024).

Definition and Importance of Co-Branding in Luxury

Co-branding denotes a strategic alliance between two companies to develop a product that capitalizes on the strengths of each (Paydas Turan, 2021; Ottoni et al., 2024). In the luxury sector, co-branding is very important as it enables firms to:

- Broaden Market Reach — Luxury businesses can engage a wider, technologically adept demographic by incorporating intelligent features.
- Elevate Brand Prestige — By collaborating with premium technology companies, luxury brands uphold their reputation as pioneers.
- Enhance Product Innovation - Merging traditional craftsmanship with contemporary smart features enhances functionality while maintaining exclusivity.

The cooperation between Apple and Hermès on the Apple Watch Hermès effectively merged Apple's technological proficiency with Hermès' artisanal craftsmanship, solidifying the watch as both a smart device and a luxury emblem.

Historical Context: Conventional Luxury Branding versus Contemporary Co-Branding

Traditionally, luxury branding emphasized scarcity, exclusivity, and artisanal skill. Brands such as Chanel, Rolex, and Louis Vuitton established their identities via heritage and exceptional craftsmanship, providing products that epitomized status and affluence (Tarquini et al., 2022; Wang, 2022). In contrast to mass-market businesses, luxury products were meticulously picked and marketed to a select clientele.

As digitization and evolving consumer tastes emerged, luxury businesses commenced their adaptation to the age of smart technology and connection. The transition was propelled by Millennials and Gen Z consumers, who prioritize innovation, digital integration, and personalized experiences in conjunction with conventional luxury attributes. This transition facilitated contemporary co-branding collaborations, exemplified as Tag Heuer x Google (smartwatches) and Lamborghini x MIT (electric automobiles), wherein technology augments the premium experience while maintaining exclusivity (Javornik et al., 2021; de et al., 2021; Varsha et al., 2021; Malarvizhi et al., 2022).

The Influence of Technology on Smart Luxury Collaborations

Technology is crucial to luxury co-branding, shaping the design, marketing, and distribution of products. Principal technologies influencing smart luxury partnerships encompass:

- AI-Driven Personalization — Tailored luxury experiences facilitated by artificial intelligence.
- Blockchain Authentication — Reliable digital certification for luxury items, guaranteeing authenticity.

- Metaverse and Virtual Luxury — Digital fashion, NFTs, and immersive e-commerce experiences.

Luxury firms must reconcile heritage with innovation to thrive in the smart luxury sector, ensuring that technology augments the luxury experience rather than dominates it.

6.2 Consumer Behavior in Premium Luxury Markets

The luxury market is undergoing a significant upheaval in consumer expectations, influenced by digital advancements, evolving generational preferences, and the incorporation of smart technologies into premium products. Traditional luxury shoppers prioritized exclusivity, heritage, and craftsmanship, whereas contemporary luxury purchasers desire a blend of innovation, digital engagement, and customisation (Kumar, 2023; Prestini and Sebastiani, 2021; Wang et al., 2022). This progression has led to smart luxury co-branding, wherein luxury brands collaborate with technology companies to develop products that appeal to both status-oriented and tech-savvy consumers.

The Transformation in Consumer Anticipations: Transitioning from Exclusivity to Innovation

Luxury has historically been linked to rarity, artistry, and permanence. Contemporary consumers require more than mere prestige; they seek innovative technology, customization, and engaging brand experiences (Al-Issa & Thanasi, 2024; Santos et al., 2024). Luxury brands must include AI, blockchain, and the metaverse, rather than depend exclusively on tradition, to maintain relevance.

The Apple x Hermès smartwatch exemplifies a successful transition from conventional exclusivity to functional, high-tech luxury, wherein buyers value both design and digital innovation. (Schmitt et al. (2022) and Le et al. (2024)). Likewise, Louis Vuitton's Horizon wireless earphones integrate fashion with superior sound technology, redefining luxury for a contemporary audience.

This transition necessitates that co-branding collaborations in the smart luxury sector reconcile history with innovation, guaranteeing that products retain their exclusivity while incorporating cutting-edge technology. High-Net-Worth Individuals (HNWIs) versus Technologically Proficient Individuals Millennials and Generation Z. The luxury market serves two main consumer sectors, each with unique expectations:

- Affluent Individuals with Significant Wealth (HNWIs)
- Value exclusivity, artisanal craftsmanship, and bespoke luxury services.

- Appreciate customized experiences and established brand prestige (e.g., Rolls-Royce Bespoke Vehicles).
- Anticipate flawless technological integration without sacrificing aesthetics (e.g., Bentley's partnership with Naim Audio for high-end in-vehicle audio systems).

Millennials and Generation Z Emphasize innovation, internet engagement, and ethical luxury.

Anticipate sustainability, technology-driven customisation, and multichannel interaction. Interact with brands via social media, influencers, and immersive retail experiences (e.g., Gucci's NFT-supported digital fashion). The Impact of Digital Engagement, Social Media, and Influencer Marketing Luxury businesses are utilizing digital storytelling, influencer partnerships, and social commerce to enhance consumer engagement. Platforms like as Instagram, TikTok, and WeChat are crucial for brand visibility, especially among younger demographics (Zhang & Erturk, 2022; Jiang, 2023).

- Influencer marketing Celebrities and technology influencers affect customer perceptions, as exemplified by TAG Heuer's smartwatch advertising.
- Live-stream shopping and virtual experiences — Luxury firms employ AR/VR technology for unique digital showrooms, such as Burberry's virtual fashion presentations.
- Social commerce and exclusive releases – Online-exclusive releases generate scarcity and urgency, hence increasing attractiveness.

By comprehending these shifting customer behaviors, luxury businesses can formulate effective smart luxury co-branding strategies that resonate with both traditional elites and digitally native consumers.

REFERENCES:

Al-Issa, N., & Thanasi, M. (2024). Mapping the Future of Tech-Infused Luxury: A Roadmap and Research Directions. [wiley.com]. *International Journal of Consumer Studies, 48*(6), e13103. DOI: 10.1111/ijcs.13103

Ashfaq, M., Jiang, Y., Yu, S., & Loureiro, S. M. C. (2020). I, Chatbot: Modeling the determinants of users' satisfaction and continuance intention of AI-powered service agents. In Telematics and Informatics (Vol. 54, p. 101473). Elsevier BV. https://doi.org/DOI: 10.1016/j.tele.2020.101473

Bian, X., & Haque, S. (2020). Counterfeit versus original patronage: Do emotional brand attachment, brand involvement, and past experience matter? In Journal of Brand Management (Vol. 27, Issue 4, p. 438). Palgrave Macmillan. https://doi.org/DOI: 10.1057/s41262-020-00189-4

Cabigiosu, A. (2020). An Overview of the Luxury Fashion Industry. In *Palgrave advances in luxury* (p. 9). Springer International Publishing., DOI: 10.1007/978-3-030-48810-9_2

Cheng, R., Zhu, J., Zhu, X., & Zhang, C. (2021). Marketing Strategies in Times of Crisis Exploring More Digital Strategies in Luxury Industry. In *Advances in economics, business and management research/Advances in Economics, Business and Management Research.* Atlantis Press., DOI: 10.2991/assehr.k.211209.365

Correia, A., & Kozak, M. (2021). Luxury product and brand purchasing behavior. https://doi.org/, https://www.tandfonline.com/doi/full/10.1080/21639159.2020.1808823DOI: 10.1080/21639159.2020.1808823

Du, X., & Li, B. (2021). Analysis of Tesla's Marketing Strategy in China. In *Advances in economics, business and management research/Advances in Economics, Business and Management Research.* Atlantis Press., DOI: 10.2991/assehr.k.211209.270

Guercini, S., Ranfagni, S., & Runfola, A. (2020). E-commerce internationalization for top luxury fashion brands: some emerging strategic issues. In Journal of Management Development (Vol. 39, Issue 4, p. 423). Emerald Publishing Limited. https://doi.org/DOI: 10.1108/JMD-10-2019-0434

Heyes, A. (2021). What is luxury hospitality? A need to move towards a scientific understanding. In Research in Hospitality Management (Vol. 11, Issue 2, p. 67). Taylor & Francis. https://doi.org/DOI: 10.1080/22243534.2021.1917919

Holmqvist, J., Wirtz, J., & Fritze, M. P. (2020). Luxury in the digital age: A multi-actor service encounter perspective. In *Journal of Business Research* (Vol. 121, p. 747). Elsevier BV., DOI: 10.1016/j.jbusres.2020.05.038

Huang, R. (2023; Vol. 47). Marketing Strategy and Operation Mode of Luxury Goods Lead to Brand Premium. In BCP Business & Management., DOI: 10.54691/bcpbm.v47i.5168

Indrakumari, R., Kumar, R., Balusamy, B., & Asirvadam, V. S. (2021). Convergence of Blockchain, AI, and IoT. In *CRC Press eBooks*. Informa., DOI: 10.1201/9781003081180

Jain, S., Mishra, S., & Mukhopadhyay, S. (2021). Critical success factors for luxury fashion brands in emerging markets: Insights from a qualitative study. In Journal of Global Fashion Marketing (Vol. 12, Issue 1, p. 47). Taylor & Francis. https://doi.org/DOI: 10.1080/20932685.2020.1845765

Jain, S., & Rathi, R. (2023). Do consumer values and perceived readiness impact secondhand luxury consumption? A goal-framing theory approach. In Journal of Product & Brand Management (Vol. 32, Issue 7, p. 973). Emerald Publishing Limited. https://doi.org/DOI: 10.1108/JPBM-10-2021-3703

Javornik, A., Duffy, K., Rokka, J., Scholz, J., Nobbs, K., Motala, A., & Goldenberg, A. (2021). Strategic approaches to augmented reality deployment by luxury brands.[]. *Journal of Business Research*, *136*, 284–292. gla.ac.uk. DOI: 10.1016/j.jbusres.2021.07.040

Javornik, A., Duffy, K., Rokka, J., Scholz, J., Nobbs, K., Motala, A., & Goldenberg, A. (2021). Strategic approaches to augmented reality deployment by luxury brands. In *Journal of Business Research* (Vol. 136, p. 284). Elsevier BV., DOI: 10.1016/j.jbusres.2021.07.040

Jeddi, Z., & Bohr, A. (2020). Remote patient monitoring using artificial intelligence. In *Elsevier eBooks* (p. 203). Elsevier BV., DOI: 10.1016/B978-0-12-818438-7.00009-5

Jiang, Y. (2023, July). The Research on the Current Situation and Trends of Social Media Marketing. In 2023 2nd International Conference on Social Sciences and Humanities and Arts (SSHA 2023) (pp. 348-355). Atlantis Press. https://doi.org/, atlantis-press.comDOI: 10.2991/978-2-38476-062-6_44

Kumar, P. (2023). Luxury consumption amidst the COVID-19 pandemic. [HTML]. *Marketing Intelligence & Planning*, *41*(1), 62–82. Advance online publication. DOI: 10.1108/MIP-02-2022-0081

Le, H. T. P. M., Park, J., Nguyen, T. T., & Yun, J. (2024). Exploring the relationship between social media use and anti-luxury behavior: Implications for interactive marketing strategies. [HTML]. *Journal of Research in Interactive Marketing*, *18*(4), 549–569. DOI: 10.1108/JRIM-04-2023-0125

Li, T. (2021). Analysis of Successful Approaches of Classic Luxury Brand's Marketing Strategies Before and After the Digital Era. In *Advances in economics, business and management research/Advances in Economics, Business and Management Research*. Atlantis Press., DOI: 10.2991/aebmr.k.210803.083

Malarvizhi, C. A., Al Mamun, A., Jayashree, S., Naznen, F., & Abir, T. (2022). Modelling the significance of social media marketing activities, brand equity and loyalty to predict consumers' willingness to pay premium price for portable tech gadgets. Heliyon, 8(8). https://doi.org/cell.comDOI: 10.1016/j.heliyon.2022.e10145

Milanesi, M., Guercini, S., & Runfola, A. (2022). Let's play! Gamification as a marketing tool to deliver a digital luxury experience. In Electronic Commerce Research (Vol. 23, Issue 4, p. 2135). Springer Science+Business Media. https://doi.org/DOI: 10.1007/s10660-021-09529-1

Nurfitriansyah, H. R., & Dirgantari, P. D. (2021). Factors That Affect the Election of Higher Education in Holistic Marketing Point of View. In Advances in economics, business and management research/Advances in Economics, Business and Management Research. Atlantis Press. https://doi.org/DOI: 10.2991/aebmr.k.210831.065

Oliveira, M. I. L. L., & Fernandes, T. (2020). Luxury brands and social media: drivers and outcomes of consumer engagement on Instagram. In Journal of Strategic Marketing (Vol. 30, Issue 4, p. 389). Taylor & Francis. https://doi.org/DOI: 10.1080/0965254X.2020.1777459

Ottoni, F. M. J., Thomé, K. M., & Porto, R. B. (2024). Advances in B2B Co-Branding: A Systematic Literature Review. [HTML]. *Journal of Business-To-Business Marketing*, •••, 1–17. DOI: 10.1080/1051712X.2024.2380683

Paydas Turan, C. (2021). Success drivers of co-branding: A meta-analysis. International Journal of Consumer Studies. DOI: , wiley.comDOI: 10.1111/ijcs.12682

Peng, Y. (2021). Marketing Strategies and New Focus for Luxury Companies under COVID-19 Pandemic. In *Advances in economics, business and management research/Advances in Economics, Business and Management Research*. Atlantis Press., DOI: 10.2991/assehr.k.211209.155

Prestini, S., & Sebastiani, R. (2021). Embracing consumer ambivalence in the luxury shopping experience. [HTML]. *Journal of Consumer Behaviour, 20*(5), 1243–1268. Advance online publication. DOI: 10.1002/cb.1933

Rathi, R., Garg, R., Kataria, A., & Chhikara, R. (2022). Evolution of luxury marketing landscape: a bibliometric analysis and future directions. In Journal of Brand Management (Vol. 29, Issue 3, p. 241). Palgrave Macmillan. https://doi.org/DOI: 10.1057/s41262-022-00273-x

Reinikainen, H., Tan, T. M., Luoma-aho, V., & Salo, J. (2021). Making and breaking relationships on social media: the impacts of brand and influencer betrayals. In *Technological Forecasting and Social Change* (Vol. 171, p. 120990). Elsevier BV., DOI: 10.1016/j.techfore.2021.120990

Reinikainen, H., Tan, T. M., Luoma-aho, V., & Salo, J. (2021). Making and breaking relationships on social media: the impacts of brand and influencer betrayals. In *Technological Forecasting and Social Change* (Vol. 171, p. 120990). Elsevier BV., DOI: 10.1016/j.techfore.2021.120990

Sahne, S. S. K., & Daronkola, H. K. (2025). Enhancing loyalty in luxury fashion with AI: the mediation role of customer relationships. In *Asia-Pacific Journal of Business Administration*. Emerald Publishing Limited., DOI: 10.1108/APJBA-08-2024-0456

Santos, V., Sousa, B., & Simões, J. (2024). Metaverse and NFT Market as AI Disruptive Technology for Digital Luxury Brands. In Business Development via AI and Digitalization: Volume 2 (pp. 1025-1036). Cham: Springer Nature Switzerland. https://doi.org/, researchgate.netDOI: 10.1007/978-3-031-62106-2_77

Schmitt, R., Rossi, A., Bensoussan, A., Schmitt, R., Rossi, A., & Bensoussan, A. (2022). Digital Transformation of the Omni-personal Customer Journey. Omni-personal Luxury: How to Transform your Luxury Business for the Digital Age, 63-139. https://doi.org/ [HTML]DOI: 10.1007/978-3-030-85769-1_5

Serdari, T. (2020). Step by Step Application of the Framework. In *Palgrave advances in luxury* (p. 169). Springer International Publishing., DOI: 10.1007/978-3-030-45301-5_11

Shang, J. (2022). Will the Covid-19 Reshape the Marketing of Luxury Brands? In *Advances in economics, business and management research/Advances in Economics, Business and Management Research.* Atlantis Press., DOI: 10.2991/aebmr.k.220307.078

Siepmann, C., & Kowalczuk, P. (2021). Understanding continued smartwatch usage: the role of emotional as well as health and fitness factors. In Electronic Markets (Vol. 31, Issue 4, p. 795). Springer Science+Business Media. https://doi.org/DOI: 10.1007/s12525-021-00458-3
Siepmann, C., & Kowalczuk, P. (2021). Understanding continued smartwatch usage: the role of emotional as well as health and fitness factors. In Electronic Markets (Vol. 31, Issue 4, p. 795). Springer Science+Business Media. https://doi.org/DOI: 10.1007/s12525-021-00458-3
Singh, B., & Kaunert, C. (2024). Redefining the Nodes of Metaverse and Luxury Hospitality: Unveiling AI and Tech-Driven Solutions in Hospitality and Tourism Industry. In Impact of AI and Tech-Driven Solutions in Hospitality and Tourism (pp. 175-202). IGI Global. DOI: [HTML]DOI: 10.4018/979-8-3693-6755-1.ch010

Tang, D. (2021). WHAT IS DIGITAL TRANSFORMATION? In EDPACS (Vol. 64, Issue 1, p. 9). Taylor & Francis. https://doi.org/DOI: 10.1080/07366981.2020.1847813

Tarquini, A., Mühlbacher, H., & Kreuzer, M. (2022). The experience of luxury craftsmanship–a strategic asset for luxury experience management. [HTML]. *Journal of Marketing Management*, *38*(13-14), 1307–1338. DOI: 10.1080/0267257X.2022.2064899

Thani, S., & Sharma, A. (2021). The Psychology behind Luxury Consumption. In *Journal of Pharmaceutical Research International* (p. 341). Sciencedomain International., DOI: 10.9734/jpri/2021/v33i57B34065

Vanhamme, J., Lindgreen, A., & Sarial-Abi, G. (2021). Luxury Ethical Consumers: Who Are They? In Journal of Business Ethics (Vol. 183, Issue 3, p. 805). Springer Science+Business Media. https://doi.org/DOI: 10.1007/s10551-021-04981-3

Varsha, P. S., Akter, S., Kumar, A., Gochhait, S., & Patagundi, B. (2021). The impact of artificial intelligence on branding: A bibliometric analysis (1982-2019). [JGIM]. *Journal of Global Information Management*, *29*(4), 221–246. DOI: 10.4018/JGIM.20210701.oa10

Wang, Y. (2022). A conceptual framework of contemporary luxury consumption. International Journal of Research in Marketing. https://doi.org/ sciencedirect.com-DOI: 10.1016/j.ijresmar.2021.10.010

Wang, Z., Yuan, R., Luo, J., & Liu, M. J. (2022). Redefining "masstige" luxury consumption in the post-COVID era. Journal of Business Research. https://doi.org/, nih.govDOI: 10.1016/j.jbusres.2022.01.057

Wei, Y., & Li, F. (2020). Omnichannel supply chain operations for luxury products with conspicuous consumers. In *Transportation Research Part E Logistics and Transportation Review* (Vol. 137, p. 101918). Elsevier BV., DOI: 10.1016/j.tre.2020.101918

Wirtz, J., Holmqvist, J., & Fritze, M. P. (2020). Luxury services. In Journal of service management (Vol. 31, Issue 4, p. 665). Emerald Publishing Limited. https://doi.org/DOI: 10.1108/JOSM-11-2019-0342

Xie, P., Shi, R., & Xu, D. (2023). Retailer service strategy on livestreaming platforms considering free riding behavior. In Annals of Operations Research. Springer Science+Business Media. https://doi.org/DOI: 10.1007/s10479-023-05201-z

Yang-yu, L., Lu, J., & Qi, G. (2021). A Summary of the Marketing Strategies of Luxury Cosmetics Brands. In *Advances in economics, business and management research/Advances in Economics, Business and Management Research.* Atlantis Press., DOI: 10.2991/assehr.k.211209.130

Yu, J., & Ko, E. (2021). UGC attributes and effects: implication for luxury brand advertising. In International Journal of Advertising (Vol. 40, Issue 6, p. 945). Taylor & Francis. https://doi.org/DOI: 10.1080/02650487.2021.1898773

Zhang, L. & Erturk, E. (2022). Potential lessons from Chinese businesses and platforms for online networking and marketing: An exploratory study. Social Sciences & Humanities Open. https://doi.org/ sciencedirect.comDOI: 10.1016/j.ssaho.2022.100274

Zniva, R., Weitzl, W., & Lindmoser, C. (2023). Be constantly different! How to manage influencer authenticity. In Electronic Commerce Research (Vol. 23, Issue 3, p. 1485). Springer Science+Business Media. https://doi.org/DOI: 10.1007/s10660-022-09653-6

Chapter 2
Challenges and Risks in Co-Branding Smart Luxury Products

Venkatesh Ganapathy
https://orcid.org/0000-0001-7500-2877
ICFAI Business School, ICFAI Foundation for Higher Education, Bangalore, India

ABSTRACT

This chapter examines the intricate dynamics of co-branding within the luxury sector, particularly in relation to smart products, which pose distinctive challenges and risks. As the market for smart luxury items expands, brands are compelled to navigate the complexities of harmonizing advanced technology with their premium image. A primary concern is the potential for brand dilution; luxury brands, renowned for their exclusivity and tradition, may face threats to their identity when collaborating with tech companies—especially if the technological offerings fail to meet the high expectations of discerning consumers. Additionally, reputational risks are paramount; any ethical, technical, or public relations issues faced by a partner brand can adversely impact the luxury brand's reputation. The operational intricacies of co-branding ventures further complicate matters. Maintaining a delicate balance between innovation and tradition is essential, as brands strive to attract younger consumers without alienating their established clientele.

DOI: 10.4018/979-8-3373-3010-5.ch002

INTRODUCTION

The luxury brand market has experienced significant growth over the past two decades, but the current economic challenges make it difficult for brand owners to convey sufficient value to justify high prices. Recent research indicates a shift from a firm- and product-centric value creation approach to one focused on personalized brand experiences, where customers and suppliers co-create value (Tynan et al., 2010).

Luxury brand collaborations reflect a significant strategy for businesses to grow their customer base. The partnership between Louis Vuitton and Stephen Sprouse serves as an example. Co-branding and partnerships are increasingly popular in the luxury sector as a means to expand market share and enhance customer awareness. These collaborations span various industries, including hospitality, automotive, and fashion.

The impact of brand names on consumer behavior in luxury fashion purchases has been analyzed (Tarnanidis et al., 2025). Luxury brand management is complex and requires a profound understanding of the target audience's needs. Maintaining exclusivity, focusing on quality, and fostering emotional connections are essential. Trends in affluent consumer behavior must be monitored closely.

Consumer engagement with co-branded luxury products is shaped by factors such as perceived value and social influence, enabling the creation of emotionally resonant experiences. Understanding how co-branding impacts consumer perception of the brands involved can help luxury brand managers develop effective co-branding strategies to enhance brand equity and drive consumer engagement (Chen & Li, 2022).

The Fall '17 collection of Supreme and Louis Vuitton marked the beginning of an era of collaborations at the center of luxury. Moncler Genius brought different perspectives to the label through collaborations with various designers. The Ritz-Carlton and Asprey International Limited created a tailor-made drink trolley. Hilton Worldwide and WWF partnered to reduce their environmental impact. Four Seasons Hotels and Resorts and Johns Hopkins Medicine International implemented the "Lead with Care" program during the COVID-19 pandemic. The co-branding collaborations (Oeppen & Jamal, 2014) fall into three categories:

1. Pairings between luxury brands
2. Partnerships with celebrities
3. Collaborations with non-luxury entities

Co-branding of smart luxury products is showcased at major fashion events so that consumers can purchase limited-edition items (Jia & Ran, 2021). The COVID-19 pandemic has transformed the fashion industry and consumer behavior, accelerating trends towards digital strategies and reshaping preferences for comfort, sustainability, and practicality. Fashion companies are adapting to these changes by embracing digital initiatives and responding to increased consumer demand for transparency and sustainable practices. Brands prioritizing sustainability can attract environmentally conscious consumers and enhance their reputation. While implementing sustainable practices may require initial investment, it offers long-term benefits, including cost savings and improved supply chain resilience. Overall, the push for sustainability fosters innovation in materials and processes, benefiting both the industry and the environment (Tarnanidis, 2024)

The right choice of partners enhances authenticity, desirability, and pricing. The success of luxury brand collaborations stems from their ability to generate heightened desirability and exclusivity, driving significant consumer interest and sales. By strategically partnering with complementary brands, celebrities, or unexpected collaborators, luxury houses can tap into new audiences and rejuvenate their image. These collaborations often create limited-edition products or experiences, fostering a sense of urgency (Gupta et al., 2024). Moreover, the inherent storytelling aspect of these partnerships, amplified by social media and high-profile events, creates buzz and reinforces the brands' positions as cultural icons. Ultimately, the success of these collaborations lies in their capacity to create a unique and memorable experience that resonates with both existing and potential customers, solidifying brand loyalty and driving revenue.

CO-BRANDING INFLUENCE IN STRATEGIC MARKETING PARTNERSHIPS

Luxury co-branding has become a particularly potent marketing strategy within China's dynamic consumer landscape (Sun, 2023). Driven by the receptiveness of Chinese consumers—especially younger generations—who view co-branded products as status symbols, and amplified by the country's vibrant digital ecosystem, these collaborations offer significant advantages. Luxury brands in China strategically partner with diverse entities, from fellow luxury houses to lifestyle brands and cultural institutions, to enhance brand image, broaden their audience, and generate excitement. A key aspect of successful co-branding in China is cultural relevance, ensuring that collaborations resonate with local sensibilities. This approach allows brands to effectively localize their marketing efforts and forge deeper connections with Chinese consumers, making co-branding an indispensable tool for navigating this complex and rapidly evolving market.

Co-branding is a strategic marketing partnership where two or more brands collaborate to create a product or service that combines their brand identities and attributes (Pinello et al., 2022). The goal of co-branding is to leverage the strengths of each partner's brand to reach a wider audience, improve customer perception, and boost sales. By combining the assets of both brands, co-branding allows companies to tap into each other's established customer base and credibility, often resulting in increased brand exposure and market reach (Ahmed & Ahmed, 2024). This partnership can take various forms, including ingredient co-branding, composite co-branding, and dual-branded products, depending on the nature of the collaboration and the respective goals of the brands involved.

One of the primary advantages of co-branding is the ability to enhance brand equity (Ottoni et al., 2024) (Steffl et al., 2024). When two brands collaborate, they bring their own unique reputations and customer loyalty, which can positively influence the perception of the new product or service. For instance, if a luxury brand partners with a popular technology company, the collaboration may create an air of exclusivity and innovation that appeals to both sets of customers (Sass et al., 2024) (Carlotto, 2024)

This kind of synergy not only attracts new consumers but also strengthens the brand positioning of each partner. Moreover, the shared promotional efforts can lower marketing costs, as both brands can pool resources to advertise the co-branded product, achieving a broader reach at a fraction of the typical cost.

Co-branding also helps mitigate risks. By partnering with an established brand, a company can reduce the uncertainties of launching a new product. The established brand brings its reputation, customer trust, and market experience to the table, providing credibility and reducing the chances of failure (Paydas, 2021) (Peasley & Woodroof, 2024). For example, when a well-known snack brand collaborates with a major soft drink company to offer a combo pack, consumers are more likely to try the product due to their trust in the individual brands. The combined power of both brands creates a sense of reliability and quality assurance that may not exist with a single, unknown brand.

However, co-branding also has its challenges (Pinello et al, 2022). A misalignment of brand values, target audience, or strategic goals between partners can lead to confusion or brand dilution. For instance, if a luxury brand collaborates with a budget brand, the luxury brand's exclusivity and high-end image may be compromised. In such cases, consumers may view the collaboration as a mismatch, causing both brands to lose their distinctiveness and appeal. Additionally, the success of co-branding depends heavily on the strength of each brand's market presence and reputation. If one brand in the partnership has a tarnished image or lacks consumer trust, the other brand might also suffer damage to its reputation.

Co-branding is a powerful marketing tool that can provide substantial benefits when executed correctly. By combining the strengths of two or more brands, companies can access new markets, reduce marketing costs, and enhance their overall brand equity. However, the success of co-branding depends on selecting the right partner, aligning brand values, and ensuring that both brands bring complementary strengths to the partnership. When done right, co-branding has the potential to create innovative products that appeal to a broader customer base and generate substantial returns for both brands involved (Lu, 2023).

SMART LUXURY: BLENDING PRESTIGE WITH INNOVATION

Smart luxury products are a blend of high-end craftsmanship and cutting-edge technology, offering affluent consumers a fusion of prestige, functionality, and innovation. These products, which include smartwatches, connected vehicles, high-tech fashion, and home appliances, cater to those who seek exclusivity and convenience. Unlike traditional luxury items that emphasize craftsmanship, rarity, and aesthetic appeal, smart luxury integrates advanced technology to enhance user experience without sacrificing design or luxury (Tarnanidis, 2024).

As consumers become increasingly knowledgeable, tech-savvy, and environmentally conscious, the concept of luxury has evolved. Today's affluent individuals seek experiences that are not just opulent but also intuitive, sustainable, and innovative, aligning with their values. This shift has given rise to the idea of "Smart Luxury," which transcends traditional notions of luxury by incorporating sustainability, personalization, and technological innovation (Tarnanidis & Manaf, 2024). Smart Luxury is increasingly recognized as a distinctive trend, appealing to a new generation of consumers who are shaping the future of luxury travel, design, and fashion (Joy et al., 2022).

Smart Luxury is now a defining trend not only among consumers but also brands and investors, who are drawn to its focus on sustainability and long-term financial and environmental benefits. This movement has particularly influenced the hospitality industry, which is reshaping the expectations of high-end travelers. Modern luxury travelers are seeking personalized, smarter, and more sustainable experiences, moving away from traditional opulence (Prevolsek & Golja, 2024).

Hotels and resorts embracing Smart Luxury now offer tailored, technology-enhanced, and eco-conscious experiences. The design philosophy emphasizes intelligent integration of technology and sustainability, with properties offering minimalist, functional spaces that reflect modern luxury aesthetics. These properties aim to provide seamless, intuitive experiences through smart technologies, such as AI-powered concierge services, and sustainability-focused designs using eco-friendly materials and energy-efficient systems (Santos et al., 2024). Notably, brands like Hilton and Marriott are leading this transformation by introducing brands like "Canopy" and "EDITION" that prioritize personalized luxury and sustainability (Sanli, 2024; Cain et al., 2024).

Additionally, the appeal of smart luxury extends beyond accommodations to products like luxury smartwatches and vehicles, which integrate advanced features such as health tracking, autonomous driving, and connectivity. These products not only elevate convenience but also cater to a lifestyle that blends sophistication with practicality, making them integral to the future of luxury consumption (Cain et al., 2024). As Smart Luxury continues to evolve, it reflects a broader shift in consumer preferences toward technology-driven, personalized, and responsible luxury experiences, offering a new paradigm in the luxury sector that merges tradition with modernity.

CO-BRANDING SMART LUXURY PRODUCTS

Co-branding of smart luxury products represents a growing trend where luxury brands collaborate with technology companies to combine the sophistication and exclusivity of luxury with the innovation and convenience of smart technology (Catoire, 2024). This partnership allows both brands to tap into the evolving consumer demand for high-tech, stylish products that offer both functionality and prestige. Smart luxury products often include items like connected watches, high-end headphones, smart home devices, and even automobiles that integrate cutting-edge technology with traditional luxury aesthetics. These co-branded collaborations aim to cater to a niche market that values both the status associated with luxury brands and the technological advancements that make life more convenient (Zancanella et al, 2024).

One of the most notable examples of co-branding in the smart luxury product market is the collaboration between luxury watchmakers, like TAG Heuer, and technology companies like Google and Intel (Minkkinen, 2024). The role of co-branding in brand loyalty. This partnership led to the creation of the TAG Heuer Connected watch, which combines the luxury watchmaking expertise of TAG Heuer with the smart features powered by Google's Android Wear platform. The result is a high-end smartwatch that retains the craftsmanship and design integrity of a traditional luxury watch while offering consumers the benefits of connectivity, fitness tracking, and app integration. This fusion appeals to affluent consumers who are not only looking for a luxurious product but also desire the convenience and functionality of smart features.

Similarly, the co-branding between high-end automobile brands and technology companies has produced smart, luxury cars that integrate the latest in automotive technology. Brands like Tesla and BMW have partnered with companies such as Apple to incorporate smart technology features like Apple CarPlay into their vehicles. This integration allows luxury car owners to seamlessly connect their smartphones to their vehicles, providing enhanced entertainment, navigation, and connectivity features. Such collaborations not only increase the desirability of the product but also enhance the overall user experience by combining the best of both worlds—luxury performance and cutting-edge technology (Arcidiacono, 2024) (Bohm et al, 2024).

Co-branding in smart luxury products also allows for expanded market reach. Luxury brands traditionally appeal to an affluent demographic, but by collaborating with tech companies, they can attract a younger, more tech-savvy audience (Górska-Warsewicz, 2024). Tech companies, on the other hand, gain access to the prestige and credibility of luxury brands, helping them expand into higher-end markets where consumers are willing to pay a premium for exclusive, high-quality products. For instance, when a luxury fashion house collaborates with a tech giant to create a limited-edition smart device, it creates a sense of exclusivity and desirability that drives demand and allows both brands to reach new consumers.

However, co-branding of smart luxury products also comes with challenges (Dong & Huanug, 2024). The integration of technology with luxury must be done with great care to ensure that the product remains true to the brand's heritage while offering innovative features that resonate with the target audience. Luxury consumers value craftsmanship, exclusivity, and design, so the introduction of technology should not overshadow the aesthetic or premium experience. Moreover, the technology must work flawlessly to meet the high standards expected from luxury products. Any malfunction or poor user experience could damage the brand's reputation and consumer trust (Singh & Shukla, 2024).

Co-branding smart luxury products allows luxury brands and tech companies to merge innovation with exclusivity, creating products that appeal to the growing demand for both status and functionality (Dallabona, 2024). By carefully aligning brand values, co-branding partners can offer a unique proposition that attracts a broad yet discerning customer base. When executed well, this collaboration can result in a product that stands out in the marketplace, offering consumers the best of both worlds—luxury and technology.

CO-BRANDING IN THE INDIAN CONTEXT

Co-branding smart luxury products in India is an emerging trend where high-end brands collaborate with technology companies to merge traditional luxury with advanced technology. With a growing affluent consumer base that values prestige and innovation, Indian consumers are increasingly seeking products that seamlessly integrate style, performance, and exclusivity (Chaurasia et al., 2024; Shahid et al., 2025).

Examples include luxury watch brands partnering with tech firms to produce smartwatches that offer fitness tracking and notifications while maintaining elegance. Similarly, luxury automobile brands like Mercedes-Benz and BMW are incorporating smart features such as advanced infotainment systems and voice control, appealing to tech-savvy consumers (Rodrigues et al., 2025).

The convergence of luxury and technology is evident in the fashion industry as well, with collaborations yielding smart textiles and accessories that combine sophisticated design with functionality. This trend caters to Indian consumers who prioritize aesthetics alongside practicality.

However, challenges remain, including meeting high expectations for quality and ensuring that technological integration does not compromise the essence of luxury (Chaurasia et al., 2024; Shahid et al., 2025). If executed well, co-branding can redefine the Indian luxury market, creating unique products that resonate with modern, affluent consumers willing to invest in innovative luxury offerings.

EVOLUTION OF LUXURY FASHION COLLABORATIONS

Elsa Schiaparelli and Salvador Dalí had an iconic 1930s-40s partnership in the 1930s and 40s (Chala, 2018). After a period of limited collaborations in the 1960s-80s, including Emilio Pucci's work with Ermenegildo Zegna and Karl Lagerfeld's initial association with Chanel, the 1990s marked the beginning of modern, consistent luxury collaborations. Nike's partnership with the Japanese venture Comme des Garçons demonstrated the power of blending high fashion with sportswear. The success of these co-branding efforts has led luxury brands to integrate them into their annual marketing strategies. There is greater awareness now about their commercial value and consumer appeal (Michel & Willing, 2020).

DEVISING A CO-BRANDING STRATEGY FOR LUXURY PRODUCTS: PRECAUTIONS

When creating a co-branding strategy for luxury products, careful planning is essential to ensure that the collaboration enhances both brands' images while aligning with their values. Luxury brands, built on exclusivity, quality, and prestige, must maintain these core attributes in any co-branding initiative while leveraging the strengths of the partnership. Below are key precautions to consider when developing a co-branding strategy for luxury products (Shan, Lu, & Cui, 2022).

The first step is ensuring that both brands share similar values, goals, and target audiences. A mismatch in brand positioning can confuse consumers and potentially damage the luxury brand's perception. For instance, if a luxury brand partners with a mass-market or inexpensive brand, it risks diluting its image and alienating its high-end clientele. Therefore, the co-branding partner should complement the luxury brand's reputation by reflecting qualities such as exclusivity, craftsmanship, and sophistication (Górska-Warsewicz, 2024).

Exclusivity is a cornerstone of luxury branding, and consumers often choose luxury products for their rarity and status. When planning a co-branding strategy, it's vital to ensure that the collaboration does not erode either brand's sense of exclusivity. Limiting availability through limited editions, selective distribution, or higher price points can help maintain the premium perception. Broad distribution channels should be avoided, as making the product too accessible can diminish its luxury appeal (Wang, 2023).

High quality is a fundamental aspect of luxury products. When entering a co-branding partnership, both brands must align their quality standards to ensure that the end product meets the expected level of excellence. If one partner has a weaker reputation for quality, or if the technology in a smart luxury product is subpar, the reputation of both brands may be compromised. Rigorous quality control and attention to detail are essential throughout the co-branding process to preserve the luxury brand's prestigious image (Mars, 2023).

Luxury consumers are highly discerning and sensitive to brand associations. Any misstep in co-branding could result in a loss of trust, which is difficult to rebuild. Before launching a co-branded product, market research, focus groups, or surveys should be conducted to gauge consumer perception. The collaboration must enhance the luxury experience rather than detract from it. The marketing message should emphasize the added value of the partnership, focusing on benefits rather than mass appeal or technology, which could undermine the luxury status (Mars, 2023).

In a co-branding partnership, it's important to maintain clear and consistent branding that highlights the strengths of both brands. The luxury brand should remain the focal point, ensuring that the collaboration does not confuse consumers or weaken its identity. There should be a clear synergy between the two brands, with careful balancing of logos, design aesthetics, and messaging. Clear guidelines for brand element usage should be established to avoid over-exposure or brand confusion (Shan, Lu, & Cui, 2022).

A co-branding partnership should not be viewed as a short-term promotional tactic. For luxury products, co-branding can be a long-term strategy that enhances both brands over time. It's important to plan for sustainability, ensuring that the partnership remains relevant and effective in the long run. This could involve evolving the product offering with each season or exploring new markets and opportunities to expand the collaboration. A one-off collaboration might generate initial excitement, but a well-thought-out, long-term strategy can help establish a lasting relationship between the brands and keep consumers engaged (Abedsoltan, Taleizadeh, & Sarker, 2022).

Luxury brands are known for their rich heritage, craftsmanship, and tradition. When incorporating technology into co-branded luxury products, it's vital to innovate while respecting the brand's traditional values. For instance, luxury watchmakers venturing into smartwatches must ensure that the innovation aligns with the craftsmanship, design, and aesthetics of the brand's heritage. The technology should enhance the product without overshadowing the core identity of the luxury brand. This delicate balance between tradition and innovation is crucial for maintaining authenticity (Yu, Rothenberg, & Moore, 2021).

Luxury customers have high expectations for product quality and the overall experience. A co-branded luxury product should not only meet but exceed these expectations. Every consumer touchpoint, from packaging to customer service, must reflect the product's luxurious nature. Offering personalized services, exclusive benefits, or unique experiences can enhance the perception of both brands. Additionally, a partnership between luxury brands and co-branding partners requires clear legal agreements to protect intellectual property, brand assets, and financial terms. This includes defining roles and responsibilities related to product development, marketing, distribution, and sales, ensuring both brands are safeguarded against asset misuse (Xue & Chawdhary, 2023).

Especially in global markets, it is crucial to consider the cultural and social implications of a co-branding strategy. Different regions may have varying attitudes toward luxury and technology. In the Indian market, for example, luxury products are often linked to heritage and prestige, while technology might represent modernity and innovation. The co-branding strategy must align with cultural perceptions and consumer behaviors in specific regions to ensure that it resonates with the local audience without compromising the integrity of the luxury brand (Jia & Ran, 2021).

Co-branding luxury products requires careful consideration of brand alignment, exclusivity, quality control, and consumer perception. By taking these precautions, luxury brands can successfully partner with technology companies or other luxury players to create innovative, high-end products that cater to the modern, affluent consumer. When done right, co-branding can enhance brand equity, increase market reach, and elevate the consumer experience, ultimately benefiting both brands in the long run (Murtas, Pedeliento, Mangiò, & Andreini, 2022).

THE STRATEGIC POWER OF SMART BRAND COLLABORATIONS

Smart brand collaborations harness the strengths of multiple entities to create innovative offerings that captivate consumers. By merging distinct brand identities, these partnerships enhance visibility and appeal across diverse markets. They foster creativity, drive engagement, and unlock new revenue streams while sharing resources and expertise. Ultimately, such collaborations amplify brand loyalty and elevate consumer experiences, making them a powerful strategic tool in today's competitive landscape (Refer Table 1)

Table 1. Smart Brand Collaborations

Factor	Details
Brand Alliances / Co-Branding	Strategic tools used by luxury brands to enhance brand equity and create unique value propositions.
Benefits of Co-Branding	Extends reach, accesses new markets, and reinforces brand image. Collaboration with brands specializing in technology, art, or design. Creates innovation, exclusivity, and desirability.
Co-Branding in the Fashion Industry	Increases annual revenues by up to 30%. Brands share resources, expertise, and marketing budgets to develop unique products. Expands customer segments and strengthens loyalty.
Successful Co-Branding Execution	Helps tap into new markets, create desirable products, and generate buzz. Exclusivity and limited availability increase appeal, particularly among younger consumers.
Alignment with Core Values	Partnerships must align with the brand's core values and strategic goals to preserve exclusivity and prestige.
Effective Co-Branding Example	Hotels partnering with high-end cosmetic brands to offer premium amenities. Reinforces brand strengths and deepens connection with clientele.
Co-Branding Benefits for Smaller Brands	Taps into each other's customer bases and reaches new demographics. Shared costs for marketing and product development. Enhances trust and generates anticipation and media coverage.
Strategic Alignment in Co-Branding	Clear strategic alignment ensures shared values, target audiences, and objectives. Provides mutual benefits, expands reach, and enhances brand equity.
Collaboration Outcomes	Fosters innovation and creativity, leading to unique products, services, or experiences. Elevates brand image, positioning, and attracts new customers.
Long-Term Brand Building	Smart collaborations contribute to long-term brand building and sustainable growth.
Maintaining Authenticity and Credibility	Ensure collaboration feels genuine and resonates with consumers. Clear, consistent communication aligned with both brands' identities.

Source: Author

MAXIMIZING BRAND EQUITY THROUGH STRATEGIC CO-BRANDING

Leuthesser, Kohli, and Suri (2003) present a framework to emphasize the importance of complementarity between parent brands and their target markets in determining the success of co-branding initiatives. The authors identify two types of complementarities: core complementarity, where both brands contribute essential benefits to the product, and extended complementarity, where one brand primarily lends its name. They differentiate between two types of co-branding: core co-branding, where both brands contribute core competencies to create innovative products, and image co-branding, where brands collaborate to create a product associated with a specific image or lifestyle. The framework highlights how co-branding can also provide access to new markets by combining brands with distinct market franchises. While core co-branding fosters innovation, image co-branding is effective in building a strong brand identity. The authors conclude that successful co-branding can enhance brand equity, increase market reach, and introduce innovative products, but it requires careful consideration of the associated risks and benefits.

Abedsoltan, Taleizadeh, and Sarker (2022) explore co-branding and collaboration strategies for remanufacturing luxury products, aiming to optimize production, expand market share, and maintain brand integrity. They suggest that luxury brands collaborate with companies like fast fashion brands to overcome limited market presence. The study stresses using distinct brand names for remanufactured items to prevent customer confusion and protect the luxury brand's image. Additionally, it highlights the importance of eco-friendly practices to attract environmentally conscious consumers, enhancing market share and brand loyalty while preserving integrity.

Yu, Rothenberg, and Moore (2021) explore factors influencing young consumers' preferences in luxury co-branding, focusing on brand combinations, retail channels, exclusivity, and price. Their research finds that collaborations between luxury and sportswear brands, when priced similarly to sportswear, are highly favored, provided they are perceived as exclusive. They suggest that luxury brands should partner with retail formats aligned with their strategy, preferring sportswear and premium streetwear over fast fashion. The study concludes that while uniqueness is important, its impact is limited without ensuring exclusivity.

Jia and Ran (2021) analyze the strategic motivations behind the Louis Vuitton (LV) and Supreme collaboration, emphasizing how such partnerships enhance brand influence and expand market reach in the luxury goods sector. They identify several driving factors for the collaboration, including boosting sales, appealing to younger consumers, and broadening market presence. The study underscores the importance of innovative product design, coordinated marketing efforts, and strategic pricing adjustments as key elements contributing to the success of luxury brand collaborations.

Paydas Turan (2021) conducts a meta-analysis of co-branding research to identify key success drivers. The study reveals that the relationship between partner brands is more crucial than individual brand characteristics, with brand image fit being more significant than product category fit or brand equity. The findings are applicable across industries and co-branding strategies, enhancing the theoretical understanding of co-branding by resolving conflicts in previous studies and providing a future research agenda.

Kania, Sukoyo, and Wibisono (2021) explore co-branding strategies between less complementary sectors, focusing on an Indonesian cosmetic brand collaborating with food brands. Their study finds that the perceived brand fit plays a crucial mediating role in influencing consumer attitudes and purchase intentions, even when attitudes towards the co-branded food brand do not directly affect attitudes towards cosmetics. The study emphasizes the "spill-over effect," where a well-known co-brand enhances the appeal of the lesser-known brand.

Ferraz, Nobre, and Barbosa (2021) explore a co-branding strategy between a luxury hotel and premium wine brands, focusing on benefits, reasons, and selection attributes from managers' perspectives. They identify that enhancing brand image is a key advantage of such collaborations and propose hypotheses for future research. The study offers insights into how co-branding can elevate brand value and reputation in the high-end market.

Wu (2022) explores Gucci's digital marketing strategies for engaging young consumers in the new media environment. The study highlights how Gucci uses social platforms to communicate its brand values and engage with young luxury consumers, addressing the challenges luxury brands face in adapting their marketing strategies to the digital landscape. The research suggests that luxury brands should target key demographics and tailor digital marketing approaches to diverse cultural backgrounds to strengthen their brand connection with young consumers.

Lin (2023) examines the impact of Intellectual Property (IP) co-branding strategies on consumer purchase intentions, using MINISO as a case study. The research reveals that MINISO has effectively leveraged IP co-branding to expand its market presence in China, positively influencing consumer behavior and driving sales. The study uses the 4Ps of Marketing Mix and SWOT analysis to assess how IP marketing strategies boost consumer engagement and offers recommendations for optimizing IP co-branding efforts.

Sun, Cryder, and Rick (2023) investigate why consumers tend to underuse co-branded credit cards outside of the featured brand. The study identifies two key mechanisms behind this behavior: consumers assume the reward structure is tied to the featured brand and view purchases outside the brand as a "bad fit." The findings suggest that this reluctance could harm both consumers and issuers by limiting the benefits of co-branded cards.

Shan, Lu, and Cui (2022) explore how co-branding can enhance brand "masstige," or "prestige for the masses." They focus on the interaction between brand personality fit and product category fit, finding that these factors significantly influence consumers' perceptions of masstige in co-branded products.

Catoire (2024) examines the impact of fast-fashion co-branding collaborations on luxury fashion brands. The study finds that while such partnerships can increase accessibility and boost short-term sales, they risk diluting the luxury image of the brand. The success of these collaborations depends on brand compatibility, with consumer involvement and brand experience moderating the outcome. The research suggests that luxury brands should prioritize collaborations with other luxury brands to preserve exclusivity and strengthen consumer perceptions of luxury.

Al-Issa, Kwiatek, and Dens (2024) explore the profiles, perceived luxury values, and purchase intentions of masstige buyers in their study of 601 UK consumers. They identify four distinct consumer segments: Premium Luxury Buyers, Masstige Buyers, Buyers of Both Premium Luxury and Masstige, and Non-luxury Buyers. The study provides a value-based positioning framework for brands to align with their target audiences' core values and better position themselves in the market.

Liu (2024) examines the marketing strategy and effectiveness of sportswear brand co-branding with entertainment celebrities, focusing on Nike's collaborations. The study identifies five main types of collaboration targets for Nike and analyzes its approach to selecting partners, developing co-branded products, and promoting them. The research offers insights into how companies can leverage celebrity partnerships to enhance brand visibility and appeal to a broader consumer base.

Barrenechea (2025) explores the Omega-Swatch "MoonSwatch" collaboration, which successfully blends Omega's luxury heritage with Swatch's mass-market appeal. The study identifies key factors contributing to the collaboration's success: brand compatibility, strategic storytelling, and value co-creation. It highlights the potential of co-branding to create competitive differentiation and value for brands with contrasting market positions while offering managerial insights on how to navigate co-branding partnerships.

Chen (2024) investigates the co-branding strategy between Heytea and Fendi, which significantly boosted sales and popularity in May 2023. The study attributes this success to value strategy, variation strategy, and vibration strategy, with key factors including enhancing product value, offering a variety of products, and creating excitement. The study provides recommendations for other companies looking to replicate this success through co-branding partnerships.

Górska-Warsewicz (2024) explores the relationship between entrepreneurial orientation (EO), innovative co-branding partnerships, and business performance. The study finds that innovativeness, proactiveness, and competitive aggressiveness positively influence business performance, and that innovative co-branding partnerships enhance business outcomes. It provides practical recommendations for managers seeking to enhance their business performance through strategic co-branding.

CO-BRANDING IN THE LUXURY INDUSTRY: OPPORTUNITIES, CHALLENGES AND RISKS

Co-branding in the luxury industry presents opportunities for enhanced brand visibility and access to new consumer segments but also poses challenges such as maintaining brand integrity and aligning values. The risks include potential dilution of brand identity and managing consumer perceptions. Successful collaborations require careful strategic planning and alignment of both brands' goals and aesthetics. Table 2 presents more details.

Table 2. Opportunities and Risks

Aspects	Details
Co-Branding Strategies	**Core Co-Branding:** Both brands contribute core competencies to foster innovation. **Image Co-Branding:** Focuses on strengthening the brand's overall identity.
Market Reach & Brand Integrity	Co-branding with larger or mainstream brands, like fast fashion, helps expand reach. Must maintain brand integrity, especially with environmentally conscious or broader demographics.
Innovation & Exclusivity	Balancing innovation with exclusivity is crucial for preserving luxury appeal. Some co-branded products are designed to meet the demand for eco-friendly options.
Consumer Preferences	Younger demographics value exclusivity, pricing, and brand combinations. Co-branding partnerships in luxury and sportswear work well if they maintain exclusivity while offering perceived value similar to affordable brands.
High-Profile Collaborations	Driven by innovative design and targeted marketing. Expands market presence and appeals to younger, trend-conscious consumers.
Co-Branding with Celebrities	Boosts visibility and builds consumer loyalty.
Opportunities of Co-Branding	Expands market reach, fosters innovation, and enhances brand image. Attracts new customer segments and creates unique products that differentiate brands in a competitive market.
Risks of Co-Branding	

continued on following page

Table 2. Continued

Aspects	Details
Brand Dilution Risk	Occurs when values, target audience, or brand image don't align. Non-luxury or misaligned brands can weaken exclusivity, leading to confusion and loss of trust.
Exclusivity Risk	Co-branded products seen as too widely available or mass-market can undermine the luxury brand's image. Overexposure risks diminishing the luxury brand's special appeal.
Guilt by Association Risk	Negative publicity or ethical issues from one brand can damage both brands' reputations, especially for luxury brands dependent on their image.
Operational Conflicts	Co-branding requires shared decision-making on product development, marketing, and distribution. Differences in values, cultural sensitivities, or target audiences may cause friction, weakening the collaboration's effectiveness.
Financial & Operational Risks	Significant investments in co-branding can result in substantial financial losses if expectations aren't met. Coordinating production, supply chains, and marketing across multiple brands is complex and prone to disruption.
Quality Control & Legal Considerations	Ensuring consistent quality standards is vital for maintaining the luxury brand's craftsmanship reputation. Legal planning is required to safeguard intellectual property rights, trademark usage, and liability.
Strategic Alignment	Both brands must share values, target audiences, and brand image that align with the luxury brand's identity. Collaboration must feel authentic to consumers; inauthentic partnerships can damage brand equity.
Innovation Management	Co-branding can foster innovation but requires careful management. Shared control over product development and marketing ensures a cohesive vision and customer experience. Coordination of operations to avoid disruptions is key.
Overexposure Caution	Luxury brands must avoid overexposure to maintain exclusivity. Collaborations should create genuine value for consumers.
Co-Branded Product Quality	Products that lack innovation or fail to meet high standards can negatively impact both brands' equity.

Source: Author

CHALLENGES IN CO-BRANDING SMART LUXURY PRODUCTS

Co-branding smart luxury products presents certain challenges, including the need for seamless integration of technology and design, as well as ensuring a shared vision between collaborating brands. These challenges must be addressed through clear communication and strategic alignment to fully capitalize on co-branding efforts and create compelling offerings that resonate with discerning consumers. Table 3 presents these challenges.

Table 3. Challenges in Co-Branding Smart Luxury Products

	Description
Maintaining Brand Integrity	Luxury brands focus on exclusivity, craftsmanship, and heritage. Integrating advanced technology into products must enhance, not overshadow, the brand's core identity. For example, a smartwatch should embody the same level of elegance and craftsmanship as a traditional luxury timepiece.
Aligning Target Audiences	Luxury brands cater to an exclusive clientele, while tech companies often target a broader, mass-market audience. Mismatched target audiences can confuse consumers and dilute the luxury brand's exclusive image, damaging its reputation.
Balancing Technology with Aesthetics	Technological features, while desirable, may clash with the traditional aesthetics and values of luxury. Merging innovation with design without compromising luxury's traditional elements presents a challenge in co-branding.
Quality Control	Luxury consumers expect the highest quality. Any failure in the technology component, such as a malfunctioning smartwatch or unreliable features, can harm the reputation of both brands. Rigorous quality checks and seamless integration are essential.
Legal and Intellectual Property Concerns	Co-branding with tech companies requires careful agreements on patents, trademarks, and proprietary technologies. Legal and intellectual property issues can complicate the partnership and lead to potential conflicts.
Sustainability Issues	As sustainability becomes more important, tech companies must ensure their smart luxury products align with eco-conscious trends. Luxury consumers increasingly demand eco-friendly products, adding complexity to co-branding initiatives.

Source: Author

SUCCESSFUL FASHION COLLABORATION BETWEEN LUXURY BRANDS

A successful collaboration between luxury brands requires careful planning and alignment of core values. This strategic partnership creates synergy, which elevates both brands. For the collaboration to be authentic, it needs to resonate with consumers, generating excitement and desirability. Limited-edition offerings and innovative designs, often launched at high-profile events, help to capture attention. Pricing and distribution must balance exclusivity with wide enough reach to maximize the appeal of smart luxury products.

Luxury branding also revolves around creating an aura of exclusivity. Scarcity, limited production, rare materials, and high pricing all contribute to the brand's luxury appeal. Luxury brands emphasize their heritage and craftsmanship, evoking a sense of timelessness. Additionally, they position their products as symbols of success and aspiration. Collaborations with artists are increasingly common, merging high-end fashion with fine art to create exclusive experiences.

To build a successful luxury brand, key elements include:

1. **Exclusivity** - Achieved through limited availability.
2. **High Quality & Craftsmanship** - Justifying premium pricing.
3. **Compelling Brand Story** - Connecting emotionally with the target audience.
4. **Distinctive Visual Identity** - Making the brand easily recognizable.
5. **Exceptional Customer Service** - Enhancing the luxury experience and fostering loyalty.
6. **Consistent Brand Image** - Maintaining the same identity across all aspects.

SMART LUXURY PRODUCTS AND CO-BRANDING

Smart luxury represents a new trend that merges high-end craftsmanship with cutting-edge technology. It's no longer just about luxury as a symbol of wealth and excess; modern consumers now value sustainability, practicality, and innovation. The term "Smart Luxury" reflects this shift, as it emphasizes not just material wealth but intelligent, thoughtful choices.

Today's luxury consumers are more informed and tech-savvy. They care about sustainability and ethical practices, and their demand for innovative products has driven the rise of smart luxury. For example, in the hospitality industry, brands like Hilton's Conrad and Canopy have incorporated smart technology to create more personalized and sustainable experiences. These hotels feature AI-powered services, smart room controls, and modern minimalist designs, catering to consumers seeking tailored, meaningful experiences.

Co-branding has become a powerful tool in the smart luxury sector. Luxury brands collaborate with tech companies to integrate advanced features while maintaining their iconic style and exclusivity. Partnerships like the one between Apple and Hermès for the Apple Watch Hermès showcase how co-branding can combine technological innovation with luxury. Similarly, car brands like Mercedes-Benz have partnered with Google to create vehicles that feature smart technology, such as autonomous driving and advanced infotainment systems.

Key Steps for Successful Smart Luxury Co-Branding

1. **Strategic Alignment**: Ensure both brands share similar values and target audiences.
2. **Innovation Balance**: Maintain luxury appeal while integrating advanced technology.
3. **Exclusive Offerings**: Create products that are both innovative and scarce, preserving the luxury brand's exclusivity.

4. **Smart Product Design**: Ensure that smart features do not overpower the brand's core craftsmanship.
5. **Legal & Ethical Considerations**: Establish clear agreements on intellectual property and brand usage.
6. **Sustainability Focus**: Align co-branded products with sustainable practices to appeal to eco-conscious consumers.
7. **Targeting the Right Demographics**: Focus on tech-savvy, modern consumers who value both innovation and luxury.

DISCUSSION & CONCLUSION

Co-branding has become a prevalent strategy in the luxury industry, offering numerous opportunities for growth and innovation. It allows luxury brands to expand their reach, create innovative products, and strengthen brand equity. However, co-branding also presents significant challenges and risks that must be carefully navigated to ensure the preservation of brand equity and customer trust.

Successful co-branding strategies in the luxury sector hinge on a deep understanding of the psychological factors that drive consumer responses. The personalities of collaborating brands influence consumer perceptions and engagement, highlighting the necessity of strong alignment between these personalities. Luxury brands can strategically partner with other brands to create synergistic relationships that enhance brand equity and offer unique value to consumers. Co-branding offers luxury brands the chance to expand their reach, create innovative products, and foster brand loyalty. By managing risks such as brand dilution, reputational damage, and logistical challenges, luxury brands can strengthen their market position.

Co-branding, while offering numerous opportunities, comes with several significant risks that must be carefully managed. One of the primary risks is brand dilution. Luxury brands rely on exclusivity and uniqueness to maintain their prestige, and collaborating with mass-market or non-luxury brands can tarnish this carefully cultivated image. To mitigate this risk, luxury brands must select co-branding partners whose values and brand images align with their own. Reputational risk is another major concern.

Negative publicity or ethical controversies involving one brand can quickly spread to damage the reputation of all parties involved in the co-branding venture. To minimize this risk, thorough due diligence is essential, as well as clear contractual agreements that outline the responsibilities and expectations of each partner. Consumer confusion is also a significant risk when it comes to co-branding. If the co-branded products or campaigns are not communicated clearly, it can lead to misunderstandings among consumers, resulting in a dilution of the brand's core message. To avoid this, maintaining a consistent brand identity across all collaborations is crucial. Ensuring clarity in communication will help maintain consumer trust and prevent any confusion about the brand's values and offerings.

Operational complexities are another challenge that comes with co-branding. These ventures require careful coordination and alignment between partners, and disagreements over creative direction, marketing strategies, or operational processes can arise. If these issues are not effectively addressed, it could lead to delays, inefficiencies, or even the termination of the partnership. The success of a co-branding venture depends on the ability of both partners to work together smoothly and harmoniously.

Financial risks must also be considered when engaging in co-branding. Collaborations often require significant investments, and if the partnership fails to resonate with consumers, the resulting financial losses can be substantial. Therefore, thorough market research and financial planning are necessary to ensure that the collaboration aligns with consumer preferences and is likely to succeed in the marketplace.

Another potential risk is over-reliance on co-branding. While collaborations can be beneficial, excessive dependence on them for growth can hinder a brand's ability to innovate and create unique value propositions independently. Maintaining a balance between leveraging co-branding opportunities and continuing to innovate on one's own is crucial for the long-term viability of the brand.

To mitigate these risks and ensure the success of co-branding ventures, luxury brands must adopt a strategic and selective approach. Choosing the right partners is crucial; they should share similar values, image, and target audiences to create an authentic and impactful collaboration. Clear communication is essential to avoid confusion and maintain brand consistency. Additionally, conducting thorough market research ensures that the co-branded venture resonates with consumers. Comprehensive agreements that outline the responsibilities and risks associated with the collaboration must be drafted to safeguard both parties. Finally, it's important to strike a balance between co-branding and innovation, ensuring that the brand remains capable of creating unique value propositions independently.

Co-branding is an increasingly popular strategy in the luxury industry, particularly in China, where it helps luxury brands connect with younger, digitally-savvy consumers and enhance brand loyalty. This often involves collaborations with brands from other sectors, such as coffee and tea, which benefit from the enhanced reputation associated with luxury brands.

However, this strategy also presents challenges. Some consumers may view the high cost of co-branded items as wasteful, while others see them as a more accessible way to own a luxury product or express their identity. There is also a risk of encouraging blind consumption driven by social pressure, leading to irresponsible spending. Therefore, consumers need to find a balance between personal choice and social responsibility when purchasing luxury goods.

Co-branding is a powerful tool for luxury brands, providing opportunities for growth and innovation. However, it must be approached with caution and careful consideration. By strategically selecting partners, ensuring clear communication, and managing potential risks such as brand dilution and reputational damage, luxury brands can leverage co-branding to enhance their brand equity and create unique value propositions for their customers.

SMART LUXURY PRODUCTS – THE FUTURE OF HIGH-END INNOVATION

Smart luxury products represent the intersection of opulence and cutting-edge technology, redefining the future of high-end innovation. These items not only showcase exquisite craftsmanship but also integrate advanced features such as connectivity, personalization, and enhanced functionality. As affluent consumers seek unique experiences, smart luxury products offer a seamless blend of aesthetics and practicality, elevating everyday life. Brands that embrace this trend are not only enhancing their value proposition but also fostering deeper emotional connections with consumers. By prioritizing sustainability and ethical practices, the evolution of smart luxury products reflects a growing demand for responsible consumption in the luxury market (Refer Table 4)

Table 4. Characteristics of Smart Luxury Products

DEFINITION	**Smart Luxury: Combining high-end craftsmanship with cutting-edge technology to create practical, sophisticated, and personalized experiences.**
KEY FEATURES	**Innovation + Elegance**: Merging high-tech with traditional luxury aesthetics. **Sustainability**: Eco-friendly and ethical practices. **Personalization**: Tailored experiences, tech customization, unique features.
CATEGORIES OF SMART LUXURY	**Smart Wearables**: Ex. Apple Watch Hermès (Luxury + Technology). **Smart Vehicles**: Ex. Mercedes-Benz x Google (Autonomous driving, infotainment). **Smart Hotels**: Ex. Hilton's Conrad and Canopy Hotels (AI concierge, smart room controls). **Smart Fashion**: Fashion brands integrating tech, such as smart fabrics or wearables.

continued on following page

Table 4. Continued

CONSUMER TRENDS DRIVING SMART LUXURY	**Tech-Savvy Consumers**: Young, modern buyers craving high-tech, exclusive products. **Sustainability Concerns**: Eco-conscious consumers. **Customization**: Desire for personalized and tailored experiences.
BENEFITS OF SMART LUXURY	**Functionality Meets Elegance**: Advanced tech, no compromise on luxury aesthetics. **Exclusive Experiences**: Products and services offering unique experiences. **Enhanced Brand Prestige**: Maintaining exclusivity with innovation.
CHALLENGES IN SMART LUXURY CO-BRANDING	**Brand Dilution**: Ensuring technology doesn't overshadow craftsmanship. **Consumer Perception**: Balancing tech with exclusivity. **Operational Complexities**: Managing production, marketing, quality across industries.
KEY EXAMPLES OF SMART LUXURY PARTNERSHIPS	**Apple & Hermès**: Apple Watch Hermès combining fashion and technology. **Mercedes-Benz & Google**: Smart cars with advanced features. **Hilton & Canopy Hotels**: Smart rooms, AI concierge.
FUTURE OF SMART LUXURY	**Growing Market**: Rising demand among younger generations. **Increased Collaboration**: More partnerships between tech and luxury brands. **Seamless Integration**: Combining the best of luxury and tech for a unique consumer experience.

Source: Author

FUTURE RESEARCH AREAS

It will be interesting to investigate the psychological factors driving consumer behavior when they buy co-branded luxury products. Luxury markets are spread across the globe. The cultural nuances and regional differences in co-branding effectiveness can lead to a more comprehensive and contextual understanding regarding the reach of co-branded luxury products. Studying the long-term impacts of co-branding strategies on brand equity, consumer loyalty, and market dynamics would be beneficial for sustainable brand development. Given that luxury items often represent trends, it's worth investigating the long-term evolution and effectiveness of co-branding strategies. Further research could delve deeper into the comprehensive effects of co-branding policies on firms, consumers, and the economy, especially in light of changing consumer behaviors and economic conditions.

REFERENCES

Here are the provided references formatted in APA 7th edition style:Abedsoltan, H., Taleizadeh, A. A., & Sarker, R. B. (2022). Optimal production of remanufactured products with collaboration and co-branding of luxury brands. *Computers & Industrial Engineering*, *172*, 108533. DOI: 10.1016/j.cie.2022.108533

Ahmed, A., & Ahmed, M. (2024). Strategic alliances and partnerships: Expanding brand reach and influence. Strategies for Brand Communications and Management, 257.

Al-Issa, N., Kwiatek, P., & Dens, N. (2024). Masstige buyers: Profile, perceived luxury values and purchase intentions. *International Journal of Consumer Studies*, *48*(1), e13005. DOI: 10.1111/ijcs.13005

Arcidiacono, D. (2024). Industry 4.0 and platforms: The new convergence. In Disruptive digitalisation and platforms (pp. 185-199). Routledge.

Barrenechea, F. (2025). Co-branding as a value-creating strategy: The MoonSwatch case study of Omega and Swatch. Available at *SSRN*, 5134152. DOI: 10.2139/ssrn.5134152

Böhm, N., Watkowski, L. M. L., & Buck, C. (2024). Establishing platform-based ecosystems: How automotive manufacturers adapt their value creation to a digital end. In *Handbook on digital platforms and business ecosystems in manufacturing* (pp. 147–163). Edward Elgar Publishing. DOI: 10.4337/9781035301003.00016

Cain, L. N., Mistry, T. G., Douglas, S., Rahman, I., & Moreo, A. (2024). Luxury hotel technology trends: A multigenerational analysis. *International Journal of Contemporary Hospitality Management*, *36*(12), 4031–4054. DOI: 10.1108/IJCHM-10-2023-1597

Carlotto, F. (2024). *Luxury brand and art collaborations: Postmodern consumer culture*. Taylor & Francis. DOI: 10.4324/9781003274094

Catoire, P. (2024). *The feedback effect of co-branding collaborations on perceived brand luxury and attitude toward luxury fashion brands: Repercussions of the "fast-fashion co-branding" trend*. Louvain School of Management.

Chala, V. (2018). Surrealism in the design of clothes: Elsa Schiaparelli and Salvador Dali (Doctoral dissertation).

Chaurasia, M., Kumar, A., & Panda, R. K. (2024). Two decades of research on "masstige" marketing: A systematic literature review and future research agenda. *International Journal of Consumer Studies*, *48*(5), e13077. DOI: 10.1111/ijcs.13077

Chen, L. (2024). Using marketing mix theory to analyze the successful reasons for Heytea's co-branding strategy. Advances in Economics. *Management and Political Sciences*, *110*, 209–216.

Chen, X., & Li, Z. (2022). Investigating the consumer evaluation of the co-branding of luxury brands. *International Journal of Research in Business & Social Science*, *11*(4), 1–14. DOI: 10.20525/ijrbs.v11i4.1808

Dallabona, A. (2024). *Luxury fashion marketing and branding: A strategic approach*. Taylor & Francis. DOI: 10.4324/9781003264811

Dong, Y., & Huanug, X. (2024). The rejuvenation of luxury brands: How to attract young consumers in China? Advances in Economics. *Management and Political Sciences*, *70*, 287–293.

Ferraz, L., Nobre, H., & Barbosa, B. (2021). When luxury vinous-concept hotel meets premium wine brands: An exploratory study on co-branding. In *Handbook of research on applied data science and artificial intelligence in business and industry* (pp. 366–380). IGI Global. DOI: 10.4018/978-1-7998-6985-6.ch017

Górska-Warsewicz, H. (2024). Relationship between entrepreneurial orientation, innovative co-branding partnership, and business performance. Journal of Entrepreneurship. *Management and Innovation*, *20*(2), 139–159. DOI: 10.7341/20242022

Górska-Warsewicz, H. (2024). *Trust and brand management: The role of brand heritage*. Taylor & Francis. DOI: 10.4324/9781032707754

Gupta, V., Hushain, J., & Mathur, A. (2024). The future of luxury brand management: A study on the impact of new technology and relationship marketing. In *AI in business: Opportunities and limitations* (Vol. 1, pp. 57–68). Springer Nature Switzerland. DOI: 10.1007/978-3-031-48479-7_6

Jia, W., & Ran, H. (2021, December). The analysis of brand co-branding strategy. In 2021 3rd International Conference on Economic Management and Cultural Industry (ICEMCI 2021) (pp. 1567-1574). Atlantis Press. DOI: 10.2991/assehr.k.211209.254

Joy, A., Zhu, Y., Peña, C., & Brouard, M. (2022). Digital future of luxury brands: Metaverse, digital fashion, and non-fungible tokens. *Strategic Change*, *31*(3), 337–343. DOI: 10.1002/jsc.2502

Kania, R., Sukoyo, K. S., & Wibisono, N. (2021). Effect of co-branding strategy elements on consumer attitude and purchase intention: A study in Indonesia. [JMI]. *Journal of Marketing Innovation*, *1*(01), 30–50. DOI: 10.35313/jmi.v1i01.17

Leuthesser, L., Kohli, C., & Suri, R. (2003). 2+2=5? A framework for using co-branding to leverage a brand. *Journal of Brand Management*, *11*(1), 35–47. DOI: 10.1057/palgrave.bm.2540146

Lin, S. (2023, June). The influential factors on consumer purchase intention: A case study on MINISO. In Proceedings of the 6th International Conference on Economic Management and Green Development (pp. 1265-1277). Springer Nature Singapore. DOI: 10.1007/978-981-19-7826-5_121

Liu, X. (2024). Research on the marketing strategy and effect of sportswear brand co-branding with entertainment celebrities. Advances in Economics. *Management and Political Sciences*, *84*, 212–217.

Lu, C. (2023). Co-branding marketing strategy and optimization of Perfect Diary from the perspective of brand equity. *Academic Journal of Business & Management*, *5*(25), 64–69.

Mars, M. M. (2023). Looking behind the label: Firm-centric brand storytelling and the development of local consumption spaces. *Local Development & Society*, *4*(2), 370–392. DOI: 10.1080/26883597.2023.2175713

Michel, G., & Willing, R. (2020). *The art of successful brand collaborations: Partnerships with artists, designers, museums, territories, sports, celebrities, science, good cause... and more*. Routledge. DOI: 10.4324/9781351014472

Minkkinen, T. (2024). The role of co-branding in brand loyalty. Unpublished manuscript.

Murtas, G., Pedeliento, G., Mangiò, F., & Andreini, D. (2022). Co-branding strategies in luxury fashion: The Off-White case. *Journal of Strategic Marketing*, •••, 1–20. DOI: 10.1080/0965254X.2022.2153258

Oeppen, J., & Jamal, A. (2014). Collaborating for success: Managerial perspectives on co-branding strategies in the fashion industry. *Journal of Marketing Management*, *30*(9-10), 925–948. DOI: 10.1080/0267257X.2014.934905

Ottoni, F. M. J., Thomé, K. M., & Porto, R. B. (2024). Advances in B2B co-branding: A systematic literature review. *Journal of Business-To-Business Marketing*, •••, 1–17. DOI: 10.1080/1051712X.2024.2197098

Paydas Turan, C. (2021). Success drivers of co-branding: A meta-analysis. *International Journal of Consumer Studies*, *45*(4), 911–936. DOI: 10.1111/ijcs.12682

Peasley, M. C., & Woodroof, P. J. (2024). Managing conflicting stakeholder interests: A co-branding perspective for how managers can mitigate negative shareholder responses to cause-related marketing. *Journal of Nonprofit & Public Sector Marketing*, •••, 1–27. DOI: 10.1080/10495142.2024.2408556

Pinello, C., Picone, P. M., & Mocciaro Li Destri, A. (2022). Co-branding research: Where we are and where we could go from here. *European Journal of Marketing*, *56*(2), 584–621. DOI: 10.1108/EJM-02-2021-0073

Rodrigues, P., Sousa, A., & Borges, A. P. (2025). Evaluation of consumer masstige brand relationship between generations. Review of Marketing Science. DOI: 10.2139/ssrn.5134152

Sanli, M. (2024). Digital transformations in the hotel industry (Doctoral dissertation, Long Island University, CW Post Center).

Santos, V., Costa, C., Au-Yong, M., & Sousa, M. (2024). Innovations in luxury tourism: Developing a novel theoretical framework. In Sustainable Tourism: Frameworks, Practices, and Innovative Solutions (pp. 59-76). Springer International Publishing.

Sass, F., Bergen, H., & Keller, M. M. (2024). *Diamonds & disruption: Evaluating brand identity fit in "Inside the Box" vs. "Outside the Box" co-branding strategies—A case study of Tiffany & Co.* LBMG Strategic Brand Management-Masters Paper Series.

Shahid, S., Husain, R., Islam, J. U., & Hollebeek, L. D. (2025). Brand love and customer brand engagement for masstige: A cross-cultural perspective. *Journal of Product and Brand Management*, *34*(4), 545–561. DOI: 10.1108/JPBM-07-2024-5370

Shan, J., Lu, H., & Cui, A. P. (2022). 1+1>2? Is co-branding an effective way to improve brand masstige? *Journal of Business Research*, *144*, 556–571. DOI: 10.1016/j.jbusres.2022.01.058

Singh, J., & Shukla, P. (2024). *Brand management: Principles and applications for effective branding*. Kogan Page Publishers.

Steffl, J., Ganassali, S., & Emes, J. (2024). Hybrid product branding strategies for brand value creation: Combining and comparing green product innovations, limited editions, and co-branding. *Journal of Product and Brand Management*, *33*(8), 1073–1087. DOI: 10.1108/JPBM-04-2023-4472

Sun, C., Cryder, C., & Rick, S. (2023). A co-branding conundrum: Consumers underuse co-branded credit cards outside of their featured brands. Available at *SSRN*, 4494708. https://doi.org/DOI: 10.2139/ssrn.4494708

Sun, X. (2023). An investigation into the impact of luxury co-branding on the purchasing behavior of Chinese millennial consumers. *Art and Society*, *2*(6), 81–107. DOI: 10.56397/AS.2023.12.11

Tarnanidis, T. (Ed.). (2024). *Consumer experience and decision-making in the metaverse*. IGI Global. DOI: 10.4018/979-8-3693-4167-4

Tarnanidis, T., & Manaf, A. H. A. (2024). Research notes on the future of marketing and consumer insights. In *New trends in marketing and consumer science* (pp. 324–336). IGI Global. DOI: 10.4018/979-8-3693-2754-8.ch017

Tarnanidis, T., Papachristou, E., Karypidis, M., & Manda, V. K. (2025). Luxury fashion consumption and branding in the affluent market. In *Cross-Cultural Perspectives on Luxury Marketing* (pp. 161–172). IGI Global.

Tarnanidis, T. K.Theodore K. Tarnanidis. (2024). Fashion consumers' response to the COVID-19 pandemic and today's future. *International Journal of Science and Research Archive*, *12*(02), 648–655. DOI: 10.30574/ijsra.2024.12.2.1286

Tynan, C., McKechnie, S., & Chhuon, C. (2010). Co-creating value for luxury brands. *Journal of Business Research*, *63*(11), 1156–1163. DOI: 10.1016/j.jbusres.2009.10.012

Wang, R. (2023). Analysis of marketing strategies of Louis Vuitton based on the marketing theory of 4Ps. Advances in Economics. *Management and Political Sciences*, *18*, 140–145.

Wu, Y. (2022, April). Research on the factors influencing the purchasing behavior of young luxury consumers: a case. In 2022 7th International Conference on Social Sciences and Economic Development (ICSSED 2022) (pp. 1338-1343). Atlantis Press. DOI: 10.2991/aebmr.k.220405.223

Xue, M. T., & Chawdhary, R. (2023). Co-branding as a masstige strategy for luxury brands: Desirable or not? *Journal of Business Research*, *158*, 113704. DOI: 10.1016/j.jbusres.2023.113704

Yu, Y., Rothenberg, L., & Moore, M. (2021). Exploring young consumers' decision-making for luxury co-branding combinations. *International Journal of Retail & Distribution Management*, *49*(3), 341–358. DOI: 10.1108/IJRDM-12-2019-0399

Zancanella, G., Mazzù, M. F., & Baccelloni, A. (2024). Co-branding masstige within luxury brands: An empirical investigation based on grounded theory. In Atti Convegno.

ADDITIONAL READING

Eiras, B., & Azevedo, A. (2019). Cobranding between fast fashion brands and luxury brands: A case study approach. In *Finding New Ways to Engage and Satisfy Global Customers: Proceedings of the 2018 Academy of Marketing Science (AMS) World Marketing Congress (WMC) 21* (pp. 771-784). Springer International Publishing. DOI: 10.1007/978-3-030-02568-7_208

Fernández Hidalgo, C., Mikano, L., & Vermeersch, T. (2016). The Effect of Co-Branding on the Fashion Luxury Consumer's Brand Equity: Comparison between the Generations Y and X.

Luck, E., Muratovski, G., & Hedley, L. (2014). Co-branding strategies for luxury fashion brands: Missoni for Target. *Global Fashion Brands: Style. Luxury & History*, *1*(1), 41–56.

Murtas, G., Pedeliento, G., Mangiò, F., & Andreini, D. (2022). Co-branding strategies in luxury fashion: The Off-White case. *Journal of Strategic Marketing*, •••, 1–20.

Wang, S. C., Soesilo, P. K., Zhang, D., & Di Benedetto, C. A. (2012). The impact of luxury brand-retailer co-branding strategy on consumers' evaluation of luxury brand image: The case of Taiwan. In *Interdisciplinary approaches to product design, innovation, & branding in international marketing* (pp. 67–89). Emerald Group Publishing Limited. DOI: 10.1108/S1474-7979(2012)0000023007

Xue, M. T., & Chawdhary, R. (2023). 'Co-branding as a masstige strategy for luxury brands: Desirable or not? *Journal of Business Research*, *158*, 113704. DOI: 10.1016/j.jbusres.2023.113704

Yu, Y., Rothenberg, L., & Moore, M. (2021). Exploring young consumers' decision-making for luxury co-branding combinations. *International Journal of Retail & Distribution Management*, *49*(3), 341–358. DOI: 10.1108/IJRDM-12-2019-0399

KEY TERMS AND DEFINITIONS

Co-Branding: A marketing strategy involving two or more brands collaborating to create a product that leverages each brand's strengths, reputation, and market presence. In the luxury sector, co-branding can enhance perceived value and attract consumers by combining the unique attributes of both brands.

Experiential Marketing: A marketing approach focused on creating memorable experiences for consumers through direct engagement with a brand. In the context of luxury products, experiential marketing aims to immerse consumers in the brand's lifestyle, allowing them to connect emotionally and socially with the product.

Luxury Products: High-end goods characterized by superior quality, craftsmanship, exclusivity, and often a prestigious brand image. These products typically command higher prices and are associated with status and sophistication in the consumer market.

Smart Luxury Products: Luxury items that incorporate advanced technology, connectivity, and innovative features to enhance user experience. These products blend opulence with functionality, offering consumers not only aesthetic appeal but also convenience and enhanced performance through smart capabilities.

Sustainability in Luxury Products: The commitment of luxury brands to ethical practices and environmentally friendly production processes. This includes sourcing materials responsibly, reducing waste, and ensuring fair labor practices. Sustainability has become an essential aspect of luxury branding as consumers increasingly prioritize eco-conscious choices in their purchasing decisions.

Value in Smart Luxury Products: The perceived worth of smart luxury products, which encompasses both their tangible benefits (such as quality and functionality) and intangible benefits (such as brand prestige and emotional connection). This value proposition is crucial for attracting discerning consumers who seek both luxury and innovation.

Chapter 3
Co-Creation, Co-Marketing, and Co-Branding Strategies for Smart Luxury Digital Ecosystems

Anuradha Yadav
https://orcid.org/0000-0001-5607-2343
REVA University, India

Vijaya Kittu Manda
https://orcid.org/0000-0002-1680-8210
PBMEIT, India

Theodore Tarnanidis
https://orcid.org/0000-0002-4836-3906
International Hellenic University, Greece

Ahmad Salman Farid
STAIN Mandailing Natal, Indonesia

ABSTRACT

The convergence of luxury and technology has redefined consumer expectations. This necessitated innovative, collaborative strategies in the smart luxury industry. This chapter examines how co-creation, co-marketing, and co-branding strategies drive value creation, consumer engagement, and competitive differentiation in this sector. The role of digital platforms, AI-driven personalization, and cross-industry

DOI: 10.4018/979-8-3373-3010-5.ch003

partnerships in shaping modern luxury branding is analyzed. A synthesis of theoretical frameworks, industry case studies, and empirical research is done. Key findings highlight the importance of strategic alignment, brand compatibility, and technology adoption for successful collaborations. The chapter also explores crisis management and future trends, such as sustainability and metaverse integration. By bridging theory and practice, this work offers actionable insights for academics and luxury brand managers navigating the evolving digital landscape. The chapter demonstrates how smart luxury brands influence collaborative strategies, innovation, and sustainability initiatives.

INTRODUCTION

Background and Context

The concepts of Luxury Marketing and Digital Innovation are coming together. This has resulted in a new marketing paradigm. Addressing modern marketing needs calls for a new set of brand collaborative strategies. Co-creation, Co-marketing, and Co-branding have evolved beyond traditional marketing instruments. They become strategic imperatives in the Smart luxury industry. A shift in consumer expectations drives this change. Globally, the number of affluent buyers is increasing. Demand for both exclusivity and functional advancement from luxury purchases is increasing. Innovative luxury products involving wearable technology, connected home devices, and digitally enhanced fashion are rising. Such products' design, development, and marketing require brand alliances that can retain aesthetic value and provide performance-driven utility.

Luxury is an evolving interdisciplinary field. Understanding luxury fashion consumption patterns among affluent consumers is essential for brands seeking to implement co-creation and co-branding strategies (T. Tarnanidis et al., 2024). Brands must remain dynamic to maintain viability, profitability, and timelessness as narratives change over time (Fletcher & Stephen, 2025). Co-creation, Co-marketing, and Co-branding Strategies in the Smart luxury Industry fit the emerging marketing discourse called Transformative Luxury Research (TLR). Figure 1 highlights key aspects of TLR.

Figure 1. Key Aspects of Transformative Luxury Research (TLR)

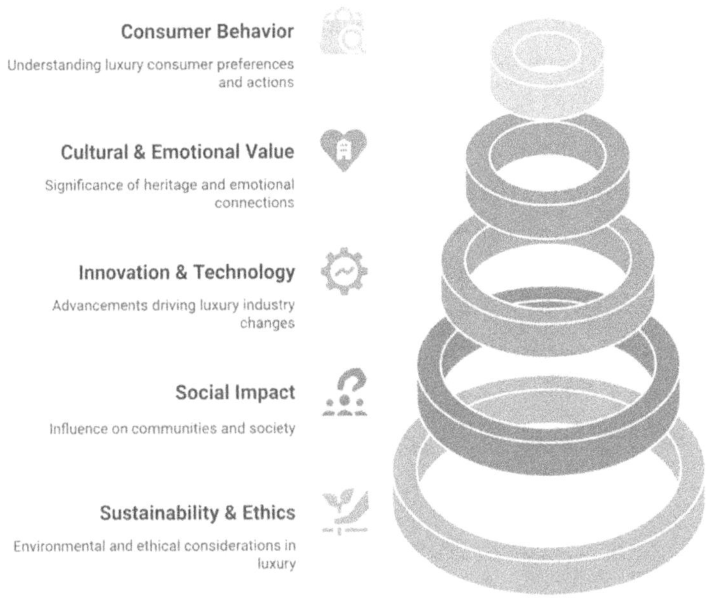

Source: Author creation

This chapter aims to provide a comprehensive analysis of how co-creation, co-marketing, and co-branding strategies. These are enabled by technology and shaped by sustainability and crisis management imperatives. They are expected to be transforming the way the Smart luxury industry works. The work incorporates theoretical perspectives with current industry practices and case examples. It aims to inform academic research and managerial decision-making in the evolving luxury landscape.

Structure of the Chapter

As Figure 2 shows, co-creation, co-marketing, and co-branding are three pillars of Smart Luxury's strategic framework considered in this study. The "Introduction" section briefly explains the importance of the three pillars in the Smart luxury segment. The background and the context of the study are explained. The second section, "Co-Creation Strategies," focuses on co-creation. It discusses strategies used and digital platforms' role in enhancing consumer engagement. The third section, "Co-Marketing Strategies," discusses the concept of co-marketing. It discusses synergistic marketing campaigns, personalization, AI-driven targeting, and the importance of cross-industry collaborations. The fourth section focuses on

"Co-branding," the concept, theoretical foundations, and key success factors. The following two sections discuss technology-driven transformations in the luxury sector and crisis management. The following section is on "Future directions for luxury brands," which previews what can shape the future of luxury. The chapter ends with a "Conclusion," giving an overview of the work.

Figure 2. Three Strategic Pillars of Smart Luxury

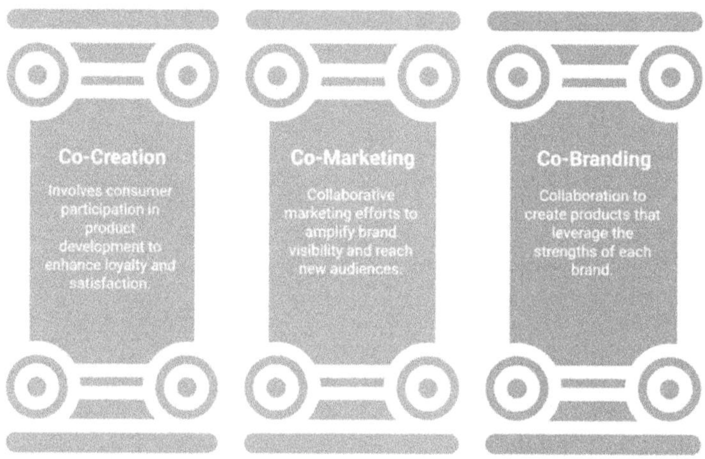

Source: Author creation

With this foundation in place, we now turn to the first core strategy shaping the Smart luxury industry - co-creation. The following section explores how collaborative approaches between brands and consumers are redefining value creation in luxury markets.

The following matrix summarizes the strategic relationships and distinctions between co-creation, co-marketing, and co-branding as applied to the Smart luxury ecosystem.

Table 1. Strategic Relationships Matrix: Co-Creation, Co-Marketing, and Co-Branding in Smart Luxury

Dimension	Co-Creation	Co-Marketing	Co-Branding	Strategic Relationships / Links
Definition	Collaborative value creation with consumers	Joint marketing and promotional activities with partners	Joint creation of a new product/service by two or more brands	Co-creation can underpin both co-marketing and co-branding
Primary Actors	Brand & consumers (sometimes other stakeholders)	Two or more brands/ companies (can include influencers)	Two or more brands, often from different sectors	Co-branding often follows co-marketing partnerships
Objective	Enhance consumer engagement, innovation, personalization	Amplify reach, share audiences, improve campaign effectiveness	Combine brand equities to create unique offerings	Co-marketing can lead to co-branding alliances
Typical Activities	Product/service ideation, user-generated content, personalization	Joint campaigns, cross-promotions, influencer partnerships	Co-designed products, joint launches, dual-branded services	All may leverage digital platforms and AI
Value Proposition	Customization, emotional connection, brand loyalty	Increased awareness, credibility, efficiency	Unique, differentiated products/services and market expansion	Co-creation boosts the authenticity of both
Role of Technology	Digital platforms, AI, Metaverse, social media	Data analytics, AI, omnichannel platforms	Technology-driven innovation (e.g., Smart products, NFTs)	Digital tools enable integration across all
Consumer Involvement	High (active participation and feedback)	Moderate (engaged as target audience)	Variable (may be involved if co-created product)	Co-creation enhances outcomes in both others
Brand Relationship	Direct, participatory	Collaborative but more transactional	Strategic, long-term alliance	Progression possible: co-creation → co-marketing → co-branding
Example in Luxury	Customizable handbags; user-designed NFT collectibles	Louis Vuitton x League of Legends campaign; influencer events	Tag Heuer x Intel smartwatch; LVMH x Kering Paris Art Museums	Often, co-marketing/ creation precedes co-branding

Source: Author creation

DISCUSSION

Co-Creation Strategies

Co-creation is a collaborative process in which companies and consumers contribute to value creation. They promote a more interactive and participatory relationship between brands and their customers. It is a strategic approach that extends beyond the traditional value creation process, which is one-way value delivery.

Consider the case of luxury travel. Many major luxury travel brands originally started as luggage manufacturers. Their products allowed travelers to move parts of their wardrobes. However, the traveler's wardrobe, jewelry, footwear, or perfumes changed as time progressed. The current traveler trend is to enrich their wardrobe after reaching the destination and to buy local items. This way, they can explore the place and go shopping. Tourists purchase dedicated accessories, "destinationwear," and fashion-related items. Thus, the landscape of luxury travel has changed (WTO & UdSi, 2025). This change made luxury travel brands need to collaborate and co-create with local stores.

Consider another example of luxury traveling. Modern luxury travelers prioritize unique, meaningful experiences over traditional opulence. The key trends in this area are personalization, wellness, authenticity, and regeneration. The post-pandemic travel era accelerated these trends because of revenge travel (Choudhary & Vijaya Kittu, 2021). Research emphasizes the need for hospitality brands to increase their collaboration among stakeholders to drive sustainable value to meet the needs of luxury travelers (Peragallo, 2025). There is a shift in luxury from product-centric to experience-centric value. This is a core idea under co-creation.

Concept of Co-Creation

The concept of Co-Creation is an extension of the traditional transactional business models. It sets the stage for understanding the value of collaborative approaches. It shifts the value generation process by involving consumers who are not merely passive recipients but get actively involved. It shifts the traditional producer-consumer dynamic by establishing reciprocal relationships.

Consumers participate in iterative interactions with the brands in the product ideation, design, and development of luxury offerings. Co-creation uses the consumer's expertise, aesthetic preferences, creativity and expertise, and technological fluency. These traits help make novel products or service experiences. It integrates user insights, knowledge, and creativity into brand innovation. It enhances value by bringing collective intelligence and allows brands to align offerings with evolving consumer preferences. This participatory approach enhances emotional connection.

It promotes a sense of ownership among luxury consumers and intensifies brand attachment and loyalty. It is instrumental in luxury contexts because such goods and services bring relevance, exclusivity, and personalization. The output will be intrinsic to brand identity. They help brands to meet evolving consumer expectations while maintaining brand prestige. Luxury brands increasingly adopt co-creation to differentiate themselves. Brands are developing integrated co-creation ecosystems. These ecosystems blend in-store (physical) and digital experiences (online) to enhance consumer involvement and innovation.

This strategy aligns with the exclusivity demanded by affluent consumers. Luxury brands must balance democratized input with exclusivity maintenance. The approach can integrate cultural preferences and technological advancements. It allows brands to align with the demands of globalized markets. The emergence of digital ecosystems has increased opportunities for brands to collaborate with consumers on product innovation and storytelling. Digital transformation accelerates co-creation possibilities using real-time feedback loops and collaborative design processes. Firms like Louis Vuitton use digital interfaces for personalization options while preserving brand heritage. The effectiveness of co-creation depends on transparent communication channels and meaningful consumer involvement. Marketers are allowing users to post their experiences on digital platforms. Brand communities undoubtedly help create and spread brand-related content and its effect on luxury appeals (Basile et al., 2024). Users doing that are found to have reinforcement of the brand's experiential and symbolic dimensions. This influences the perception of value and attitude towards the product.

Prahalad and Ramaswamy pioneered the concept of co-creation (Prahalad & Ramaswamy, 2005). Vargo and Lusch's Service-Dominant (S-D) Logic emphasizes that value is co-created through the integration of resources by various actors (including firms and consumers) rather than being embedded in products and delivered unilaterally (Vargo et al., 2008). Füller discussed how consumers' roles and levels of influence in co-creation differ depending on the structure and objectives of the engagement model (Füller, 2010). The works of Ramaswamy and Ozcan focused on how engagement platforms enable and shape co-creation experiences (Ramaswamy & Ozcan, 2014).

Digital Platforms and Consumer Engagement

Digital platforms are crucial in facilitating co-creation. These platforms facilitate seamless communication and serve as enablers of co-creation strategies. They provide interactive spaces for consumer-brand collaboration. They enable firms to engage with consumers, gather feedback, and promote collaborative ideation and innovation. These platforms enable real-time interaction, allowing brands to crowdsource ideas, test prototypes, and gauge consumer sentiment efficiently. They also allow real-time tracking of consumer preferences. Engagement metrics guide iterative improvements. The digital ecosystem amplifies consumer participation. Figure 3 gives a list of popular digital platforms used in co-creation.

Figure 3. Digital platforms used in co-creation

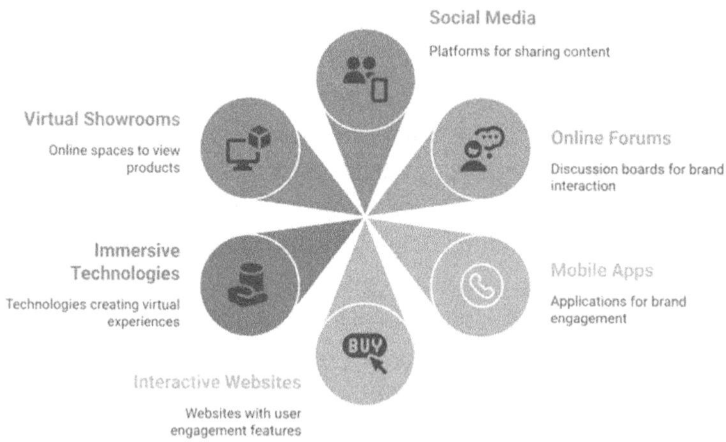

Source: Author creation

Experiential engagement refers to creating immersive, interactive, and memorable interactions with a brand. Gamification and virtual showrooms are two popular methods of enhancing experiential engagement.

Gamification uses game-design elements and game principles in non-game contexts. They can be used in the context of luxury to create subtle elements of playfulness. Some popular gamified elements are:

1. Virtual treasure hunts for exclusive content
2. Interactive quizzes about brand heritage
3. Challenges to unlock early access to collections

These elements create a sense of excitement and discovery. Rewards can be given through points systems, virtual badges, or tiered access.

Virtual showrooms are immersive digital environments. They replicate or reimagine physical retail spaces and offer a powerful way to engage luxury consumers online. They provide a sensory experience with the help of high-quality 3D renderings. They help visualize how items might look in their own space, such as virtual furniture placement or trying on virtual jewelry.

Consumer engagement is critical, as it determines the success of co-creation initiatives. So, digital platforms are essential for dynamic co-creation. Firms must encourage active participation, build trust, and provide incentives to motivate consumers. High levels of consumer engagement are observable on platforms that support two-way communication and reward creative participation. Platforms like Instagram, WeChat, and dedicated brand forums are digital arenas for luxury consumers to influence product narratives and features. Brands run interactive campaigns on these platforms. Co-creation with consumers can also be seen as a great community engagement tool. Social media can build trust and even help rebuild trust during a crisis (Venturini, 2023).

User-Generated Content and Brand Communities

User-generated content (UGC) is a key driver and is central to co-creation in luxury branding. UGC can range from product reviews to unboxing videos. Consumers contribute content through reviews, social media posts, videos, and collaborative design initiatives. It shapes brand narratives within the digital ecosystems and serves as authentic testimony to brand value and user experience. They are a form of peer influence in luxury branding. They influence brand perception and market positioning. Platforms like Instagram, TikTok, and X are examples where consumers are encouraged to co-narrate brand stories. Luxury brands curate and amplify select user contributions to reinforce narratives of exclusivity and innovation while maintaining accessibility. These strategies strengthen emotional bonds and drive organic reach. Research indicates that UGC-driven communities increase consumer retention and advocacy.

Brand communities emerge as spaces/hubs for shared experiences. They can both be open or invite-only. Both facilitate the aggregation of UGC. They cultivate a sense of belonging and bring a shared identity among members. These communities often act as incubators for co-creation, where members share insights and co-develop product ideas. Consumers use the communities to exchange feedback, propose enhancements, and collaborate on limited-edition releases. They provide emotional connections and strengthen loyalty. Interactive co-creation campaigns encourage consumers to take ownership of brand evolution. They drive credibility and engagement. Brand communities can cultivate advocacy through shared values and exclusive experiences. Community-driven content also accelerates new products and services diffusion across digitally connected luxury audiences.

Digital Luxury Experiences

Luxury brands integrate digital experiences to enhance exclusivity and personalization in co-creation strategies. This became possible because of technological advancements in immersive technologies and other disruptive technologies.

The Metaverse offers an immersive environment for consumers. They allow interactive product launches and bring brand engagement through virtual storefronts, customizable avatars, and interactive luxury spaces. Virtual showrooms and digital twins simulate real-world interactions. Consumers can co-design digital assets, such as bespoke garments for virtual avatars or custom digital collectibles. So, they are ideal co-creation spaces for virtual luxury experiences. Virtual land in the Metaverse is emerging as a new asset class, with global companies and investors buying virtual land to explore business potentials (Leonhard et al., 2025). These virtual lands present new opportunities for luxury investment and have the potential for digital assets to be appreciated.

Non-fungible tokens (NFTs) introduce scarcity in digital assets and brand differentiation. So, they bring exclusive ownership of virtual luxury items (Jang & Kang, 2025b). They enable digital ownership. Brands such as Gucci, Louis Vuitton, and Balenciaga have pioneered NFT-based collaborations. Brands like Gucci and Balenciaga experiment with virtual fashion and tokenized collectibles.

The integration of Blockchain brings transparency, provenance, and authenticity verification and ensures value preservation in digital luxury markets (Manda et al., 2024). These digital frontiers redefine exclusivity and are appealing to tech-savvy luxury consumers. These technologies blur physical and digital boundaries. Artificial Intelligence (AI) is increasingly used in marketing luxury products. AI-driven customization further enhances hyper-personalized digital engagements. However, the research emphasizes balancing AI implementation with human oversight in luxury retail settings (Li & Kang, 2025). Luxury consumers often view AI technology as morally intended during crises. It can heighten feelings of betrayal when interacting with human-centered AI. It suggests training service employees to maintain warm, approachable demeanors to counteract potential AI-related backlash. The integration of blockchain and metaverse technologies is reshaping customer experiences in luxury, enabling new forms of digital ownership and immersive engagement (Khaliq & Manda, 2023).

While co-creation emphasizes the joint value development between brands and consumers, luxury brands increasingly use collaborative marketing initiatives with external partners. The following section delves into co-marketing strategies, highlighting how alliances can amplify brand reach and impact.

Co-Marketing Strategies

Concept of Co-Marketing

Co-marketing involves partnering with other brands or organizations to promote each other's products or services. It is often done when between brands to promote complementary products or services. This builds on collaboration and highlights the benefits of joint marketing efforts. It represents a strategic alliance focused on joint promotional activities. Partners use mutual resources and expertise while minimizing resource expenditure to achieve shared marketing objectives. Its primary aim is to amplify market reach and enhance brand visibility. This differs from co-branding's focus on product creation. It can take various forms, including joint product launches and marketing campaigns, shared promotional activities, and coordinated branding efforts. In Smart Luxury, co-marketing takes advantage of multiple firms' resources, audiences, and brand equity to maximize campaign efficiency. It integrates luxury aesthetics and merges technological sophistication with exclusive brand narratives.

This strategy extends beyond traditional collaborations, involving more collaboration, integration, and risk-sharing between the participating brands. The strategic alliances enhance differentiation, reinforcing exclusivity and brand positioning through co-branded initiatives. It integrates well with the digital ecosystems and consumer touchpoints. It creates a seamless experience for target audiences. Data-driven insights underpin the selection of partners to ensure brand congruence and cultural resonance. It allows partners to access complementary customer segments effectively. Shared campaigns can significantly reduce individual marketing expenditures. The strategy builds credibility through association with reputable partners. Strategic alignment in messaging ensures consistency across partner brands. Trust and complementary value propositions are critical for long-term success. Success lies in clearly defined (often common) goals and mutual benefit. Co-marketing promotes shared innovation, brand storytelling, and hybrid value propositions. It enhances the perceived value proposition for target consumers. Co-marketing can enhance brand awareness, credibility, and loyalty by using the strengths and reputations of the partner brands. It does not always need to be done with other firms. Non-firm-controlled touchpoints, like influencers and peer networks, can be key drivers of engagement. So, this could imply opportunities for co-marketing initiatives with influencers or community-driven collaborations (Holmqvist et al., 2025).

The consumption of luxury fashion brands is an investment, especially for millennials from the Association of Southeast Asian Nations (ASEAN) countries like Indonesia (Sun et al., 2024). This perception requires different marketing strategies because the goods and services should now also have symbolic and financial values. Culture, fashion, tourism, media, and publishers (including technologies and influencers) are strongly connected. They make a good case/opportunity to co-market their services collaboratively (WTO & UdSi, 2025).

Synergistic Marketing Campaigns

Synergistic marketing campaigns are the operational core of co-marketing partnerships. They integrate complementary brand strengths to enhance visibility and consumer appeal. These campaigns rely on shared communication strategies, unified brand messaging, and co-developed promotional efforts across multiple channels. Partners pool financial resources and marketing budgets for greater impact. This can include co-branded advertising, social media campaigns, and event sponsorships. In luxury, co-marketing is characterized by aligning objectives, resources, and communication strategies. Co-marketing partnerships between luxury and technology brands employ influencer collaborations, digital storytelling, and immersive brand experiences. The object is to drive consumer interaction. Luxury-tech collaborations often blend exclusivity with innovation to attract premium consumers. For example, a luxury fashion brand may partner with a high-end tech brand to launch a co-branded smartwatch. The campaign can highlight the fusion of style and technology. The Smartwatch brings in functionality and prestige. The fashion brand brings in style and aesthetics. Such campaigns appeal to customers who value both luxury and innovation. Louis Vuitton x League of Legends is a popular campaign demonstrating how storytelling enhances engagement.

Markets use campaigns to enhance perceived value. They employ integrated digital platforms, immersive experiences, and real-time consumer engagement to accomplish this. Cross-promotions can be run on partner platforms. They can extend audience engagement significantly. Strategic synchronization of promotional timing, messaging, and influencer engagement significantly boosts returns for marketing efforts when applied across multiple market segments. Joint messaging combines distinct brand identities into a cohesive narrative. Some popular examples of this are:

1. Shared digital advertising
2. Collaborative content creation
3. Co-hosted events.

Effective campaigns require meticulous planning and coordinated execution. Measuring joint campaign success demands shared metrics and transparent reporting. Metrics such as sentiment analysis, conversion rates (engagement rates), ROI, and social listening inform the post-campaign evaluation. Data-driven insights optimize cross-brand synergies, ensuring brand narratives align with consumer expectations and market trends. The insights optimize campaign resonance across diverse demographics. Luxury brands are using omnichannel strategies to blend physical and digital efforts. This helps extend campaign life and increase consumer engagement. The ultimate objective is to create marketing outcomes unattainable independently.

Personalization and AI-Driven Targeting

Personalization and Artificial Intelligence (AI)-driven targeting are critical components of co-marketing strategies in the Smart luxury industry. AI algorithms analyze combined customer datasets for deeper insights. Brands use AI and data analytics (DA) to refine their offerings. These tools help understand their cultural nuances and cultivate deeper customer loyalty (Khamoushi Sahne & Kalantari Daronkola, 2025). Integrating DA helps build engagement strategies by identifying influential co-creators and mapping their impact on brand value. Effective engagement enhances brand loyalty, strengthens market positioning, and ensures relevance across diverse markets. Machine learning algorithms analyze behavioral, demographic, and psychographic data to predict consumer preferences accurately. Segmentation models identify niche audiences effectively. Dynamic segmentation and real-time content adaptation allow brands to make relevant marketing messages for different audience clusters. Predictive analytics can refine segmentation, identify high-value consumers, and optimize campaign effectiveness. They can facilitate anticipatory service, drive higher engagement and purchase intent, and enhance experiential interactions. Further, they anticipate market trends and ensure timely campaign adjustments.

AI-powered personalization enables luxury brands to offer hyper-individualized experiences at a scale that resonates with individual customers (Khamoushi Sahne & Kalantari Daronkola, 2025). This can include tailored content, product recommendations, and targeted advertising. Joint marketing messages can be hyper-personalized based on shared data analytics. Consider a co-branded luxury e-commerce platform as an example. The platform may use AI to analyze customer behavior and preferences (including cultural preferences). It can then offer personalized product suggestions and promotions. AI can refine the campaign messaging for global audiences. AI optimizes channel selection and budget allocation for co-marketing initiatives.

AI-driven targeting also supports ethical data management, ensuring compliance with privacy regulations and enhancing consumer trust within the luxury segment. For instance, a luxury car brand can partner with a technology company. The two can develop a co-branded AI-powered voice assistant. It can provide personalized recommendations and services to car owners based on their preferences and habits. This way, the offering will align well with the luxury expectations of the customer. Personalization enhances consumer loyalty and brand resonance.

Cross-Industry Collaborations

Cross-industry collaborations involve partnerships between brands from different sectors to create innovative products or services. In the Smart luxury industry, fashion-tech partnerships merge luxury fashion with advanced technology. This form of partnership is becoming increasingly popular.

For example, consider a luxury fashion brand interested in partnering with a tech company. The two can develop a co-branded Smart clothing line. Their collaboration can combine the brand's fashion expertise with the tech company's innovative technologies. The two brands can create a unique product that appeals to customers seeking style and functionality. Another example is Smart wearables that blend aesthetics with functionality. These collaborations target generally tech-savvy luxury consumers. The Tag Heuer x Intel smartwatches are a testimony to this. Co-marketing campaigns might feature wearable technology, Smart/intelligence fabrics, and connected devices. These devices are displayed in stylish, high-fashion settings. Luxury automotive brands partnering with smart home technology providers is another example. Their collaboration can create a lifestyle narrative that can involve multiple luxury products.

Cross-industry collaborations redefine the boundaries of luxury co-marketing. It enhances creativity through interdisciplinary expertise. Such collaborations can bridge gaps between aesthetics and functionality. Some hybrid offerings that they can make are smart wearables, augmented reality experiences, and digital collectibles, expanding traditional product portfolios. The convergence of fashion and technology facilitates narrative-driven campaigns emphasizing innovation and exclusivity. Collaborative projects often involve co-branded digital experiences, limited-edition releases, and exclusive online communities. Joint ventures use complementary expertise to drive innovation. Strategic alliances use unique technological competencies and creative capital, accelerating market differentiation in the rapidly evolving Smart luxury ecosystem. They expand market presence, especially in emerging regions, while reinforcing brand prestige. Such partnerships redefine luxury experiences in the digital era. Cross-industry alliances also mitigate risks by diversifying consumer bases.

Building on the collaborative spirit of co-marketing, some luxury brands take partnership a step further through strategic co-branding. The following section examines how co-branding combines the strengths of multiple brands to deliver innovative products and experiences in the Smart luxury sector.

Co-Branding In The Smart Luxury Industry

Concept of Co-Branding

Co-branding involves a strategic partnership between two or more independent brands to collaborate and jointly create a new product, service, or experience that combines the strengths of both brands. This can be seen as a more advanced form of collaboration, building on the concepts of co-creation and co-marketing. Within the Smart luxury sector, it represents a strategic alliance. It typically involves a traditional luxury and prestige marque and a technology-focused entity.

This partnership aims to develop and market technologically advanced Smart consumer products or services. Doing so enhances perceived value and innovation. These innovative products cater to the evolving preferences of discerning consumers. The resulting products often blend sophisticated design with cutting-edge functionality. These distinctive offerings retain the core attributes of luxury, such as exclusivity and superior craftsmanship. They appeal to a global audience. Such partnerships often align with experiential marketing paradigms, emphasizing hedonic and symbolic consumption. The principles of experiential marketing say that the boundaries between products, services, and experiences are increasingly blurred.

In this context, luxury brands contribute intangible assets such as brand equity, prestige, exclusivity, and emotional appeal. At the same time, technology partners bring in functional expertise, innovation capabilities, and digital infrastructure. Together, they create unique offerings that resonate with affluent consumers. While creating a strong brand is important, as Figure 4 shows, managing it in the short- and long-term is even more important (Shams et al., 2025a). Co-branding can be a powerful tool for managing and growing a brand, providing a strong brand fit, values, and messaging to ensure a successful partnership (Shams et al., 2025a).

Figure 4. Balancing Short-term and Long-term goals in co-branding is important

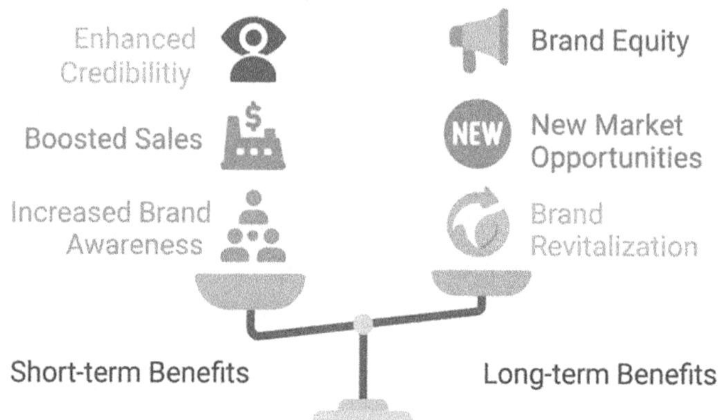

Source: Author creation

Theoretical Foundations of Co-Branding

The theoretical foundation of co-branding is rooted in several academic theories. Table 2 presents some popular theories connected to Co-branding.

Table 2. Theories connecting with Co-branding

Theory	Connection with Co-branding
Associative Brand Theories	These theories examine how brand collaborations shape consumer perceptions, enhancing credibility and desirability.
Signaling Theory	Explains how co-branded products communicate enhanced quality or attributes, heritage, and innovation to target audiences. For example, an association with a reputable tech brand can signal innovation. Conversely, partnering with a luxury brand generally signals premium quality. This dual signaling enhances market differentiation in multicultural contexts.
Consumer-based Brand Equity (CBBE)	The core brand equity theory emphasizes shared value creation and highlights how brand associations enhance perceived quality. Co-branding uses complementary brand strengths to elevate consumer perceptions. CBBE posits that well-executed co-branding strategies reinforce brand recall, loyalty, and perceived value, optimizing competitive positioning in luxury markets.
Resource-Based View (RBV)	This theory posits that firms can gain a competitive advantage by strategically bundling complementary resources. RBV views alliances as resource-sharing mechanisms. Here, luxury provides brand equity, while tech offers innovation capabilities.
Brand Alliance Theory	This theory examines how partnerships influence consumer evaluations of parent brands. Successful integration enhances the perceptions of both partners.

continued on following page

Table 2. Continued

Theory	Connection with Co-branding
Network Theory & Associative Network Theory (ANT)	This theory treats alliances as nodes that create value within a larger ecosystem. ANT explains how co-branding influences consumer perceptions by linking two brand schemas, shaping attitudes and expectations.
Service-Dominant Logic	This posits that both brands act as operant resources, jointly co-creating value with the consumer.
Social Identity Theory (SIT)	SIT says consumers align with brands that reflect their self-concept and aspirational identities. When applied to Smart luxury, this theory explains how a joint offering can echo more deeply with target audiences. This is achieved by meeting both style preferences and performance needs.

Source: Author creation

Key Success Factors for Co-Branding

Researchers have put efforts towards identifying key factors that determine the success of a co-branding strategy (Chen & Choi, 2025). Successful co-branding in Smart Luxury depends on several factors. Strategic alignment helps partnering brands coordinate their efforts and resources to achieve a common objective. Clarity in partnership objectives minimizes brand dilution risks. Effective governance structures ensure seamless collaboration, balancing creative autonomy with strategic alignment. A long-term strategic vision ensures that the partnership remains relevant and productive.

Mutual brand enhancement is important. The collaboration must elevate both participants. High perceived fit between partners enhances consumer acceptance and reduces cognitive dissonance. Empirical studies suggest the need for deep conceptual integration rather than superficial endorsements.

Clear articulation of the complementary value proposition for the target consumer is essential. Consumers must perceive distinct benefits from the combined offering. Shared commitment to quality, both in technology and finishing, is non-negotiable. The operational synergy between partners allows for smoother product development and marketing execution. Effective communication strategies must convey the rationale and benefits of the partnership. Careful selection of distribution channels aligning with both brands' exclusivity is crucial. Managing intellectual property and contractual obligations can prevent future conflicts. Ongoing monitoring of consumer response allows for necessary adjustments.

A clear example of co-branding is the collaboration between Parisian city leaders and luxury groups like LVMH and Kering—the collaboration aimed at developing Parisian private art museums (PAMs) (Green, 2024). PAMs fuse the cultural prestige of Paris with the exclusivity and sophistication of luxury brands. They create a shared brand identity that enhances both entities. AMs reinforce Paris' identity as a global culture, art, and luxury capital. It aligns with its legacy as a hub for haute couture and refined lifestyles. For LVMH and Kering, PAMs serve as cultural extensions of their brands, embedding their names in iconic architectural and artistic landmarks (e.g., LVMH's Fondation Louis Vuitton). This elevates their prestige beyond commercial products to cultural patronage.

Brand Compatibility and Strategic Alignment

Brand compatibility is fundamental to successful Smart luxury co-branding. Brand compatibility refers to the degree to which the partnering brands share similar values, attributes, and philosophies. It involves assessing the fit between the partners' brand identities and images. It requires congruence in brand personality, prestige, and consumer perceptions. Values congruence ensures the partnership feels authentic to consumers. Perceived quality levels must be comparable to avoid brand dilution.

Strategic alignment requires shared objectives and KPIs, a long-term vision, and integrated marketing efforts. Goals might include market entry, innovation showcases, or reaching new demographics. Partners must agree on the positioning of the co-branded product or service. Alignment extends to marketing messages and overall communication tone. Long-term visions for both brands should accommodate the collaborative venture. Target audience overlap, or logical extension, is necessary for market acceptance. Luxury brands must ensure that technological partners uphold craftsmanship and exclusivity standards. Cross-industry collaborations demand meticulous positioning to avoid over-commercialization.

Mismatched brand identities can trigger skepticism or erode trust. Misalignment in strategy or compatibility can damage the reputation of one or both brands. It risks confusing consumers and undermining the perceived value of the collaboration. Sometimes, co-branding can bring tension. This is especially true when the association is between public good and private influence (Green, 2024). For example, while the PAMs contribute to Paris' cultural vibrancy, their private funding raises concerns. Specific concerns about prioritizing marketing-driven iconicity over authentic public cultural enrichment were raised. The exclusivity of PAMs may cater more to elite audiences, potentially limiting accessibility for broader Parisian communities.

Consumer Perception and Market Fit

In luxury marketing, consumer perception is crucial. The value of luxury goods often lies in the brand's image and storytelling as much as in the product itself. Consumers seek high-quality craftsmanship, emotional resonance, exclusivity, and status. Their perception of a brand is critical to its success. Achieving market fit means aligning a brand's offerings with its target audience's evolving desires and expectations. This involves delivering products and experiences that feel both aspirational and personally meaningful. Brands that monitor and respond to shifts in consumer perception can maintain relevance and command premium pricing. They are also better positioned to endure loyalty in the competitive luxury market.

Co-branding is a mutually beneficial partnership and requires strategic alignment between the brands. It uses the strength of both brands to positively affect customer attitudes to increase the perceived value, thereby leading to higher purchasing intention (Nurrosidah et al., 2025). Co-branding appeals to affluent customers (Semwal et al., 2024) by attracting their attention and influencing their purchasing intention. The success is higher when the brands are compatible products.

Consider a case of a prestigious travel brand licensing its name to a luxury hotel operator. Luxury hotel operators benefit from increased brand recognition and prestige by associating with prestigious travel brands. They can attract higher-end clientele and get an opportunity to drive higher revenue through premium positioning. The Travel brand benefits from generating additional income through licensing fees and expands brand presence into new markets (e.g., hospitality). They can enhance brand equity by aligning with the luxury experience.

Co-branding success depends on product, brand, and promotional compatibility (Nurrosidah et al., 2025). Not all co-branding efforts result in a successful marriage. Therefore, there is a need to have a deeper understanding of the factors responsible (Chen & Choi, 2025). For example, high popularity may cause confusion and competition in consumers' minds. So, partner brand popularity can reduce the price premium of the co-branded product, as is observed in sportswear. This goes against the idea that more popular brands always add more value (Lu et al., 2025). Celebrity endorsements usually increase price premiums but can be less effective if the partner brand is linked to a celebrity (Lu et al., 2025).

There are some geographic locations where luxury is essential, such as the UAE. Also, co-branding is common and increasingly used in certain products, such as credit cards (Sathakathulla, 2025).

Marketing strategies can apply to commonly and regularly used products to give them luxury status. Chinese tea brand Heytea did something similar to position itself as the "Louis Vuitton of Tea" in China and beyond (Wang, 2024). To accomplish this, Heytea took a series of steps:

1. Using Natural and Premium Ingredients & Craftsmanship
2. High-end branding and market positioning that included an iconic logo that signifies exclusivity in fashion

3. Created scarcity and made the act of purchasing a drink an event rather than just a transaction
4. It turned its product into a form of "social currency," particularly among Gen Z consumers
5. Turned into the first tea brand to collaborate with a luxury house by partnering with FENDI to create exclusive beverages.
6. Heytea entered cross-industry collaborations with premium brands like Baileys, Fenty Beauty, and Stanley. Heytea was able to upscale brand perception with this.
7. Expanded into prestigious international markets such as Singapore, New York, London, and Sydney, reinforcing its global luxury appeal.
8. Heytea created a mobile app named Heytea GO to promote the concept of digital luxury. The app allows customers to pre-order and customize their drinks. Customers get the feeling of a VIP concierge service with the app.

Another example is Luckin Coffee and Guizhou Maotai's co-branding strategy for the Moutai-flavored latte. Their strategy has enabled both brands to expand across different market segments (Zhang, 2025). They combined a luxury brand with a fast-moving consumer goods (FMCG) model. The Moutai-flavored latte created a significant buzz. It attracted attention from coffee enthusiasts and those interested in the premium liquor brand. The collaboration also turned into an example of bringing the strengths of both brands - blending luxury and mass-market appeal.

Researchers examine co-branding and crossover brand extension as potential innovation strategies (Zhang et al., 2025). A study that captured insights from heritage brand managers finds that Brand extension has an advantage over co-branding. The mindset of the consumer affects the perception. Fixed mindset consumers appear to maintain a negative bias against co-branding. However, those with a growth mindset neutralize the disadvantage of co-branding.

In 2023, Louis Vuitton and Yayoi Kusama collaborated. They brought a concept called "appreciation task" to the fore (Kłeczek & Hajdas, 2025). This concept denoted a shift from decoration-centric to appreciation-centric practices. It enriches the consumer experience by attracting a deeper engagement with art. The luxury brand was responsible for actively communicating the artistic intent behind collaborations. It also showed that luxury brand–artist collaborations go beyond commercial success when executed thoughtfully. The collaboration promotes cultural appreciation and social transformation.

Technological innovation has become an essential enabler as these collaborative strategies evolve, transforming how luxury brands operate and engage with consumers. The following section investigates the pivotal role of digital technologies in driving the Smart luxury industry forward.

TECHNOLOGY-DRIVEN TRANSFORMATION IN LUXURY BRANDING

Technology is leveraged to modernize luxury retail while maintaining exclusivity and appeal (Surià, 2024). Brands are increasingly adopting innovation and technology to provide value to consumers. Adoption of augmented reality (AR) and virtual reality (VR) to offer personalized and interactive customer experiences. Studies show that Western and Eastern designs aim to create immersive environments that promote consumer engagement and brand loyalty (Surià, 2024). At the same time, physical spaces are strategically designed to enhance emotional connections with luxury brands.

Brands ignoring or less interested in it will be pressured to adopt technology and incorporate sustainability practices. Thus, technology and sustainability are not just for public relations but a key part of brand strategy. They are a strategic priority for innovation for products and services (Shams et al., 2025b). Luxury hospitality, for example, inspires innovative strategies that align profitability with environmental and social responsibility (Peragallo, 2025).

Studies also discussed integrating lean techniques (like Lean Six Sigma and Value Stream Mapping) with Industry 4.0 technologies. The integration improves competitiveness and boosts sustainability in luxury fashion. The practices can improve quality and reduce waste (Bilancia et al., 2025). Blockchain is a leading and disruptive Industry 4.0 technology with several applications in the luxury industry.

Non-fungible tokens (NFT) are based on Blockchain technology. NFT-based authentication has increased the warranting value and structural assurance, especially among luxury retailers (Jang & Kang, 2025b). Blockchain-enabled NFT authentication enhances trust and purchase intention in online luxury retail. Technology-based authentication of the product increases warranting value (product-level trust) for luxury goods. The structural assurance (platform-level trust) reassures consumers about the credibility of online luxury platforms like Farfetch or The RealReal. NFTs distinctly impact new and pre-loved luxury goods (Jang & Kang, 2025a). The Metaverse provides digital experiences, and marketers must address fundamental luxury attributes such as exclusivity and authenticity in the "new luxury" era (Cardoso et al., 2025).

The sustainable luxury products market has increased over the years thanks to changing consumer demand for high-end luxury products with a conscience (Husain et al., 2025). Research says sustainability and circular economy (CE) align with luxury exclusivity. For example, CE principles such as closed-loop systems, recycled materials, and eco-friendly practices can be applied to luxury apparel (Bhardwaj et al., 2025). Slow fashion, biodegradable materials, and clean supply chains appeal to consumers valuing sustainability and craftsmanship (Bhardwaj et al., 2025).

Second-hand luxury fashion (Holmqvist et al., 2025) and the sharing economy are a strong example of changing traditional economic dynamics (Husain et al., 2025). Circular economy adoption enhances brand prestige through sustainability. The customer engagement journey has three factors - price sensitivity, uniqueness, and sustainability-driven by specific motives and touchpoints. Consumer motivations evolve, with sustainability becoming more prominent as familiarity with circular practices grows (Holmqvist et al., 2025). As confirmed by research, luxury and sustainability are strongly compatible, especially in the hotel industry, with sustainability communication and customer engagement playing a vital link (Schuck et al., 2025).

Digital Product Passports (DPPs) are used even in the secondary luxury market. DPPs enhance the authenticity verification of secondary luxury goods (Slaton & Pookulangara, 2025). Blockchain can store high-quality, traceable information. It can reduce perceived financial, social, time, contamination, and performance risks associated with online second-hand luxury shopping (Youn et al., 2025). Such steps include improving customer experience and developing marketing innovations. It also supports a circular economy by enabling luxury product repurchase and promoting sustainability. Studies show that consumers with lower price consciousness value product diagnosticity (detailed product information) more. However, those with higher price consciousness prioritize trustworthiness when using Blockchain (Youn et al., 2025).

However, the same digital advancements that fuel innovation also introduce new risks, particularly in reputation management. The subsequent section explores crisis management in the luxury sector, highlighting strategies for safeguarding brand prestige in an increasingly connected world.

CRISIS MANAGEMENT AND REPUTATION RECOVERY IN LUXURY BRANDING

Balenciaga's 2022 Scandal is a case study on luxury brand management in the digital age (Venturini, 2023). Luxury brands are increasingly expected to take stances on social/political issues such as sustainability and child protection. Failure to align with consumer values risks reputational damage (e.g., Balenciaga's 2022 scandal). Social media amplifies consumer empowerment, enabling rapid backlash (e.g., #CancelBalenciaga). Delayed responses (e.g., Balenciaga's apology) worsen backlash. So, luxury brands must address controversies swiftly and authentically to mitigate long-term damage. They need to monitor social media to preempt or manage firestorms.

Learning from both the opportunities and challenges outlined, luxury brands must chart a course for the future. The following section outlines emerging trends and strategic directions shaping the luxury industry in the coming years.

FUTURE DIRECTIONS FOR LUXURY BRANDS

Luxury brands will continue to evolve with time. Figure 5 gives a glimpse of key future directions for luxury brands.

Figure 5. Future Directions for Luxury Brands

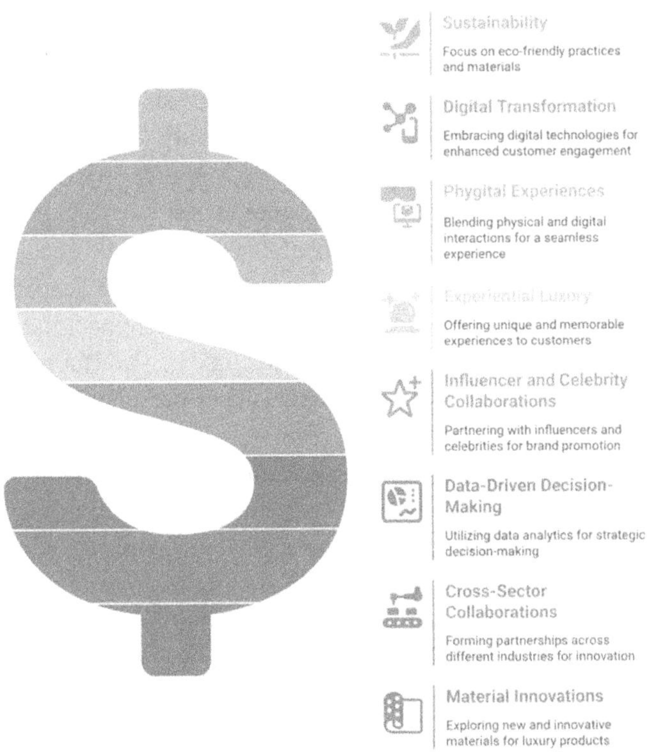

Source: Author creation

Sustainability and Luxury: Luxury brands will have to prioritize sustainability. They must incorporate eco-friendly materials and practices (Wu et al., 2025). Such brands can appeal to environmentally conscious consumers and improve their brand image. The emergence of slow fashion and sustainability-focused consumption patterns among wealthy consumers is influencing luxury brands to adopt circular economy principles and eco-conscious branding (T. Tarnanidis et al., 2025). Consumers demand verifiable ethical sourcing and production processes. Blockchain adoption may increase supply chain transparency. Circular business models like resale and rental will gain further traction. Authentic communication regarding environmental impact becomes crucial. Brands must adopt circular business models such as product take-back schemes and luxury resale platforms. Upcycled and ethically sourced materials are increasingly integrated into core product lines. They reflect evolving consumer values and regulatory pressures. Brands like Stella McCartney and Gucci have already adopted sustainable practices. Increased researcher focus is seen on the once-exclusive luxury tourism and its transformation into a regenerative model that is more sustainable. Regenerative tourism is close to biodiversity conservation, cultural appreciation, and community inclusion. It also offers green branding opportunities, allowing luxury brands to be closer to UN SDGs (Arantes et al., 2025).

Digital Transformation: Luxury brands will invest in digital technologies. Businesses quickly adopt virtual and augmented reality (VAR) and artificial intelligence (AI). Brands' adaptation of technologies requires strategic foresight and agility. These technologies will enhance customer experiences and provide personalized services. For instance, virtual try-ons and personalized recommendations will become increasingly popular. Hyper-personalization will deepen through AI and advanced data analytics. Predictive modeling may anticipate individual client needs proactively. Bespoke digital configurations could become more sophisticated. The integration of artificial intelligence enables hyper-personalized experiences in product development and consumer interaction. Algorithmic curation of offerings, informed by real-time data analytics, enhances individual engagement and optimizes conversion rates.

Phygital experiences: Phygital experiences require seamless integration across all touchpoints. Augmented reality can enrich both online discovery and physical retail. Virtual reality offers the potential for immersive brand storytelling. Smart store technologies might enable highly personalized in-person service. Metaverse strategies and NFT utility will continue to evolve.

Several future research possibilities exist in this segment. Empirical validation is needed to test strategies across luxury sectors like fashion, hospitality, and automotive. Sector-specific studies can investigate how consumers perceive phygital experiences and their impact on behavior. These studies can also examine the long-term effects on brand loyalty and equity. Comparative analyses between markets or cultures can examine how local preferences shape tech adoption. The studies can include augmented reality, virtual reality, and NFTs. Longitudinal studies that look into how phygital experiences evolve and influence customer journeys can also be taken up. Partnerships with the industry can provide the technical know-how and infrastructure to conduct real-world experiments. Advanced analytics and AI-driven personalization can also be used in phygital contexts.

Experiential Luxury: Luxury brands aim to create immersive experiences. Exclusive events and personalized services are two examples. These experiences will bring brand loyalty and emotional connections with customers. Direct-to-avatar commerce in virtual worlds has come up recently. It can become a new brand engagement space and a novel revenue stream. Brands like Louis Vuitton and Burberry have successfully created experiential luxury experiences.

Influencer and Celebrity Collaborations: Luxury brands will continue to partner with influencers and celebrities. The buzz and appeal to new audiences that they create cannot be underrated. They are impactful and can drive brand awareness and sales. For example, consider Mercedes-Maybach from the automotive luxury-tech sector. They collaborated with Dolby Atmos and created an immersive in-car audio experience. Celebrities can be used to support it with some promotional content. Their involvement can reinforce the brand's image as a leader in opulence and sound engineering. The brand appeals to affluent consumers who value heritage and innovation.

Data-Driven Decision-Making: Luxury brands will use data analytics to inform their marketing strategies and product development. Data will help brands understand consumer preferences and behaviors. This will enable luxury brands to make informed decisions and optimize their marketing efforts. Data-driven insights enable predictive demand planning, reducing overproduction and minimizing waste. Advanced CRM systems integrate omnichannel touchpoints, creating unified consumer profiles and enhancing loyalty program effectiveness. Geotargeted marketing campaigns leverage location-based data to deliver contextually relevant messaging and drive in-store traffic. Adaptive brand architectures allow luxury firms to address local cultural nuances while maintaining global consistency and prestige.

Cross-sector collaborations: Collaboration between luxury brands and technology firms accelerate innovation in smart products and digital services. Strategic partnerships with fintech companies facilitate seamless, secure payment solutions and enhance the luxury purchase journey. Future empirical research can add considerable value. Researchers can explore how cross-sector collaborations between luxury brands and technology firms (including fintech) impact innovation, customer experience, and brand perception. Possible avenues are mentioned in Table 3.

Table 3. Future Empirical Research Directions on Cross-Sector Collaborations

Research Area	Potential studies that future researchers can take up
Consumer Perceptions and Adoption	• Conduct quantitative studies or experiments to measure how luxury consumers perceive and adopt Smart products and services from such collaborations. • Survey or interview customers to assess perceived trust, exclusivity, and value in luxury-tech or luxury-fintech co-created offerings.
Impact on Purchase Behavior	• Analyze how fintech-enabled seamless payment solutions influence purchase intent, conversion rates, and customer loyalty in digital and physical luxury retail environments. • Use behavioral data to examine if and how secure, tech-driven payment methods alter luxury shoppers' journeys or basket sizes.
Brand Equity and Co-Branding Outcomes	• Longitudinal studies could track the effects of technology collaborations on luxury brand equity, reputation, and customer retention. • Comparative case studies can investigate which types of technology partners (e.g., fintech vs. smart home tech) yield the most positive brand outcomes.
Innovation and Value Co-Creation	• Empirically test how joint product development processes influence the speed and quality of innovation in luxury goods and services. • Examine the role of customer co-creation in cross-sector partnerships (e.g., luxury brands and tech start-ups involving end-users in product design).
Sector- and Culture-Specific Effects	• Explore variations in collaboration outcomes across luxury sectors (fashion, automotive, hospitality) and geographic markets, considering cultural differences in technology adoption and luxury consumption.
Trust and Risk Management	• Assess consumer perceptions of data security, privacy, and risk in luxury-fintech collaborations, especially in high-value transactions.

Source: Author creation

Material innovations: Material innovation presents significant opportunities for differentiation. Investment in sustainable, high-performance, or Smart textiles will grow. The selection of construction materials for event booths during exclusive high-profile promotions (HPP) is vital (Wu et al., 2025). Opting for sustainable alternatives can dramatically reduce the environmental burden of these prestigious events, sometimes by approximately 73%. Collaborations with technology firms can drive material science breakthroughs. Materials themselves may become key storytelling elements.

CONCLUSION

This chapter explored how co-branding, co-creation, and co-marketing strategies transform the Smart luxury industry. The analysis shows that these strategies, enabled by technology, are crucial for creating value and enhancing consumer engagement. Co-creation creates a collaborative environment between brands and consumers. Co-marketing and co-branding use this partnership to amplify the brand reach and create innovative offerings. They allow luxury brands to meet the evolving consumer preferences for unique, personalized experiences and to integrate technological advancements. However, successful implementation requires careful strategic alignment and brand compatibility. More importantly, understanding consumer perceptions is required. Technologies such as AI and Data Analytics can help to some extent. This study is limited by its focus on current trends and may not fully capture the long-term evolution of the strategies. Hence, future research should investigate the long-term impact of these strategies on brand equity and consumer loyalty in the Smart luxury sector.

REFERENCES

Arantes, L., Vaz, L., & Costa, S. B. (2025). From Exclusive to Regenerative: The Transformative Power of Luxury Tourism. In Sousa, B. B., Arantes, L., & Bhartiya, S. (Eds.), *Regenerative Tourism for Social Development* (pp. 93–118). IGI Global., DOI: 10.4018/979-8-3373-0235-5.ch005

Basile, V., Brandão, A., & Ferreira, M. (2024). Does user-generated content influence value co-creation in the context of luxury fashion brand communities? Matching inclusivity and exclusivity. *Italian Journal of Marketing*, *2024*(4), 419–444. DOI: 10.1007/s43039-024-00103-9

Bhardwaj, P., Priya, P. S., & Sharma, S. (2025). Exploring the Influence of Circular Economy Paradigm on Convergence of Sustainability and Luxury Apparel: A Systematic Review. *Circular Economy and Sustainability*, *5*(3), 2295–2316. Advance online publication. DOI: 10.1007/s43615-024-00489-w

Bilancia, A., Costa, F., & Staudacher, A. P. (2025). Achieving sustainability and circular economy in the luxury fashion industry through lean practices: A systematic literature review. *Computers & Industrial Engineering*, *111107*, 111107. Advance online publication. DOI: 10.1016/j.cie.2025.111107

Cardoso, C., Casteleiro-Pitrez, J., & Lobo, H. (2025). Embracing the Luxury Paradox: New Luxury, Digitalization, and Digital Experiences in the Metaverse. In Raposo, D., Neves, J., Silva, R., Correia Castilho, L., & Dias, R. (Eds.), *Advances in Design, Music and Arts III* (Vol. 49, pp. 449–466). Springer Nature Switzerland., DOI: 10.1007/978-3-031-73705-3_34

Chen, Y., & Choi, J. (2025). The Effectiveness of Collaboration: The Features of Co-Branded Products Affecting Brand Equity and Consumer Sentiment. *The Journal of Society for E-Business Studies*, *30*(1), 1–22. DOI: 10.7838/jsebs.2025.30.1.001

Choudhary, N., & Vijaya Kittu, M. (2021). Revenge Travel or the Road Less Traveled? Opting for a suitable option in Pandemic Times. *International Journal of Advance & Innovative Research*, *8*(3). Advance online publication. DOI: 10.17605/OSF.IO/83BCG

Fletcher, K.-A., & Stephen, C. (2025). Luxury Brands in Dynamic Times. In Gbadamosi, A. (Ed.), *Brands, Branding, and Consumerism* (pp. 311–344). Springer Nature Switzerland., DOI: 10.1007/978-3-031-80859-3_8

Füller, J. (2010). Refining Virtual Co-Creation from a Consumer Perspective. *California Management Review*, *52*(2), 98–122. DOI: 10.1525/cmr.2010.52.2.98

Green, A. (2024). Private Art Museums in Paris: How France's Luxury Titans Shape Urban Branding Through Iconic Architecture. *Luxury*, *11*(1–3), 69–91. DOI: 10.1080/20511817.2025.2466345

Holmqvist, J., Berger, C., De Keyser, A., & Verleye, K. (2025). Luxury in the Circular Economy: An Engagement Journey Perspective. *Journal of Consumer Behaviour*, cb.2460. DOI: 10.1002/cb.2460

Husain, R. V. S., Kamath, R., & Moorthy, J. (2025). Navigating the Sustainability–Luxury Paradox: Bibliometric Insights and Research Directions. *Journal of Consumer Behaviour*, cb.2469. DOI: 10.1002/cb.2469

Jang, J., & Kang, J. (2025a). Blockchain meets luxury: The role of NFT authentication in luxury retail platforms. *Journal of Retailing and Consumer Services*, *84*, 104262. DOI: 10.1016/j.jretconser.2025.104262

Jang, J., & Kang, J. (2025b, January 18). From New to Pre-Loved: The Impact of Blockchain-enabled NFT Authentication on Warranting Value and Assurance in Luxury Markets. *Making Waves Toward A Sustainable and Equitable Future*. Making Waves Toward A Sustainable and Equitable Future. DOI: 10.31274/itaa.18818

Khaliq, L. N., & Manda, V. K. (2023). Customer Experience in the Web 3.0 Era: The Meeting of Blockchain and the Metaverse. In Majeed, M., Ofori, K. S., Amoako, G. K., Alolo, A.-R., & Awini, G. (Eds.), (pp. 258–274). Advances in Marketing, Customer Relationship Management, and E-Services. IGI Global., DOI: 10.4018/978-1-6684-7649-9.ch015

Khamoushi Sahne, S. S., & Kalantari Daronkola, H. (2025). Enhancing loyalty in luxury fashion with AI: The mediation role of customer relationships. *Asia-Pacific Journal of Business Administration*. Advance online publication. DOI: 10.1108/APJBA-08-2024-0456

Kłeczek, R., & Hajdas, M. (2025). The appreciation task for luxury brand–artist collaborations. *Journal of Brand Management*, *32*(3), 203–226. Advance online publication. DOI: 10.1057/s41262-025-00378-z

Leonhard, H., Nagl, M., & Schaefers, W. (2025). Virtual land in the Metaverse? Exploring the dynamic correlation with physical real estate. *Journal of European Real Estate Research*, *18*(1), 121–151. DOI: 10.1108/JERER-04-2024-0025

Li, J., & Kang, J. (2025). Human-centred AI vs machine-centred AI: Navigating AI ethical crises in luxury retail. *International Journal of Retail & Distribution Management*, *53*(7), 639–653. Advance online publication. DOI: 10.1108/IJRDM-11-2024-0600

Lu, C., Zhang, Y., Chen, Z., Sang, Z., & Liu, Z. (2025). How do marketing practices impact price premium of co-branded sportswear? The moderating effect of product category fit and partner brand name. *International Journal of Sports Marketing & Sponsorship*. Advance online publication. DOI: 10.1108/IJSMS-12-2023-0248

Manda, V. K., Sagi, S., & Yadav, A. (2024). Blockchain in Advertising and Marketing: Revolutionizing the Industry Through Transparency and Trust. In Tarnanidis, T. K., & Sklavounos, N. (Eds.), (pp. 89–112). Advances in Marketing, Customer Relationship Management, and E-Services. IGI Global., DOI: 10.4018/979-8-3693-2754-8.ch005

Nurrosidah, A., Djatmika, E. T., & Dewi, T. S. (2025). Exploring the Role of Perceived Value and Customer Attitude in Co-Branding: Impact on Purchase Intention for Somethinc Cosmetics X Bittersweet by Najla. *Journal of Applied Business. Taxation and Economics Research*, *4*(3), 445–458. DOI: 10.54408/jabter.v4i3.365

Peragallo, N. A. (2025). *Empowering Place: Unlocking Value for Investors by Integrating Indigenous Values in Luxury Hospitality* [Program in Real Estate Development., Massachusetts Institute of Technology]. https://hdl.handle.net/1721.1/158878

Prahalad, C. K., & Ramaswamy, V. (2005). *The future of competition: Co-creating unique value with customers* (Nachdr.). Harvard Business School Press.

Ramaswamy, V., & Ozcan, K. (2014). *The Co-Creation Paradigm*. Stanford University Press., DOI: 10.1515/9780804790758

Sathakathulla, A. A. (2025). The Influence of Credit Cards on Consumer Buying Behavior among Expats in the UAE. *International Journal of Advances in Engineering and Management*, *7*(1), 347–361. DOI: 10.35629/5252-0701347361

Schuck, K. A., Derqui, B., Grimaldi, D., & Mehn, A. (2025). Sustainability and Luxury in Hospitality: Achieving Compatibility for a Better Future. *International Journal of Hospitality & Tourism Administration*, •••, 1–24. DOI: 10.1080/15256480.2025.2476405

Semwal, R., Tripathi, N., Tyagi, P., & Singh, A. (2024). Synergising Brands: Licensing and Collaboration Strategies in Tourism. In Albattat, A., Azman, N., Valeri, M., Talib, Z. M., & Tyagi, P. K. (Eds.), *Advances in Hospitality, Tourism, and the Services Industry* (pp. 163–180). IGI Global., DOI: 10.4018/979-8-3693-4171-1.ch007

Shams, S. M. R., Brown, D. M., & Hardcastle, K. (2025a). Brand Management, Marketing Communication, and Promotion Management for People, Planet, and Profit. In S. M. R. Shams, D. M. Brown, & K. Hardcastle, *Sustainable Marketing* (pp. 227–298). Springer Nature Switzerland. DOI: 10.1007/978-3-031-79130-7_4

Shams, S. M. R., Brown, D. M., & Hardcastle, K. (2025b). Product and Service Development for People, Planet, and Profit. In S. M. R. Shams, D. M. Brown, & K. Hardcastle, *Sustainable Marketing* (pp. 103–167). Springer Nature Switzerland. DOI: 10.1007/978-3-031-79130-7_2

Slaton, K., & Pookulangara, S. (2025). Exploring Consumer Use of Digital Product Passports for Secondary Luxury Consumption: A Conceptual Study. *Advances in Consumer Research. Association for Consumer Research (U. S.)*, *2*(1). https://acr-journal.com/article/exploring-consumer-use-of-digital-product-passports-for-secondary-luxury-consumption-a-conceptual-study-902/

Sun, Y., Wang, R., Bai, H., Whyatt, G., & Ohandi, T. T. (2024). Perceived values of luxury fashion brands—Evidence from Indonesian Millennials. *SN Business & Economics*, *4*(12), 169. DOI: 10.1007/s43546-024-00755-x

Surià, M. H. (2024). *Architecture and Design of Luxury Fashion Brands' Flagships and Boutiques: A Comparison Between Western and Eastern Culture* [Master in Fashion Management Thesis, ISEM Fashion Business Schoo, Universidad de Navarra]. https://dadun.unav.edu/entities/publication/2a3eec15-f55b-4b64-9b38-98b24bd983ac

Tarnanidis, T., Karypidis, M., Manda, V. K., Bilali, S., & Papachristou, E. (2025). Discovering the Patterns of Slow Fashion Consumption in the Rich Market. In *Sustainable Practices in the Fashion and Retail Industry* (p. 20). IGI Global., https://www.igi-global.com/gateway/chapter/372015 DOI: 10.4018/979-8-3693-9959-0.ch007

Tarnanidis, T., Papachristou, E., Karypidis, M., & Manda, V. K. (2024). Luxury Fashion Consumption and Branding in the Affluent Market. In Tarnanidis, T. K. (Ed.), (pp. 161–172). Advances in Marketing, Customer Relationship Management, and E-Services. IGI Global., DOI: 10.4018/979-8-3693-7535-8.ch008

Vargo, S. L., Maglio, P. P., & Akaka, M. A. (2008). On value and value co-creation: A service systems and service logic perspective. *European Management Journal*, *26*(3), 145–152. DOI: 10.1016/j.emj.2008.04.003

Venturini, A. (2023). *Understanding the Consumer-Brand Relationship in the Hyperconnected Environment* [PhD program in Communication, Markets and Society, IULM]. https://apeiron.iulm.it/bitstream/10808/64689/2/venturini_brands_over_heels.pdf

Wang, Z. (2024). Heytea: Revolutionizing the Beverage Industry With Innovation and Customer Engagement. In Elhaoussine, Y., Appelgryn, H., & Wang, L. (Eds.), (pp. 19–42). Advances in Business Strategy and Competitive Advantage. IGI Global., DOI: 10.4018/979-8-3693-2921-4.ch002

WTO & UdSi (Ed.). (2025). *Fashion and Cultural Tourism – Connecting Creators, Businesses and Destinations*. UN Tourism., DOI: 10.18111/9789284426287

Wu, D., Ren, Z., Lei, H., & Zhao, S. (2025). Life cycle assessment of a luxury fashion brand's HPP event booth in the context of circular economy. *Scientific Reports, 15*(1), 5153. DOI: 10.1038/s41598-024-84528-4 PMID: 39934184

Youn, S., Hwang, J., & Ju, N. (2025). Blockchain transparency in reducing second-hand luxury shopping uncertainty and the role of price consciousness. *Fashion and Textiles, 12*(1), 10. DOI: 10.1186/s40691-025-00420-3

Zhang, Y. (2025). Cross-border co-branding influencing consumer purchases: Case of Luckin Coffee's Moutai Flavored Latte. In Chew, F. P. (Ed.), *Global Dialogue on Media Dynamics, Trends and Perspectives on Public Relations and Communication* (p. 6). CRC Press., DOI: 10.1201/9781003564966

Zhang, Y., Guo, X., & Qu, Y. (2025). Brand extension or co-branding: The consumer's attribution of responsibility to the crossover strategies of heritage brands. *Journal of Retailing and Consumer Services, 82*, 104133. DOI: 10.1016/j.jretconser.2024.104133

KEY TERMS AND DEFINITIONS

Co-Creation: Co-creation is the collaborative process where multiple stakeholders work together to design, develop, or innovate products, services, or experiences. It focuses on shared input, creativity, and mutual value creation. It uses diverse perspectives to achieve outcomes that benefit all parties involved.

Co-Marketing: Co-Marketing refers to a strategic collaboration between two or more brands to jointly plan, execute, and promote marketing campaigns or initiatives. Co-marketing centers on synchronized promotional efforts that maximize each partner's audience reach. It improves brand equity and marketing capabilities. In the Smart luxury industry, co-marketing increases visibility. It enhances consumer engagement and aligns brand narratives across digital and experiential platforms.

Co-Branding: Co-branding is a strategic marketing partnership between two or more brands to create a combined product, service, or campaign. The partners take advantage of each partner's strengths, equity, and consumer perception. The collaboration enhances value proposition, differentiates offerings, and accesses new markets. They blend complementary attributes such as luxury prestige with technological innovation while maintaining brand integrity.

Luxury: Luxury refers to high-end products or services that offer exceptional quality, craftsmanship, and exclusivity. Luxury goods and experiences often convey status, sophistication, and refinement. They typically possess unique characteristics, such as high-quality materials, scarcity, attention to detail and precision, and unique design and aesthetics. Such products generally have premium pricing.

Smart Luxury: Smart luxury uses advanced digital technologies, data-driven personalization, and intelligent services in the traditional luxury sector. Smart luxury aims to enhance customer experience, product value, and brand engagement. It uses technologies such as Artificial Intelligence, Augmented Reality (AR), Virtual Reality (VR), Smart Devices, and the Internet of Things (IoT).

Chapter 4

The Psychology of Smart Consumer Products:
Behavioral Insights for Co-Branding Success

Jomon Jose M.
https://orcid.org/0009-0003-4341-9936
Sahrdaya Institute of Management Studies, Kodakara, India

Jyothis R. Mathews
https://orcid.org/0009-0003-8279-6080
Ahalia School of Management, Palakkad, India

ABSTRACT

Co-branding in the luxury space is a travelled and lofty highway. Luxury brands have historically guarded their brand and market zealously, enduring only through exclusivity, legacy, and craftsmanship. Even the stodgiest of luxury players have been forced to adopt strategic partnerships in the push for personalization and digitisation of consumer lives, which means even faster. Imagine co-branding once in a high-end round table, with heritage extended to monogamy with innovation — the strategic marriage is for real. Smart tech in luxury commodities has added functionality, data driven design, and consumer engagement. The chapter will probe the biography and obituary of co-brands in luxury territories, dig beneath somekey drivers and innovative trends we can expect playing a pivotal role for brand binding alliances, and how this is shapingup as a future (sensory) landscape.

DOI: 10.4018/979-8-3373-3010-5.ch004

1. INTRODUCTION

The contemporary luxury market is being radically reshaped by the intersection of technological innovation and changing expectations of consumers. No longer driven by timeless craftsmanship and exclusivity alone, luxury is now being ever more characterized by digital sophistication and smart functionality (Shamim et al., 2024); (Tasci & Guillet, 2011); (Joy et al., 2022); (Hassan et al., 2015). Intelligent consumer goods—from AI-driven fashion accessories to connected luxury vehicles—are rewriting the rules of owning and enjoying luxury (Machado et al., 2022); (C. Chen et al., 2022); (Suzuki & Kanno, 2022). In the new environment, co-branding is at its forefront by facilitating collaboration between heritage luxury brands and innovation leaders to blend their respective strengths (P. Sharma et al., 2021); (Shamim et al., 2024); (Chiang, 2023).. More than a marketing tool, co-branding for the smart luxury space engages the deeper psychological drivers of identity, aspiration, trust, and emotional connection (Machado et al., 2022); (P. Sharma et al., 2021); (Chiang, 2023).. This chapter examines the cornerstones of smart luxury branding and the behavioral drivers of co-branded product success (Tasci & Guillet, 2011); (Hassan et al., 2015); (Joy et al., 2022); (Suzuki & Kanno, 2022).

1.1 Smart Luxury Product Overview And The Technological Role In Reshaping Luxury Branding

In this digital age, luxury has shifted from its traditional values of rarity, heritage and hand-made-work to a new level of creativity; intelligence and interactivity (Pantano et al., 2018); (Gao & Yang, 2022); (Choi & Kim, 2016); (Majeed et al., 2024). The convergence of cutting-edge technology and high-end aesthetics in smart luxury products is paving the way for a new era in consumer luxury branding (Chiang, 2023); (C. Chen et al., 2022). Merging exclusive demands of luxury with the practical prowess of contemporary tech, these goods set a new precedent for owning and enjoying luxury (Gao & Yang, 2022); (Choi & Kim, 2016). Gone are the days of mere handcrafted perfectionism or family heritage; today's high-end consumers expect a seamless blend of sophistication and smart functionality (C. Chen et al., 2022); (Pantano et al., 2018). This is most clearly seen through collaborations such as the Apple-Hermès smartwatch, pairing Silicon Valley know-how with Parisian sophistication, or Louis Vuitton's Tambour Horizon, a luxury watch rebooted with digital connectivity and worldwide travel capabilities (Gao & Yang, 2022); (Choi & Kim, 2016); (Majeed et al., 2024). These are not simply symbols of wealth or prestige—they define a superior lifestyle that is globally relevant and technologically enabled (Chiang, 2023); (C. Chen et al., 2022); (Pantano et al., 2018).

Technology is not a secondary aspect of the luxury storyline anymore—it is now a lead player. Artificial Intelligence (AI), the Internet of Things (IoT), Augmented Reality (AR), and blockchain are transforming the luxury product lifecycle, from creation and production to personalization, customer experience, and after-sales service (Joy et al., 2022); (Zhou & Zhao, 2024); (Hoyer et al., 2020); (Zuo, 2024). Intelligent luxury goods are not only meant to impress the eye but to stimulate the mind and anticipate the needs of the consumer by being embedded with intelligence (Hoyer et al., 2020); (Zuo, 2024). For instance, luxury cars now feature AI-driven driving, biometric readers, and voice-controlled digital assistants that present a distinctly intuitive user interface, transforming a high-end car into a personalized command center (L. Xu & Mehta, 2022); (De Boissieu et al., 2021). In a similar vein, high-end fashion establishments are using AR for virtual try-ons and interactive catwalk shows, enabling customers to experience luxury within highly customized, tech-driven environments (L. Xu & Mehta, 2022); (De Boissieu et al., 2021). These technologies mirror a new language of luxury that values user experience, responsiveness, and smart customisation as much as historical signals of price, material, and heritage (Joy et al., 2022); (Zhou & Zhao, 2024); (Hoyer et al., 2020).

It is the psychological connection which smart luxury goods evoke among today's consumers that makes this transfiguration so intriguing. These goods satisfy not only extrinsic aspirations for social standing and esteem, as well as for convenience, control, and self-expression, but they satisfy intrinsically as well (De Kerviler & Rodriguez, 2019); (Dubois et al., 2020); (Das et al., 2021). The capability to keep track of one's health using a diamond-studded smart ring, for example, or to personalise light and mood through a voice-controlled home luxury system, is not only appealing to the senses but to the mind as well (Chan & Northey, 2021); (Das et al., 2021). Smart luxury goods address the twenty-first-century consumer's quest for overall well-being, connectedness, and digital empowerment, thus positioning material commodities within contemporary values (Dubois et al., 2020); (Chan & Northey, 2021). In effect, they move the locus of luxury branding away from ownership of objects to orchestrating experiences (De Kerviler & Rodriguez, 2019); (Dubois et al., 2020); (Chan & Northey, 2021).

In addition, technology allows luxury brands to balance exclusivity with mass personalization. With AI-driven analytics and large data analytics, luxury houses can create hyper-personalized experiences and solutions that engage on a deep, personal level with uniquely individual consumers (Rahman et al., 2023); (Park & Ahn, 2024); (Sahne & Daronkola, 2025); (Song & Bonanni, 2024). This ability shifts the dynamic between buyer and brand from transactional to relational. It's not just consumers who buy luxury, but also those who co-create, participate, and experience it (Sahne & Daronkola, 2025); (Song & Bonanni, 2024); (N. V. P. S et al., 2021). Through this collaborative approach, not only does it foster greater brand loyalty but also incorporates high-end products into everyday life, catering to both practical and aspirational needs (Rahman et al., 2023); (Park & Ahn, 2024); (N. V. P. S et al., 2021). In the end, technology blending with luxury branding represents a tectonic shift in the very essence of luxury (Creevey et al., 2021); (Joy et al., 2022). It's no longer static, it'll be indestructible; it has become dynamic, intelligent and absorbed in the rhythms of life. Smart technologies are being used by top-tier brands to redefine the contours of desire, experience, and prestige for 21 century consumers (Al-Issa & Thanasi, 2024); (Creevey et al., 2021); (Joy et al., 2022).

1.2 Importance of Co-Branding in Luxury Markets

In luxury marketing, co-branding has become a key strategy rather than a peripheral branding strategy. With consumers of luxury seeking not only heritage- and craftsmanship-defined exclusivity but innovation, experience, as well as technological resonance, co-branding creates a rich terrain for richer value propositions (Quamina et al., 2023); (Suzuki & Kanno, 2022). Co-branding within luxury markets is the strategic collaboration between two or more dominant brands—an often-disparate set of brands across different sectors—that unite to create a synergistic offering worth more than the sum of its parts(Zhao, 2024); (Shen et al., 2017); (X. Chen & Li, 2022). These collaborations harness the strengths, cachets, and consumer bases of each partner brand to develop distinctive, extremely differentiated products with commanding premium attention and pricing power(Shen et al., 2017); (X. Chen & Li, 2022). In the process, co-branding achieves brand storytelling that surpasses single-brand stories, often giving rise to a novel culture and commerce phenomenon(Suzuki & Kanno, 2022); (Zhao, 2024)..By collaborating with brands such as Gucci, Nordstrom and Montblanc or another famous brand like Arnaud Bergha from H&M, iconic collaborations can not only drive sales but also create lasting brand equity and market resonance among target audiences (Quamina et al., 2023); (Suzuki & Kanno, 2022); (Shen et al., 2017).

In addition to being a fresh tactic, co-branding is essential in luxury markets as it achieves strategic branding objectives such as market expansion, cross segmentation, and positioning changes (Zhao, 2024); (Shen et al., 2017). In a world where expectations of consumers change at a rapid pace and that of brands is growing weaker by the day, co-branding helps luxury brands remain relevant by drawing on the knowledge leveraged by their partners within a given area of expertise, innovation potential, or cultural capital (Quamina et al., 2023); (Suzuki & Kanno, 2022); (X. Chen & Li, 2022). For example, a conventional luxury apparel brand partnership with a technological innovator signifies adaptability and vision for the next generation without compromising on its core high-end base of clients. The double-appeal widens the market for the brand while maintaining its aspirational factor (Zhao, 2024); (Shen et al., 2017); (X. Chen & Li, 2022). Also, such partnerships typically yield limited-series or bespoke solutions, which align seamlessly with the luxury industry's obsession with rarity and uniqueness. These starts serving as culture-driven artifacts stimulating desire, virality, as well as scarcity—all key psychological drivers for luxury consumption (Quamina et al., 2023); (Suzuki & Kanno, 2022); (Zhao, 2024).

At a psychological level, co-branded luxury goods fulfill a multi-faceted consumer psychology. Prestige brands contribute to the consolidation of identity and status, while alternative experiences evoke a sense of exploration and innovation (Yu et al., 2020); (Shaikh & Gummerum, 2025). The co-branding approach creates a sense of intimacy and insider knowledge, as consumers experience tangible cultural events or aesthetic convergence (X. Chen & Li, 2022); (Yu et al., 2020). In addition, co-branding often involves a multiple-layered symbolism: they are not merely perceived as statements of wealth or taste but also statements about contemporary culture, innovation, and personal intelligence. This multifaceted role adds experiential and emotional value to the luxury product, thus increasing attachment to a brand and premium price paid (X. Chen & Li, 2022); (Shaikh & Gummerum, 2025).

Strategically, co-branding is a vehicle through which luxury brands can address market challenges like economic recessions, generation changes, or technological disruption too. With complementary brands, they will be able to co-share risk, co-invest, benefit from one another's distribution networks, customer understanding, and brand capital (Quamina et al., 2023); (Suzuki & Kanno, 2022). For sectors where, discriminative consumers are present and competitive differentiation is extremely important, co-branding allows innovation and storytelling on a scope few brands can do independently (Suzuki & Kanno, 2022); (Abedsoltan et al., 2022); (Wu & Kim, 2023). It is not only a device for exposure and hype creation but a platform for innovation leadership, strategic rebirth, as well as long-term development of a brand (Quamina et al., 2023); (Suzuki & Kanno, 2022); (Wu & Kim, 2023). At its core, co-branding within luxury markets is no longer a periodic creative exercise—it is a deliberate strategic decision that strengthens brand storytelling, increases experiential content, and establishes new definitions of value for a discerning, constantly changing consumer audience(Quamina et al., 2023); (Suzuki & Kanno, 2022). Co-branding is a most effective driver of differentiation, aspiration, and long-term market applicability within the hypercompetitive environment of the luxury market when done with authenticity, aesthetic balance, and respect for the values of both brands involved(Quamina et al., 2023); (Suzuki & Kanno, 2022); (Shen et al., 2017).

1.3 The Psychological Impact of Branding on Consumer Behavior

In the luxury market, branding far surpasses a logo, name, or visual identity—it is a psychological construct rooted within the perception, decision process, and self-concept of the consumer (Kowalczyk & Mitchell, 2021); (Shahid & Paul, 2021). The strength of a luxury brand comes from its ability to create emotional resonance, instill trust, and convey symbolic meaning. The psychological aspect is especially powerful where high-technology-enabled smart consumer goods meet high-end luxury cues (Prentice & Loureiro, 2018); (J. Kumar, 2024). The consumer does not buy a luxury product, they internalize what it signifies—status, accomplishment, being one-of-a-kind, or cultural capital (J. Kumar, 2024); (Shahid & Paul, 2021). Branding, then, is a mental short cut or a heuristic, allowing consumers to rapidly determine product value and societal meaning amidst a saturated marketplace (Kowalczyk & Mitchell, 2021); (Shahid & Paul, 2021);(Prentice & Loureiro, 2018).

At its essence, the psychological effects of branding can be explained by reference to a number of well-established approaches to understanding consumer behavior. The Theory of Planned Behavior (TPB) suggests attitudes, subjective norms, and perceived behavioral control determine intention, which itself causes actual behavior (Philiastides & Ratcliff, 2013); (Reimann et al., 2011); (V. Kumar & Kaushal, 2021). In luxury branding, consumers develop positive attitudes towards brands on the basis of perceived prestige, craftsmanship, and heritage, which are reinforced through social approval (subjective norms) and strengthened if they feel financially or socially able to pay for the product (perceived behavioral control) (Guan, 2023); (V. Kumar & Kaushal, 2021). Maslow's Hierarchy of Needs is a second lens through which to view this, with luxury commodities placed at the esteem and self-actualization needs stage, affirming consumers' aspirations for success, status, and personal accomplishment (Philiastides & Ratcliff, 2013); (Reimann et al., 2011); (Guan, 2023).

In the age of the internet, the psychological power of branding is heightened by the ubiquity of stories about brands across media channels. Consumers are exposed continuously to staged luxury images, endorsements by influencers, experiential branding, and customized marketing—all orchestrated to shape the development of schemas about a brand in the head (Christodoulides, 2009); (Yoganathan et al., 2021); (Budnitsky & Jia, 2018). Schemas are mental frameworks that contain knowledge and feeling associations about a brand, directing attention, memory, and judgment (Kur, 2016); (Yoganathan et al., 2021). The more congruent and emotionally charged the storyline about a brand, the stronger the schema, the more likely a consumer will display advocacy and commitment behaviors (Yoganathan et al., 2021); (Budnitsky & Jia, 2018). For smart luxury goods, these stories tend to incorporate innovation, customization, and contemporary sophistication dimensions, developing a powerful blend of emotional and utility appeal (Kur, 2016); (Christodoulides, 2009).

One of the most powerful psychological effects of branding is the ability to exert influence on identity formation and self-expression. People tend to use luxury brands as "identity markers" to present their ideal selves to the world (Chernev et al., 2011); (Harmon-Kizer et al., 2013); (Kongsri & Jaroenwanit, 2024). This is summarized by the theory of self-congruity, stating that people will buy brands whose personality is aligned with their own conception of themselves or their desired conception of themselves (Kongsri & Jaroenwanit, 2024); (Harmon-Kizer et al., 2013). A luxury smartwatch branded by a famous heritage house and a prominent technology firm, for instance, enables the buyer to signal both high taste as well as technological expertise—qualities that might match their conception of themselves or their societal positioning aspirations (Chernev et al., 2011); (Kongsri & Jaroenwanit, 2024). In addition, a perception of membership among a high-end set of early movers or culturally astute buyers reinforces emotional affinity and perceived value for the brand (Chernev et al., 2011); (Harmon-Kizer et al., 2013).

Emotional branding is another important dimension, which is about establishing a relationship between the user and the brand on the basis of storytelling, experience through the senses, and symbolic meaning (Y. Kim & Sullivan, 2019); (Bandyopadhyay et al., 2021). In luxury markets, emotional branding is a differentiator, as it makes a product not an object of function but a significant possession (Bishnoi & Singh, 2021); (S. Q. U. Hassan et al., 2016). When luxury brands are able to access emotions comprised of nostalgia, aspiration, pride, or desire, they engage the brain's limbic system, which handles memory and emotion, resulting in greater recall and preference for a brand (Bandyopadhyay et al., 2021); (Bishnoi & Singh, 2021). The emotional bond is even stronger for co-branded smart products, where the coming together of beauty and innovation results in layered stimulation through the senses as well as the mind (Y. Kim & Sullivan, 2019); (S. Q. U. Hassan et al., 2016).

In addition, the psychological effects of branding also enter dimensions of perceived value and price justification. People pay a premium for luxury co-branded goods not necessarily because they are superior in quality but because they generate psychological benefits for them—social approval, enhancement of self, and richness of experience (Hamzah & Pontes, 2022); (Chiang & Jang, 2007); (Grewal et al., 1998). This is consistent with the tenets of behavioral economics, which state that decision-making is subject to biases, heuristics, and emotional reactions rather than purely logical assessment (Chiang & Jang, 2007); (Grewal et al., 1998). For example, limited edition co-branded merchandise is made more appealing by the scarcity effect, whereas the halo effect results in people inferring high quality and high standing from familiar, known, high-quality-associated brands (Hamzah & Pontes, 2022); (Chiang & Jang, 2007); (Grewal et al., 1998).

In total, the psychological effect of branding on consumer behavior is immense, far-reaching, and rooted firmly within emotional, cognitive, and social drivers (Anisimova, 2007); (Schmitt, 2011); (Sprott & Liu, 2016); (Xue, 2024). Those luxury brands who comprehend these psychological levers, as well as strategically utilize them, are more effective at establishing enduring relationships with consumers, encouraging loyal behavior, and holding a clear, delineated position within a competitive international marketplace (Anisimova, 2007); (Schmitt, 2011). With smarter and smarter technology on the rise, the dilemma—and opportunity—is to balance brand experiences that meet both the logical and emotive aspects of the luxury buyer's mindset (Anisimova, 2007); (Sprott & Liu, 2016); (Xue, 2024).

2. CO-BRANDING IN THE SMART LUXURY SECTOR

The evolution and revolution of the co-branding journey in luxury markets have been driven by changes in consumer lifestyles that have accelerated digitization and sought more personalized experiences (Suzuki & Kanno, 2022); (Zhao, 2024). Luxury brands, once comfortable with exclusivity, heritage, and craftsmanship, and maintaining their brand identity and market position jealously, have been pushed by such trends even towards more conservative luxury houses to form strategic partnerships (Quamina et al., 2023); (Suzuki & Kanno, 2022). Co-branding, which was previously a novelty in the high-end segment, is now a strategic attempt to merge tradition with innovation (Shen et al., 2017); (Chen & Li, 2022). The integration of intelligent technology into luxury products has added new layers of functionality, data-driven design, and consumer engagement (Suzuki & Kanno, 2022); (Zhao, 2024). This section follows the historical trajectory of co-branding in the luxury industry, examines the new trends evolving and redesigning brand alliances, and presents examples of successful alliances that have been able to successfully balance prestige with technological advancement (Quamina et al., 2023); (Shen et al., 2017); (Chen & Li, 2022).

2.1 Co-Branding in the Luxury Industry: A Historical Overview

The idea of co-branding—two or more brands coming together to create one product or service—has roots that go far back in marketing history, but its acceptance within the luxury industry has been more slow and subtle. Luxury branding has traditionally been characterized by exclusivity, heritage, and close management of brand stories and identity(Quamina et al., 2023); (Suzuki & Kanno, 2022). Sharing brand space, particularly with non-traditional or non-luxury partners, was initially viewed as a threat to brand purity. But as consumer expectations changed and markets globalized, co-branding came to be viewed as a strategic vehicle in the luxury arena, providing a means of accessing new demographics, injecting freshness into portfolio brand lines, and creating cultural currency(Quamina et al., 2023); (Suzuki & Kanno, 2022);(Abedsoltan et al., 2022).

Historically, the earliest forms of luxury co-branding were typically within the same industry or closely related domains—such as luxury automakers collaborating with elite fashion designers or premium interior decorators(Shen et al., 2017); (Mrad et al., 2018). Examples include BMW's partnership with Louis Vuitton to create bespoke luggage sets that matched the car's interior, or Bugatti's alliance with ParmigianiFleurier, a Swiss luxury watchmaker, which produced high-end timepieces inspired by the aesthetics of Bugatti's engineering(Shen et al., 2017); (Mrad et al., 2018). They were extensions of mutual values: craftsmanship, innovation, and prestige. They were meant not simply to expand the product's desirability but to grow the story each brand was already executing (Quamina et al., 2023); (Mrad et al., 2018); (Suzuki & Kanno, 2022).

With time, especially from the early 2000s, luxury brands started venturing into more aggressive cross-industry collaborations as digitalization, globalization, and changing consumer values called for creativity in brand narratives (Ramaswamy & Ozcan, 2018); (Shahid et al., 2024). The Hermès-Apple Watch partnership, which was introduced in 2015, was a major turning point (Hsiao, 2017); (Ramaswamy & Ozcan, 2018); (Gil, 2019). It wasn't just about looks; it represented the intersection of fashion and functionality, tradition and technology. Hermès added craftsmanship legacy and exclusivity, whereas Apple added innovation, convenience, and digital attractiveness (Ramaswamy & Ozcan, 2018); (Gil, 2019). The two combined effectively to appeal not just to historical luxury customers but also to younger, technology-enabled customers that cared about personalization and digital literacy (Hsiao, 2017); (Shahid et al., 2024).

This evolution over time was motivated by various forces. Initially, shifting consumer tastes—particularly among millennials and Gen Z—prompted luxury brands to shed their inward-facing habits (Soni & Kumar, 2024); (Ruiz & Cruz, 2023). Younger consumers more and more turned to products that could capture their diverse identities, merging the cultivated with the modern, the analog with the digital (P. Singh, 2024); (Joy et al., 2022). Second, emerging economies started to have an impact on luxury consumption behavior, especially in Asia and the Middle East, where cultural acceptance of technology meshed well with a desire for status purchases (Ruiz & Cruz, 2023); (P. Singh, 2024). Third, growth in experiential and lifestyle branding led luxury houses to extend their conceptual and commercial horizons—providing not merely products, but lifestyles (Soni & Kumar, 2024); (Ruiz & Cruz, 2023); (Joy et al., 2022).

Another turning point in the historical trajectory of luxury co-branding was the arrival of streetwear collaborations, particularly Louis Vuitton's historic collaboration with Supreme in 2017 (Ochkovskaya et al., 2021); (Xiang & Cho, 2024). Although not a technologically focused endeavor, it was a turning point toward cultural co-branding, with even the oldest luxury houses showing that they were prepared to open their codes for others to interpret and keep up with the times (Baumgarth & Wieker, 2020); (Xiang & Cho, 2024). This paved the way for increasingly innovative, tech-enriched partnerships in the years that followed (Ochkovskaya et al., 2021); (Baumgarth & Wieker, 2020).Therefore, the historical development of co-branding among luxury brands is one of a transformation from conservative, tradition-maintaining alliances to revolutionary, cross-industry collaborations that strive to capture the spirit of the times(Quamina et al., 2023); (Mrad et al., 2018); (Shen et al., 2017). While maintaining brand DNA continues to be important, the readiness to co-opt strategic alliances—particularly with technology companies—has become a necessary route for luxury brands to drive innovation, grow, and prosper in a more competitive and digitalized business environment(Quamina et al., 2023); (Mrad et al., 2018); (Shen et al., 2017).

2.2 Key Trends and Innovations in Smart Luxury Branding

The alignment of technology and luxury has ignited a revolutionary tidal wave of branding innovations that reshape the way high-end brands express value, beauty, and exclusivity. During the time of digital transformation, smart luxury branding became an advanced ecosystem that interlaces leading-edge innovation with inspirational stories (Keller, 2009); (Pantano et al., 2018); (Rachbini et al., 2023). Smart luxury branding differs from traditional luxury marketing, which dominated by scarcity, legacy, and status, is being more defined by personalization, interactivity, and digital integration—renewing the old standards of value creation (Keller, 2009); (Pantano et al., 2018); (Rachbini et al., 2023).

One of the most important trends in smart luxury branding is the marrying of wearable technology with fashionable design. This is not just a technological advance, however—it is a philosophical shift. Luxury brands are not simply taking on tech as an add-on; they are integrating it into the fabric of their being (Nieroda et al., 2018); (Yu et al., 2019). The Apple-Hermès smartwatch, for instance, demonstrates how technological functionality can blend with handmade craftsmanship, redefining luxury for a generation that demands both sophistication and practicality (Blazquez et al., 2020); (Yu et al., 2019); (Joy et al., 2022). Likewise, TAG Heuer's partnership with Google and Intel to produce the TAG Heuer Connected watch demonstrates how Swiss watchmaking can exist alongside Silicon Valley technology, heralding a new model of co-branding in which heritage and futurism enhance instead of clash (Nieroda et al., 2018); (Blazquez et al., 2020); (Joy et al., 2022).

Another revolutionary trend is the movement away from product-focused branding toward experience-based luxury. Consumers no longer just purchase a smart purse or a networked watch—they are investing in an experience ecosystem, typically developed through AI, IoT, or AR-based interfaces (Husain et al., 2022); (Atwal & Williams, 2009); (De Kerviler & Rodriguez, 2019). For example, Louis Vuitton's Tambour Horizon smartwatch provides customized travel assistance, weather information, and insider city guides, thus infusing luxury into the customer's daily life and world travels (Atwal & Williams, 2009); (Ko et al., 2016). This experiential shift mirrors a richer comprehension of luxury as a lived feeling, not an inert ownership—a fundamental psychological transformation that co-branding alliances need to tactically leverage (Husain et al., 2022); (Atwal & Williams, 2009); (Ko et al., 2016); (De Kerviler & Rodriguez, 2019).

Sophisticated luxury branding is also seeing a greater focus on sustainability and moral storytelling. With consumers increasingly becoming socially aware, luxury brands are embedding smart technologies that enable open supply chains, circular fashion initiatives, and digital product passports (Cheah et al., 2021); (Al-Issa & Thanasi, 2024); (Kunz et al., 2020). Partnerships with technology startups that focus on blockchain and AI-driven traceability—like LVMH's collaboration with Aura Blockchain Consortium—exhibit a double commitment to innovation and ethics. These partnerships not only authenticate products but also position luxury branding in line with the values of a new generation of consumers who expect both status and social responsibility (Cheah et al., 2021); (Al-Issa & Thanasi, 2024).

Another turning point is the emergence of hyper-personalization using data analytics and artificial intelligence. Sophisticated luxury brands now use customer information—from behavior patterns to biometrics—to customize their products and services with unprecedented accuracy (Davenport, 2023); (Desai, 2021); (X. Huang et al., 2023). This is often achieved through co-branding collaborations with AI technology companies and wearables firms, which allow luxury brands to craft very personalized products and services (Gungunawat et al., 2024); (X. Huang et al., 2023). The application of AI-created designs, virtual stylists, and predictive recommendation engines turns branding from a generic one-size-fits-all formula into an intelligent conversation between the brand and the consumer (Davenport, 2023); (Desai, 2021); (Gungunawat et al., 2024).

Furthermore, digital-native luxury branding has emerged as a critical trend, especially post-pandemic. Brands that were once reluctant to embrace e-commerce have now transitioned into phygital ecosystems, seamlessly blending offline luxury touchpoints with immersive online platforms (Hoang et al., 2022); (Shah & Ali, 2023); (Zhu, 2023). Collaborations with tech giants, gaming platforms, and digital artists—such as Gucci's ventures in the metaverse and Balenciaga's digital fashion launches in Fortnite—are not mere experiments but strategic branding recalibrations aimed at future-proofing relevance (Shah & Ali, 2023); (Zhu, 2023). These efforts enable luxury brands to reach younger consumers within spaces where authenticity, novelty, and interaction reconfigure status (Hoang et al., 2022); (Shah & Ali, 2023).

Overall, the development of smart luxury branding is being fueled by a series of innovations that break beyond conventional luxury models (Quamina et al., 2023); (Husain et al., 2022). From wearable technology and experiential ecosystems to sustainability, hyper-personalization, and digital immersion, these currents represent a seismic change in the way luxury is imagined, communicated, and consumed (Husain et al., 2022); (Pantano et al., 2018). Co-branding here is not just a marketing strategy but a transformative catalyst through which luxury brands are able to adopt the future without compromising on their heritage. It is this synergy of timelessness and futurism that is going to define the next wave of smart luxury branding (Quamina et al., 2023); (Pantano et al., 2018).

2.3 Examples of Successful Smart Luxury Co-Branding Collaborations

The strategic marriage of smart technology and luxury branding has yielded some outstanding co-branding partnerships that not only upended traditional product categories but have also reshaped consumer expectations within segments(Quamina et al., 2023); (Moon & Sprott, 2015). These alliances illustrate the way brand equity, when synergistically merged, can yield innovative, emotionally charged, and commercially viable products (Quamina et al., 2023); (Chapman & Dilmperi, 2022). More significantly, these instances highlight how luxury brands preserve their identity while embracing modernity—without alienating heritage customers or sacrificing exclusivity (Moon & Sprott, 2015); (Chapman & Dilmperi, 2022).

One of the most famous and oft-quoted instances is the Apple-Hermès partnership, which elegantly combined Apple's technological prowess with Hermès' tradition of French craftsmanship (Li et al., 2021); (D. Zhu, 2022). The Apple Watch Hermès, a blend of precision technology and artisanal leather craftsmanship, does more than tell time—it provides a fashion statement based on luxury heritage and digital ease (Migicovsky et al., 2021); (Li et al., 2021). This partnership prospered by maintaining the DNA of each brand: Apple did not dilute the prestige of Hermès, and Hermès absorbed Apple's customer-driven innovation (Li et al., 2021); (D. Zhu, 2022). The outcome was a product that extended Apple's appeal to prosperous lifestyle categories without losing its status while offering Hermès a pipeline to a new, youthful demographic of technology adopters—trading on each other's respective value improvements (Migicovsky et al., 2021); (Li et al., 2021).

Another significant example is the TAG Heuer Connected Modular line, a collaboration between TAG Heuer, Intel, and Google. By merging TAG Heuer's Swiss watchmaking heritage with Intel's hardware prowess and Google's Android Wear OS, the partnership brought to market a smartwatch that preserved the aesthetic depth and modular construction of a classic TAG Heuer watch, while incorporating effortless smart features (Raffaelli, 2018); (Hoffmann et al., 2018). What makes this alliance unique is its modularity, which enabled customers to toggle between smart and mechanical modules—combining the emotional nostalgia of vintage watches with the functionality of contemporary technology. This innovation not only improved product customization but also maintained the symbolic and emotional value that consumers attach to luxury watches (Raffaelli, 2018); (Hoffmann et al., 2018); (Tajeddini & Mueller, 2011).

A good example in fashion-tech is Louis Vuitton's Tambour Horizon, a smartwatch designed in partnership with Google and Qualcomm. The Tambour Horizon does not merely feature digital functionality—it is infused with the essence of travel, one of Louis Vuitton's three brand pillars (Choi & Kim, 2016); (Margariti et al., 2019); (Godey et al., 2016). With bespoke-designed dials, sole city guides, and live travel news alerts, the smartwatch becomes an extension of the brand self, presenting a luxurious and experiential brand story that deeply resonates among globe-trotting elites. It reinforces the fact that luxury in the modern era is no longer merely the product itself, but the lifestyle that it experiences (Choi & Kim, 2016); (Margariti et al., 2019).

In the world of fashion accessories, Prada's partnership with Adidas brought smart sneakers that combined high fashion and streetwear with sustainability (Escobedo et al., 2020); (Dai et al., 2024); (Mirzajani et al., 2022). The limited-edition Prada x Adidas Luna Rossa 21 not only featured both brands' unique styles but also came with NFC chips inserted in the tongue, allowing owners to check authenticity and access special content. This virtual enhancement provided an added dimension of traceability and ownership pride, illustrating how intelligent functionality can increase the perceived value and emotional attachment to co-branded luxury goods (Escobedo et al., 2020); (Dai et al., 2024). Outside of wearables, Montblanc's Summit Smartwatch line is another successful co-branding story, using Google's Wear OS to revolutionize an old writing and watch brand into a tech-savvy luxury brand (Turan, 2021); (Sarkar & Banerjee, 2019). By retaining its classic traditional design while adding smart capabilities like fitness monitoring, navigation, and voice guidance, Montblanc captured the sensibilities of today's working professionals who want classy understatement with contemporary capability (Turan, 2021); (Childs & Jin, 2020).

In the automotive industry, BMW's co-branding with Montblanc on intelligent key fobs and connected accessories also shows how co-branding can cross categories. These fobs are more than car keys—they are status symbols that embody the spirit of both high-end German engineering and European luxury craftsmanship, reinforcing brand identity even in micro-interactions(Quamina et al., 2023); (Pinello et al., 2022). These examples highlight an important insight: successful smart luxury co-branding is not merely about combining functionalities but about combining identities, narratives, and emotional experiences(Suzuki & Kanno, 2022); (Yu et al., 2020); (Tynan et al., 2009). The solution lies in respect for each other's brand heritage, a shared vision for innovation, and profound knowledge of consumer desires (Quamina et al., 2023); (Mrad et al., 2018); (Suzuki & Kanno, 2022); (Yu et al., 2020). In both instances, the partnership created something that was not just technologically advanced but also emotionally engaging and therefore commercially successful without compromising exclusivity or authenticity. These collaborations are thus a gauge by which future brand partnerships in the smart luxury space will be measured (Quamina et al., 2023); (Mrad et al., 2018); (Suzuki & Kanno, 2022); (Tynan et al., 2009).

3. THE INFLUENCE OF CONSUMER PSYCHOLOGY IN SMART LUXURY BUYING

Any successful luxury brand strategy has at its core a deep understanding of consumer psychology. The image of a brand is not only defined by its looks or price but by the emotional, cognitive, and social cues it conveys to its public. In the case of smart luxury co-branding, this is even more complex (Lin, 2023); (Ma, 2023); (J. Yu & Wang, 2024). How a consumer interprets a brand as authentic, trustworthy, and exclusive can either make or break their purchase decision. This section explores the emotional motivations of luxury consumption, the psychological significance of brand status and prestige, and the complex interaction between consumer identity and brand association (Wei, 2022); (Ma, 2023). Through discovering how psychological drivers shape brand perception, we have an understanding of why some co-branded smart products manage to create desire while others do not resonate (Lin, 2023); (Wei, 2022); (J. Yu & Wang, 2024).

3.1 How Consumer Emotions Influence Brand Perception

In the world of intelligent luxury products, consumer feeling is not incidental—it's pivotal in the building of brand attitude and, most importantly, informing purchase. The emotional reactions to branding are lodged deep within psychological models that shoppers employ in deciding value, self, and worthiness (L. Xu & Mehta, 2022); (Shahid & Paul, 2021); (Kauppinen-Räisänen et al., 2020). When a luxury brand goes into a co-branding exercise, particularly with a technology brand, the emotional connection it triggers can greatly augment or detract from the value perceived by the product. Luxury is no longer judged by consumers based on concrete qualities such as design, quality, or cost; they now judge how something makes them feel—about themselves and the world at large (L. Xu & Mehta, 2022); (Shahid & Paul, 2021).

Emotions serve as strong mental shortcuts that drive consumers' interpretation of brand signals. For example, excitement and novelty can be elicited when a consumer observes a classic luxury brand such as Dior or Gucci pairing up with a pioneering technology brand); (Suzuki & Kanno, 2022); (Tynan et al., 2009). Such an emotional response involves the brain's reward system, with the product thus becoming more attractive due to conformity with the aspiration, lifestyle, and developing self-concept of the consumer(Quamina et al., 2023); (Mrad et al., 2018). Specifically, products which capture both tradition and innovation tend to be seen as "future-proofed luxury," providing customers with emotional assurance that they are making an investment in experiences that are classic but forward-looking (Quamina et al., 2023); (Mrad et al., 2018); (Suzuki & Kanno, 2022); (Tynan et al., 2009).

Secondly, emotional branding takes advantage of consumers' need for narrative consistency. Clever luxury products tend to be positioned as part of a grander life narrative—of achievement, aspiration, distinctiveness, and standing (Bi, 2024); (Kim & Sullivan, 2019); (Loureiro et al., 2019). When co-branding initiatives are emotionally engaging, like via good storytelling, limited edition releases, or cause marketing, they develop higher-level psychological involvement. Buyers establish affective bonds, building loyalty beyond rational calculations of price-value. A customer might, for instance, select a co-branded intelligent watch not on the basis of features per se but because it represents a synthesis of heritage and innovative quality that reflects their professional or personal persona (Bi, 2024); (Kim & Sullivan, 2019).

Influencing behavior in an emotional state is also felt after purchase behavior. Those that are emotionally satisfied tend to support the brand positively by writing and sharing reviews or through word of mouth, effectively building the brand's social capital (Bonifield & Cole, 2006); (Santos & Boote, 2003); (Yassin & Soares, 2023). Emotional dissonance when a product's failure to attain emotional expectation generated due to it being luxurious fails, and customers become alienated from the brand. This is most dangerous in co-branding situations, where an imbalance of the emotional tone between the collaborating brands can confuse or frustrate the customer, undermining the perceived authenticity of the collaboration (Bonifield & Cole, 2006); (Santos & Boote, 2003).Therefore, emotional resonance is not accidental but necessary in influencing how consumers view smart luxury brands(Husain et al., 2022); (Kaur & Soch, 2025).. It is both a filter and amplifier—it amplifies attractiveness, increases brand affection, and strengthens the consumer's perception that they are making a valuable, emotionally satisfying decision. In the more complicated environment of smart luxury branding, levering emotional psychology is not only advantageous—it is essential (Han & Kim, 2020); (Husain et al., 2022); (Kaur & Soch, 2025).

3.2 The Role of Trust and Exclusivity in Smart Luxury Purchases

Trust and exclusivity are two basic pillars on which the charm of luxury is based, and their importance becomes even greater in the case of smart luxury co-branded products (Pantano et al., 2018); (Husain et al., 2022); (Suzuki & Kanno, 2022); (Tynan et al., 2009). Customers interacting with premium, technologically advanced products are not just buying a product; they are making a brand story investment that needs to connect with their values, aspirations, and social self. In a context so subtle and high-risk, trust acts as the medium of credibility and exclusivity as the symbol of desirability (Pantano et al., 2018); (Husain et al., 2022); (Suzuki & Kanno, 2022).

Trust, above all, is the psychological connection between consumers and brands. When luxury businesses align with technology businesses, especially those not in the classic luxury universe, there tends to be an element of uncertainty (Pantano et al., 2018); (Alexander & Contreras, 2016); (Mrad et al., 2018). The consumers have to trust that the collaboration will maintain the craftsmanship, prestige, and symbolic capital of the luxury brand. For instance, when TAG Heuer formed a co-branding deal with Intel and Google for its smartwatches, the product's success hinged on the consumers trusting that the technological innovation introduced by tech giants would not compromise the painstaking heritage of Swiss watchmaking. This so-called continuity of quality and heritage throughout the alliance is essential in sustaining consumer trust (Alexander & Contreras, 2016); (K. Kim et al., 2014); (Mrad et al., 2018).

In addition, the intangible aspect of technology—its quick changes, possible obsolescence, and data privacy issues—adds another level of complicating factors in the establishment of trust (Lankton et al., 2015); (Schuetz et al., 2024). Luxury buyers, being more cautious and risk-averse, require assurance that they are protecting their data, the technology remains up-to-date, and the brand will still provide support to their product after purchasing (Lankton et al., 2015); (Paska & Budnik, 2023). A co-branded intelligent product would thus have to convey not only the superior beauty and performance but also reliability, durability, and after-sales dedication. The reputation of the brand becomes a guarantee of technological reliability, and any incongruity or mismatch in this case can chip away consumer faith (Lankton et al., 2015); (Schuetz et al., 2024); (Paska & Budnik, 2023).

Exclusivity, by contrast, acts as the psychological guardian of luxury. It realizes the core human aspiration for differentiation and social rise. Intelligent luxury goods, which are widely available or mass-made, can unwittingly compromise this exclusivity (Wang et al., 2021); (Petrova & Pruzhansky, 2011); (Y. Kim, 2018). Therefore, co-branded efforts have to walk a tightrope—delivering cutting-edge innovation while maintaining the air of scarcity. Limited releases, tailored experiences, and customized services are usually used to reinforce this air of exclusiveness (Septianto et al., 2020); (Petrova & Pruzhansky, 2011). For example, Louis Vuitton's Tambour Horizon smartwatch combines advanced technology with personalized features that appeal to every individual, enhancing their uniqueness and status (Wang et al., 2021); (Septianto et al., 2020); (Y. Kim, 2018).

Exclusivity, further, is not just scarcity but signaling. Consumers utilize luxury products as social and psychological signals of taste, success, and identity. Exclusivity in co-branded smart luxury products conveys not just wealth but cultural and technological awareness as well (Ajitha & Sivakumar, 2017); (Das et al., 2021); (Y. Wang & Griskevicius, 2013). It is a dual proficiency of style and innovation—a sign of prestige in an evolving society. In executing it properly, this type of branding weaves an emotional story in which the customer is not only a consumer but a connoisseur, a member of an exclusive club who recognizes and adopts the future (Ajitha & Sivakumar, 2017); (Das et al., 2021). Overall, trust and uniqueness are not the niceties of doing business but psychological levers that make smart luxury co-branding strategies appealing. They reduce perceived risk from technological integration while heightening the symbolic potential of ownership. To address the markets with these offerings, losing these psychological factors is not only a lost opportunity—it is a danger to the very existence of the brand (Zhao, 2024); (Y. Yu et al., 2020).

3.3 Psychological Effects of Brand Prestige and Status

Brand prestige and social status have been prime drivers of luxury consumption for years, serving as psychological anchors to self-expression, identity affirmation, and social signaling (Han & Kim, 2020); (Shaikh & Gummerum, 2025). In the context of smart luxury co-branded products, prestige dynamics become all the more relevant as consumers want products that balance status symbolism and technological acumen seamlessly (Pantano et al., 2018); (Han & Kim, 2020). This intersection of traditional luxury signifiers and high-tech innovation redefines the act of consumption as a strategic psychological move—where the possession of a co-branded intelligent product says something beyond wealth; it says something about cultural capital, contemporaneity, and intellectual sophistication (Pantano et al., 2018); (Han & Kim, 2020); (Shaikh & Gummerum, 2025).

The psychological effect of brand prestige starts with the idea of self-congruence. People are inherently attracted to brands that are consistent with their ideal self-concept and goals. When a high-end brand collaborates with a well-known technology company, the co-branded product that follows has a double prestige that fulfills emotional as well as cognitive aspirations (Kumagai & Nagasawa, 2019); (C. Choi, 2024); (Kressmann et al., 2006). A Montblanc and Google smart product, for instance, conveys not just commitment to high-end craftsmanship but also an openness to advanced innovation. The consumers take on this prestige inside, feeling themselves forward-thinking yet sophisticated, elite yet connected to the future (Kumagai & Nagasawa, 2019); (C. Choi, 2024). In addition, prestige branding serves an important role in inter-personal perception. Social identity theory suggests that individuals identify themselves with their group affiliations, and prestigious brands function as status-construction tools in these groups. Co-branded smart luxury goods become strong markers of identity in professional, social, and even online environments. The owner of a TAG Heuer Connected smartwatch or a Prada x LG fashion accessory is not so much flaunting financial ability, but advertising a lifestyle that includes innovation, taste, and cultural sophistication. Such products are "badges of belonging" to aspirational social groups (Kressmann et al., 2006); (Diawara, 1995).

Neurological studies also validate the strong influences of brand prestige. Neuromarketing research shows that viewing luxury brand logos stimulates areas of the brain linked to reward, pleasure, and self-esteem—like the medial prefrontal cortex and the ventral striatum. When these logos are combined with technologically advanced features, reward circuitry is doubly stimulated, increasing the psychological pleasure of ownership (Rahmadani et al., 2022); (Schaefer & Rotte, 2007); (Cherubino et al., 2019). Thus, smart luxury co-branded products set up a positive feedback loop such that perceived prestige fuels emotional reward, which reciprocally bolsters brand loyalty and future buying intention (Rahmadani et al., 2022); (Schaefer & Rotte, 2007).

Still, smart luxury brand prestige is dynamic and must be maintained and improved constantly through exclusivity, quality craftsmanship, innovativeness, and restricted availability. A misjudged proliferation of co-branding or partnerships with a maladaptive technological counterpart may dilute prestige, leading to the disaccumulation of psychological advantages ascribed to status signaling (Phau & Prendergast, 2000); (Fionda & Moore, 2009). To illustrate, should a luxury company pursue a mass-market partnership when insufficiently differentiated, it jeopardizes alienation from core buyers seeking exclusivity and symbolic difference (Phau & Prendergast, 2000); (Fionda & Moore, 2009); (J. C. Kim et al., 2018). Finally, it is the psychological impact of brand status and prestige that acts as the unsung power driving want, interaction, and word-of-mouth in the smart luxury sector (Pino et al., 2017); (Y. J. Han et al., 2010). By recognizing these primal drivers, marketers have the ability to design co-branding programs that not only engage consumers' outward status aspirations but tap into their own personal stories—providing them with not simply a product, but a psychological projection of themselves they desire to be (Pino et al., 2017); (Y. J. Han et al., 2010); (Kapferer & Valette-Florence, 2017).

4. BEHAVIORAL ECONOMICS AND NEUROMARKETING IN SMART LUXURY

Understanding luxury consumer behavior through demographics or market trends is not enough—it takes understanding the subconscious mind. Neuromarketing and behavioral economics provide strong methods for unlocking how consumers respond to co-branded smart luxury brands (Hsieh & Chang, 2016); (J. Kumar & Nayak, 2019). From scarcity to social proof, cognitive biases significantly influence value perception. Meanwhile, sensory branding and multisensory experiences have become essential in developing the emotional atmosphere that comes with luxury products (J. Kumar & Nayak, 2019); (Leckie et al., 2016). This chapter delves into how luxury brands can leverage behavioral science to maximize pricing strategies, maximize recall, and trigger desire. Through real-world examples and case studies, we illustrate how the combination of neuroscience and psychology in co-branding can maximize consumer engagement and loyalty (Hsieh & Chang, 2016); (Leckie et al., 2016).

4.1 The Role of Cognitive Biases in Consumer Decisions

In the era of intelligent luxury co-branding, consumer decisions are anything but simply rational; instead, they tend to be governed by cognitive biases that subtly impact perception, judgment, and choice (Baghi & Gabrielli, 2018); (Q. Zhang et al., 2021). These cognitive biases—psychological heuristics-driven mental shortcuts—are the backbone of behavioral economics, and when deliberately exploited in luxury co-branding, they maximize consumer desirability, urgency, and loyalty (Quamina et al., 2023); (Q. Zhang et al., 2021). Three of the most powerful biases operating in this arena are anchoring, scarcity, and social proof—each with differing implications for the way intelligent luxury products are perceived and used (Baghi & Gabrielli, 2018); (Quamina et al., 2023).

Anchoring bias is the tendency for humans to place great weight on the initial piece of information presented—the "anchor"—when making choices. In co-branded intelligent luxury products, the anchor tends to be set through price positioning or comparative worth (Kurtz, 2020); (Weatherill, 2017); (Burger, 1986). For example, when a high-end smartwatch is launched with an extremely high baseline price because of the exclusive heritage of one brand, any subsequent co-branded product at a slightly reduced price is seen as a "bargain," even though it remains pricey by traditional standards (Weatherill, 2017); (Burger, 1986). Shoppers anchor their perception on first sight, and high-end brands take advantage of this bias to maintain a high-end image while lowering barriers to entry. In addition, technical specifications (such as better battery life or health sensors) benchmarked against a rival's inferior product can also raise the value perceived by the co-branded product (Kurtz, 2020); (Burger, 1986).

Another highly effective lever that is based on the fear of missing out (FOMO) is scarcity bias. As consumers feel that a product is scarce in terms of quantity, time, or exclusivity, they tend to act spontaneously and ascribe higher value to it. Luxury brands have long wielded scarcity as an emotional lever, but among smart co-branded products, the impact is amplified by the acceleration of development cycles and limited releases (Rosendo-Rios & Shukla, 2022); (Shi et al., 2020); (Sung et al., 2020). An example of a smart luxury co-branded partnership between Hublot and a tech company, launched in a limited global production run of 500 pieces, instantly provokes urgency and psychological ownership—even from buyers previously on the fence. Their transient nature gives them emotional value, making a transaction a trophy purchase (Rosendo-Rios & Shukla, 2022); (Shi et al., 2020). Social proof, arguably the most social of these biases, is based on the principle that individuals mimic others' behavior—particularly those who are respected or perceived as influential. In clever luxury branding, this is most often realized in the form of celebrity endorsements, influencer advertising, or user-generated content that is organic(Rosendo-Rios & Shukla, 2022); (Shi et al., 2020); (Sung et al., 2020). When a celebrity is spotted sporting a Louis Vuitton x Huawei smart accessory, it not only legitimizes the product's design and tech authenticity but also inspires aspirational buyers to do the same. In addition, social media comments, unboxing videos, and celebrity product launches generate collective energy, pushing individual interest into social movement. Buyers instinctively think, "If others are purchasing it—particularly individuals I aspire to—then it must be worthwhile(Rosendo-Rios & Shukla, 2022); (Shi et al., 2020)."

These biases, although frequently unconscious, are not negative manipulations; instead, they are psychological tools that co-branded luxury brands need to understand and responsibly deploy(Ho et al., 2019); (Hsieh & Chang, 2016); (Nadeem et al., 2021). The commercial success of an intelligent luxury product relies not just on its concrete attributes but on its capacity to induce these intuitive reactions, giving rise to narratives of value, prestige, and social belonging(Hsieh & Chang, 2016); (Nadeem et al., 2021). By designing strategies that resonate with anchoring, scarcity, and social proof, co-branding efforts can connect with consumers' cognitive architecture—engagement as psychological as it is transactional(Ho et al., 2019); (Hsieh & Chang, 2016).

4.2 Neuromarketing Insights: Sensory Branding and Its Influence on Luxury Perception

Within the world of intelligent luxury co-branding, the war for consumer attention and devotion reaches far beyond product features—it is anchored deeply in the world of the senses. Neuromarketing, a hybrid discipline of neuroscience and marketing, investigates how consumers' brains react to sensory stimulation, emotional stimuli, and brand experiences (Joshi, 2024); (Alsharif et al., 2023); (Kajla et al., 2023); (Terenteva & Vovchenko, 2023). Luxury brands collaborating with technology companies find that incorporating sensory branding techniques guided by neuromarketing not only raises product perception but also creates more powerful emotional and neurological connections with consumers (Kajla et al., 2023); (Terenteva & Vovchenko, 2023). Luxury is about experience, and sensory branding makes that experience immersive, memorable, and irresistibly desirable (Joshi, 2024); (Alsharif et al., 2023).

Vision is perhaps the most powerful sense in luxury branding. Clever co-branded products are carefully crafted to entice with visual sophistication—streamlined screens, luxurious textures, legendary logos, and balanced color schemes (Kusumasondjaja, 2019); (Giglio et al., 2019). When a high-end fashion brand collaborates with a tech company, e.g., when Montblanc teamed up with Summit smartwatches, graphical elements such as elegant watch faces, high-end straps, and elegant interfaces are not coincidental—they are deliberate design stimuli designed to trigger the reward system in the brain (J. E. Lee et al., 2018); (Kusumasondjaja, 2019). Neuromarketing research with eye-tracking and EEG methods indicates that customers experience higher brain activity and more emotional participation when confronted with visually better products, particularly when these images resonate with their self-image as a style-conscious consumer (J. E. Lee et al., 2018); (Kusumasondjaja, 2019); (Giglio et al., 2019).

Sound also has an important function in embedding brand image and consumer happiness. Think about the distinctive startup sound of a Bang & Olufsen x LG smart speaker or the tactile click of a high-end smart pen (Knoeferle et al., 2016); (Carnevale et al., 2017). These sound experiences are not just functional—auditory stimuli that are emotional triggers and encode the brand in memory (Lowe & Haws, 2016); (Huh et al., 2023). In neuromarketing tests, pleasing soundscapes have been proven to elevate dwell time, perceived value, and emotional affinity, especially when harmonized with brand themes like luxury, tranquility, or innovation (Jun & Lee, 2021); (Dogaru et al., 2024).

Touch is at the core of the tactile pleasure that is the essence of luxury. Co-branded intelligent products that contain materials such as sapphire glass, ceramic bezels, Italian leather, or brushed titanium offer sensory richness that grounds perceptions of quality and exclusivity (Peck & Shu, 2009); (Akram & Lavuri, 2023); (Shankar & Jain, 2020). The feeling of engaging with such a product—how it sits in the hand, the weight, the smoothness of the user interface—all combine to underpin a subconscious story of high-quality craftsmanship(Akram & Lavuri, 2023); (Shankar & Jain, 2020). Neuroscientific research shows that these haptic experiences activate areas of the brain linked with trust and pleasure that then impact willingness to pay and brand loyalty (Peck & Shu, 2009); (Akram & Lavuri, 2023).

Even smell, while harder to incorporate into digital products, is being tested in experiential co-branding contexts like high-end pop-up boutiques or interactive showrooms. (Castillo-Villar & Villasante-Arellano, 2020); (Wiedmann et al., 2017). Dior and Samsung have tried multisensory brand spaces where customers are not only interacting with the product but also with a carefully curated environment that reinforces emotional connection and brand narrative (Rodrigues, 2019) (Wiedmann et al., 2017). Smell, when combined with sight and sound, has been found in fMRI research to create profound emotional memory and brand recall, frequently more intensely than visual or audio alone (Rodrigues, 2019); (Castillo-Villar & Villasante-Arellano, 2020).

In addition, neuromarketing focuses on the power of emotional storytelling and narrative framing (Song et al., 2025); (Endres et al., 2024). When a co-branded smart product is introduced not just as a device but as an icon of heritage, innovation, and uniqueness, it resonates with the brain's limbic system—the hub of emotions, identity, and decision-making (Wang et al., 2016); (Song et al., 2025). The brain is not a logic processor when it comes to buying luxury; it is a narrative processor. By emotionally intelligent campaigns, experiential environments, and multi-sensory cues, intelligent luxury brands are able to develop an engaging brand ecosystem that goes beyond function and resonates in the soul (Wang et al., 2016); (Song et al., 2025); (Meyerding & Mehlhose, 2018). In short, neuromarketing provides luxury co-branding a deep competitive edge: the capacity to build sensory-rich, emotionally resonant experiences that communicate at a neurological level (Quamina et al., 2023); (Suzuki & Kanno, 2022); (Meyerding & Mehlhose, 2018). It transforms branding from page to pulse—insisting that what consumers see, hear, and feel harmonizes perfectly with what the brand represents. In an era where attention is ephemeral and loyalty is fleeting, the art of sensation might just be the key to sustained commitment in intelligent luxury markets (Quamina et al., 2023); (Suzuki & Kanno, 2022).

4.3 Pricing Psychology in Co-Branded Luxury Smart Products

Pricing in the luxury market is never an economic choice—it is always a strategic, psychological move. In the world of smart luxury co-branding, where high-end fashion and advanced technology come together, the price point speaks volumes beyond product value. It dictates status, workmanship, innovation, and emotional gratification (Parguel et al., 2015); (Royo-Vela & Sánchez, 2022). The psychology used to price such co-branded smart products accesses a rich reservoir of consumer behavior knowledge, leveraging subconscious biases and perceptions that strongly influence buying behavior (Parguel et al., 2015); (Royo-Vela & Sánchez, 2022). One of the strongest psychological pricing techniques is anchoring. In co-branded luxury categories, customers tend to judge the price of a product in terms of an initial "anchor" they are given—whether the starting price of a flagship luxury product or the comparison with an expensive tech gadget (Reis, 2022); (X. Jiang et al., 2022); (Barnes & Higgins, 2017). For instance, when Louis Vuitton makes a partnership with a technology company to launch a $2,000 smartwatch, that tag may seem rational or even appealing when compared psychologically to a $10,000 mechanical watch made by the same fashion brand. The perceived value is not measured in a vacuum but against such cognitive reference points. Anchoring assists brands to justify luxury pricing by influencing consumers' expectations and rationales insidiously (Reis, 2022); (X. Jiang et al., 2022). Scarcity and restrictiveness are also key to the pricing story. Smart product limited editions like TAG Heuer's Connected Calibre E4 or Gucci x Oura Ring are made deliberately in small numbers to raise desirability (Frew et al., 2021); (Cheshire & Sheppard, 1989). Behaviorally, scarcity activates FOMO (fear of missing out), creating urgency and perceived value. Customers are neurologically programmed to give more value to products that they perceive as being rare or temporary. This effect, exacerbated by co-branding, levers a notion of privilege and insider status well-suited to luxury positioning (Reis, 2022); (Frew et al., 2021).

Price-quality heuristics dominate in the vindication of premium labels. With luxury co-branding, such high prices even serve as hints of quality, uniqueness, and innovation even in the absence of real technical leadership (Chen & Li, 2022); (Y. Yu et al., 2020). If a smart handbag from Prada comes with an in-built NFC chip to enable luxury services and authenticate the handbag, its premium price not only announces the innovation but also primes the consumer to view it as higher status and advanced (Y. Yu et al., 2020); (Quamina et al., 2023); (Suzuki & Kanno, 2022). Behavioral economics demonstrates that consumers tend to associate price with performance, particularly in emotionally charged product categories such as luxury fashion and wearable technology (Quamina et al., 2023); (Chen & Li, 2022).

In addition, partitioned pricing—where the complete value of the product is covertly unbundled into parts—can minimize perceived cost obstacles. For example, a co-branded smartwatch can have discrete prices for design enhancements, app ecosystems, or service plans, each justified independently (Reyes-Menendez et al., 2021); (Pentina et al., 2018). This can make the luxury product feel more modular and bespoke, inducing the perception of layering value as opposed to bundled expense(Reyes-Menendez et al., 2021); (Pentina et al., 2018).Framing effects also affect perception. A high-end-tech collaboration that frames a smart product as a "next-generation heirloom" or a "limited masterpiece in wearable art" reframes the price from a transaction cost to an emotional investment. Customers don't only purchase the product—they purchase the story, the symbolism, and the identity that comes with it. Framing the experience as an experience instead of a commodity reframes the price as an access point to an elite lifestyle(Quamina et al., 2023); (Suzuki & Kanno, 2022); (Pentina et al., 2018).

Lastly, decoy pricing is an sophisticated strategy most commonly used with luxury co-branded lines. A company would release three types of a high-tech product: a standard product, a mid-range co-branded one, and an over-the-top deluxe model with personalization options (Yuan & Xiao, 2022); (Pennington et al., 2022); (Heath & Chatterjee, 1995). The fact that there is a ridiculously expensive choice available makes the mid-range alternative—still significantly priced—more palatable by comparison and naturally nudges people towards a rewarding option using a form of comparative bias (Yuan & Xiao, 2022); (Pennington et al., 2022). Overall, co-branded smart luxury pricing psychology is a subtle art of behavioral cues and perceptual maneuvers. It is not merely a matter of what people can pay, but what they perceive the product to be worth in value, experience, and rarity. With the growth of smart luxury, the art of controlling these psychology pricing levers will be necessary to maintain fascination, preserve margins, and foster brand elitism in an intensified competitive market (Heath & Chatterjee, 1995); (Quamina et al., 2023); (Suzuki & Kanno, 2022); (Pentina et al., 2018).

5. ARTIFICIAL INTELLIGENCE, PERSONALIZATION, AND ETHICAL CONSIDERATIONS

With consumer aspirations in today's times shaped by immediacy, personalization, and digital sophistication, AI is the game-changer for luxury co-branding. Artificial intelligence not only allows for hyper-personalization but also predictive analytics, customized content and real-time engagement tactics (S et al., 2021); (Maldonado-Canca et al., 2024); (Shen et al., 2017). With luxury brands collaborating with tech companies to provide wiser, more intelligent products, the value of AI in driving brand value and customer happiness becomes paramount (Maldonado-Canca et al., 2024); (D. Kumar & Suthar, 2024). Yet, as there are benefits, so too are there ethical challenges surrounding data protection, algorithmic bias, and consumer control (Maldonado-Canca et al., 2024); (Shen et al., 2017). This section looks at the double-edged sword of AI in being an enabler and a disruptor in the co-branding world. It also explores how data-driven personalization can reinforce emotional branding while maintaining the exclusivity that defines luxury (S et al., 2021); (D. Kumar & Suthar, 2024); (Shen et al., 2017).

5.1 How AI-Driven Personalization Enhances Consumer Engagement

Artificial Intelligence (AI) transformed the luxury space, but not by diminishing human touch, by intensifying personalization to heretofore unimaginable dimensions (Xu & Mehta, 2022); (S et al., 2021). Within the context of co-branded smart consumer products—where design and luxury marry functionality and technology—AI is that intermediary connecting brand identity to singular consumer need (Park & Ahn, 2024); (S et al., 2021). The ability to provide tailored experience via AI-driven algorithms not only rewrote rules of consumer engagement but emerged as a sine qua non in determining success for luxury co-branding (Xu & Mehta, 2022); (Park & Ahn, 2024). The core of AI-powered personalization is predictive analytics, which leverages customer information—like browsing patterns, previous purchases, lifestyle choices, and even biometric signals—to predict what consumers need before they can say it (Parashar et al., 2023); (Jain, 2024). For instance, in co-branded high-end fitness wearables, AI platforms can monitor user behaviors to provide personalized wellness recommendations (Porbundarwala et al., 2023); (Parashar et al., 2023). A collaboration between a high-end brand and a technology company such as Oura or Fitbit can include AI to recommend bespoke wellness programs, diet advice, or sleep improvement, all tailored to the individual. Such hyper-personalization turns the product from an instrument into a lifestyle partner, creating deeper emotional engagement and attachment (Porbundarwala et al., 2023); (Parashar et al., 2023); (Jain, 2024).

Natural Language Processing (NLP), which is a sub-discipline of AI, also plays an important role in amplifying consumer-brand conversation. Luxury chatbots and virtual assistants fueled by artificial intelligence offer live, conversational assistance and learn from every experience to make responses even better the next time around (Umarova & Jurayeva, 2024); (Babatunde et al., 2024). Take an example of a co-branded app created by a fashion brand and a technology leader: as customers ask questions regarding product availability, styling ideas, or compatibility with their intelligent devices, AI responds with accuracy, tone, and content in every individual's language and behavior (Babatunde et al., 2024); (B. Yu, 2021). This not only enhances customer service but also enhances the perception of exclusivity that is expected from luxury brands (Umarova & Jurayeva, 2024); (B. Yu, 2021).

Aside from communication, AI-based recommendation engines are crucial. They study a user's history of interactions to recommend products, activities, or content that match their interests (Y. Zhang, 2024); (Boka et al., 2024). On co-branded e-commerce websites or mobile apps, AI tailors product lists in real-time—displaying a consumer with luxury smartwatches with the precise strap style or app functionalities they have liked in the past, based on their past choices. This anticipatory customization is not only making things more convenient; it delivers a curated, boutique-like experience in a digital environment (Y. Zhang, 2024); (Boka et al., 2024); (Y. Li et al., 2019). In addition, emotional AI, which can pick up and react to human emotions via voice, facial expression, and physiological signals, is surfacing as a potent weapon (Xu & Mehta, 2022); (S et al., 2021). In luxury co-branding, this might take the form of intelligent mirrors or wearable technology that varies its function or look-and-feel based on the mood of the user (Baudier et al., 2022); (Shahzad et al., 2024). Picture a co-branded high-end fitness gadget that lowers its interface hue, provides meditative sound, or changes health recommendations when it detects stress. These interactions improve customer satisfaction by making the product intuitive, responsive, and caring (Baudier et al., 2022); (Xu & Mehta, 2022); (S et al., 2021); (Shahzad et al., 2024).

AI-powered personalization also extends to post-purchase engagement, which is critical in sustaining brand loyalty and advocacy (Rahman et al., 2023); (Sahne & Daronkola, 2025); (Chakraborty et al., 2024). Intelligent luxury products with AI can send customized care advice, anniversary reminders, or special upgrade invitations, maintaining a long-term relationship beyond the first sale (Chakraborty et al., 2024); (X. Song & Bonanni, 2024). These experiences are frequently complemented by loyalty programs that leverage AI to recommend personalized rewards or content, reinforcing a sense of belonging and premium value (Rahman et al., 2023); (Sahne & Daronkola, 2025); Chakraborty et al., 2024); (X. Song & Bonanni, 2024). Notably, AI enables co-branded luxury items to break down the boundaries between digital and physical spaces by integrating phygitals—merging physical touch points with digital moments (Xu & Mehta, 2022); (Rahman et al., 2023). For example, an AI-infused luxury handbag may provide users access to a virtual styling consultant via a tap or scan, or inform the owner of special events that align with their fashion persona. Such a harmonious, smart interaction transcends the static object status to a dynamic, adaptive experience (Xu & Mehta, 2022); (Rahman et al., 2023); (Joy et al., 2022). To conclude, AI-powered personalization is no longer a futuristic dream but a fundamental strategy in smart luxury co-branding (Xu & Mehta, 2022); (Rahman et al., 2023). It enables brands to know, predict, and respond to the subtle needs of discerning consumers. By combining the accuracy of machine learning and the emotional intelligence of luxury branding, AI supercharges not only engagement but also loyalty, advocacy, and perceived value of co-branded smart products (Xu & Mehta, 2022); (Rahman et al., 2023); (V. Kumar et al., 2019).

5.2 Data-Driven Marketing and Its Effect on Co-Branding Strategies

In the luxury segment, where branding is critically dependent on emotional connection, beauty, and rarity, data-driven marketing provides an effective vehicle for the deployment of strategies with precision and flexibility(Quamina et al., 2023); (Suzuki & Kanno, 2022); (Arrigo, 2018). With the inclusion of co-branding into the mix—especially in intelligent consumer goods—the role of data intensifies even further(Quamina et al., 2023); (Suzuki & Kanno, 2022); (Chapman & Dilmperi, 2022). In this case, two or more brands are not just coming together but also bringing systems and values along to cater to a common consumer base. Data is the guide that directs this convergence, such that the collaboration speaks meaningfully to target consumers and is aligned with changing market expectations(Quamina et al., 2023); (Suzuki & Kanno, 2022); (Chapman & Dilmperi, 2022); (Arrigo, 2018).

The power of data-driven marketing lies in the potential to turn vast reservoirs of raw consumer data into usable intelligence (D. S. Johnson et al., 2019); (D. Shah & Murthi, 2020). These sources include behavioral data (purchase history, online browsing, mobile app usage), demographic profiles, social media sentiment analysis, and real-time user comments from networked devices (Grandhi et al., 2020); (Saura et al., 2021); (D. Shah & Murthi, 2020). In co-branded luxury smart products, e.g., a premium smartwatch designed by a luxury fashion company and a technology company, data allows marketers to segment audiences with great accuracy and present personalized messaging, offers, and product suggestions that resonate with one's lifestyle and aspirations (Grandhi et al., 2020); (Saura et al., 2021); (D. S. Johnson et al., 2019); (D. Shah & Murthi, 2020). Data also has an important role in brand alignment and positioning. Prior to rolling out a co-branded product, companies look at consumer data to determine overlap in brand audiences, assess shared values, and determine overlap in lifestyle touchpoints (S. Jun & Park, 2017); (Fuchs & Diamantopoulos, 2010); (Ranfagni et al., 2021). This strategic alignment allows the partnership to feel organic and not forced. For instance, a partnership between a high-end car manufacturer and a tech-savvy company could be guided by data that indicates a shared consumer interest in innovation, sustainability, and high-end design. Such findings guide everything from product look to tone of messaging, launch dates, and channel strategy (S. Jun & Park, 2017); (Fuchs & Diamantopoulos, 2010); (Ranfagni et al., 2021).

Data-driven marketing, in practice, enables dynamic campaign optimization. Marketing groups can track which segments of the audience react to a co-branded rollout in real-time and make messaging, visuals, or channels shifts accordingly (Jihua & Lin, 2024); (Mallipeddi et al., 2018). For example, if data indicate that Gen Z shoppers interact more with video content highlighting the technology aspect of the partnership, whereas Millennials react better to influencer narratives focusing on luxury heritage, marketers can reallocate spending and creative assets to align with those findings mid-campaign. This flexibility increases relevance and optimizes return on investment (Jihua & Lin, 2024); (Mallipeddi et al., 2018); (Cutrona et al., 2019). In addition, information powers content personalization across touchpoints, a key luxury market expectation. Co-branded luxury websites, mobile apps, and platforms now adjust according to who is looking at them (Liu et al., 2019); (Mandler et al., 2019); (Xie & Lou, 2020). A logged-in user of a luxury fashion brand's app may view AI-personalized content highlighting the intelligent aspects of their wearable device in the context of a runway show (Moreau et al., 2020); (Xie & Lou, 2020). At the same time, an astute user visiting from one of a device brand's websites may be welcomed with content emphasizing functionality and innovation. The same product is described through lenses most engaging to different personas, all facilitated by data segmentation (Liu et al., 2019); (Mandler et al., 2019); (Moreau et al., 2020).

Notably, data allows co-branded initiatives to monitor the customer lifetime value (CLV) and engagement patterns throughout the product life cycle (Nadeem et al., 2021); (Solem, 2016). From pre-launch awareness to after-purchase support, intelligent products feed information back to the brand system at all times (Nadeem et al., 2021); (Solem, 2016); (Tran et al., 2022). These insights not only tell the brand what sells, but also how products are used, what features contribute to satisfaction and where improvement or new product innovation is required (Solem, 2016); (Tran et al., 2022). For example, heat maps from wearable tech usage could show that certain biometric attributes are more popular among fitness users than among fashion-oriented consumers, which could inform future product development or niche marketing (Nadeem et al., 2021); (Solem, 2016); (Tran et al., 2022).

Additionally, predictive modeling driven by machine learning allows brands to predict changes in consumer expectations and adapt co-branding strategies in advance (Adams & Abhayawansa, 2021); (X. Wang et al., 2023). For instance, if trends in the data show that consumers are increasingly demanding sustainable tech-luxury hybrids, brands can incorporate ethical sourcing, sustainable packaging, or green technology into upcoming collaborations (X. Wang et al., 2023); (Cunha et al., 2014). Predictive signals also guide price models, with insights on which customer segments are more sensitive to premium pricing, limited editions, or exclusive upgrades (Adams & Abhayawansa, 2021); (Cunha et al., 2014). Yet, although the potential for data-driven marketing in co-branding is gigantic, it also requires a judicious balance between personalization and privacy. Luxury consumers, more than others, anticipate discretion and respect for their data (Hayes et al., 2021); (Aguirre et al., 2016); (Bleier et al., 2020). Co-branded alliances need to walk this tightrope with great care, promising transparent data use policies and tight cybersecurity measures. Brands that are seen to take responsibility in handling consumer data not only protect trust but also enhance their prestige among discerning consumers (Hayes et al., 2021); (Aguirre et al., 2016); (Y. Zhu & Kanjanamekanant, 2020); (Bleier et al., 2020). In summary, data-driven marketing has emerged as a cornerstone of effective co-branding strategies within the smart luxury industry(Quamina et al., 2023); (Suzuki & Kanno, 2022). It informs partnership decisions, enhances targeting accuracy, maximizes campaign performance, and enhances consumer experience from discovery to loyalty(Chapman & Dilmperi, 2022); (S et al., 2021). As everyday life incorporates more smart products, the capability to leverage and ethically utilize consumer data will shape the future of innovation and influence in luxury co-branding(Quamina et al., 2023); (Suzuki & Kanno, 2022); (Chapman & Dilmperi, 2022); (S et al., 2021).

5.3 The Ethical Considerations of AI in Luxury Marketing

As artificial intelligence (AI) continues to redefine the boundaries of luxury marketing, especially in co-branded intelligent products, ethical concerns have become a key point of focus (Xu & Mehta, 2022); (S et al., 2021). In contrast to mass-market segments where personalization is transactional, luxury consumers expect a more subtle and respectful approach—one that maintains dignity, privacy, and exclusivity alongside innovation and relevance. Co-branding initiatives involving AI must negotiate a delicate web of consumer expectations, technological capabilities, and moral obligations (Xu & Mehta, 2022); (S et al., 2021); (Gonçalves et al., 2023). At the center of AI ethics in luxury branding is the ethical use of consumer data. Luxury brands generally develop rich emotional bonds with their clients, who likewise look for discretion and integrity (Zakiyah, 2023); (Xu & Mehta, 2022); (Rahman et al., 2023). The use of AI for personalization—be it in terms of predictive algorithms, chatbots, or facial recognition technology—is premised on the availability of sensitive personal data (Xu & Mehta, 2022); (Rahman et al., 2023). This may consist of behavioral data, wearability biometrics, and even user-interaction-generated psychographic profiles. As marvelous personalization becomes possible with AI, there's also the threat of encroaching on privacy boundaries if controlled without openness and permission (Zakiyah, 2023); (Xu & Mehta, 2022).

Smart luxury co-branded products, like fashion wearables or intelligent cars, garner real-time flows of information from users, and these can be used to offer hyper-personalized experiences (Roy et al., 2019); (Pantano et al., 2018). Yet, the opacity in how this data is handled, stored, and transferred between co-operating brands is a cause for concern. Consumers might not know who is the owner of their data, for how long it will be stored, and if it will be used for anything other than the original intent (Roy et al., 2019); (Liang et al., 2021). Brands that do not explain these terms in clear terms might undermine the same trust that they have worked so hard to establish. Conversely, individuals who do adopt ethical AI models—such as transparent opt-in structures, anonymous data use, and open data-sharing contracts—are in a stronger position to fortify loyalty among a digitally aware population (Roy et al., 2019); (Pantano et al., 2018); (Liang et al., 2021). Another ethical aspect is algorithmic bias. AI technologies, particularly those developed using imbalanced datasets, have the potential to perpetuate stereotypes by accident or generate exclusionary results (Kordzadeh & Ghasemaghaei, 2021); (Ukanwa, 2024). In the high-end market, where diversity and individuality should be the norm, an unfair algorithm privileging some consumer profiles over others can lead to exclusion. Recommendation engines, for instance, may underrepresent some demographic groups in product offerings, or facial recognition tools may work poorly with people of color. Such flaws are not technical ones—they are brand missteps with deep reputational implications. Ethical AI requires active auditing, inclusive data training, and engagement of diverse teams in model creation (Kordzadeh & Ghasemaghaei, 2021); (Ukanwa, 2024); (Norori et al., 2021).

Emotional manipulation by AI is another tightrope on ethics. Luxury advertising has long played on desire, status, and aspiration—but when AI acquires the ability to subtly take advantage of psychological triggers, it can enter the realm of manipulation (Bakpayev et al., 2020); (Campbell et al., 2021). Methods like dynamic pricing, scarcity cues, or careful scarcity can be amplified with the help of AI, which could push consumers towards impulsive decisions they might come to regret later (Campbell et al., 2021); (Bakir et al., 2024). Behavioral nudges can, in fact, increase participation but need to be used with care and justice, especially in big-ticket luxury transactions. A co-branded offering incorporating AI in the form of FOMO (fear of missing out) may drive short-term conversions, but can jeopardize brand equity if buyers are deceived or coerced (Bakpayev et al., 2020); (Campbell et al., 2021) (Bakir et al., 2024). Moreover, AI-driven job displacement is an ethical issue that is frequently neglected. As AI takes over such tasks as customer service, curation, and even content generation, jobs previously served by human specialists—such as in-store stylists or luxury brand representatives—could decline (V. Shankar, 2024);

(Modhoriye et al., 2023). This change might dismantle the high-touch, bespoke service that luxury consumers demand. Co-branding partners need to think about how to balance AI effectiveness with human warmth, perhaps through re-skilling employees for higher-order jobs like experience curators or digital concierges so that the brand's heritage of personal touch is not compromised (V. Shankar, 2024); (Modhoriye et al., 2023); (Pavashe et al., 2023).

Authenticity is also a concern. Authenticity is a highly valued currency in luxury branding. Once experiences created by AI are indistinguishable from ones created by human beings, one is left pondering the fundamental nature of luxury itself (Xu & Mehta, 2022); (Park & Ahn, 2024). Will an AI-tailored fashion experience be equivalent in emotional impact to a wise human stylist? Can an algorithm ever grasp the subtlety of cultural heritage, artistic reference, or mastery-level craftsmanship underlying a luxury brand? Co-branded collaborations that depend greatly on AI need to tread very carefully—utilizing the technology to enhance, not substitute, the artistry and heritage that luxury represents (Xu & Mehta, 2022); (Park & Ahn, 2024); (Shahzad et al., 2024).

Finally, there is the issue of sustainability and the environmental impact of AI. Though not much talked about in luxury marketing communities, the data centers that run AI models burn enormous energy resources (Falk et al., 2024); (S. Shankar, 2024). As luxury brands become more aligned with environmental stewardship, they need to consider whether their AI-based marketing infrastructures are in line with these principles (Falk et al., 2024); (Chauhan, 2024). Ethical co-branding must not only take into account the user experience but also the larger social and planetary impact. Collaborations that embrace green AI strategies—like carbon-neutral data centers or energy-saving algorithms—will set the industry on the right path (Falk et al., 2024); (S. Shankar, 2024). In conclusion, the embedding of AI in co-branded smart luxury marketing is a range of ethical issues—from privacy and bias to manipulation, displacement, authenticity, and sustainability. While the potential of AI is great, so is its responsibility (Xu & Mehta, 2022); (Hermann, 2021). The luxury industry, through its aspirational positioning and high consumer trust, has a special duty to be an advocate of ethical practices. Co-branding partnerships need to infuse these values into all aspects of strategy, technology, and communication. Only then can they create a future where innovation and integrity exist side by side in the service of a truly enhanced consumer experience (Xu & Mehta, 2022); (Hermann, 2021).

6. CHALLENGES AND RISKS IN SMART LUXURY CO-BRANDING

The charm of smart luxury co-branding is that it can combine the sophistication and tradition of premium brands with the advanced technology of tech companies. Yet, this partnership is not without major obstacles (Suzuki & Kanno, 2022); (Prandelli et al., 2024). As it is that brands strive to balance their unique profiles and provide technology-enabled experiences without sacrificing their identity, they land on a complicated ground with strategic pitfalls and perceptual issues (Suzuki & Kanno, 2022); (Chapman & Dilmperi, 2022). Brand value misalignments, consumer distrust toward technological incursions into established luxury spaces, and the risk of exclusivity compromise all threaten success for such tie-ups (Chapman & Dilmperi, 2022); (Shen et al., 2017). Within this section, we explore the central issues with which brands have to contend—varying from identity clashes and credibility issues to the sensitive protection of brand value and prestige amidst a changing digital-luxury environment (Suzuki & Kanno, 2022); (Prandelli et al., 2024).

6.1 Threats to Brand Identity and Positioning

Of the biggest challenges facing smart luxury co-branding is the retention of individual brand identity coherence and integrity. Luxury brands are based on strong historical roots, scarcity, and emotional storytelling that take decades or centuries to build(Suzuki & Kanno, 2022); (Prandelli et al., 2024). They carefully construct brand positions addressing a small elite audience of wealthy individuals who relate luxury with authenticity, rarity, and handmade quality(Prandelli et al., 2024); (Tynan et al., 2009). When a luxury brand signs on for co-branding with a technology company that might represent a more frenetic, more innovation-motivated, often mass-market culture, the threat to brand identity dilution is great(Prandelli et al., 2024); (Tynan et al., 2009). These two universes, complementary as they are to the possibility of generating new consumer value, generally possess different cultural DNA, communication styles, and symbolic meaning. Misalignment in tone, messaging, or values can confuse customers and alienate the loyal customer base of both or either involved brand(Suzuki & Kanno, 2022); (Prandelli et al., 2024)..

A heritage luxury brand, for example, might own tradition, timeless appeal, and artistic heritage, whereas a technology partner might own functionality, cutting-edge performance, and iterative-based innovation. If these positioning approaches are not balanced well, the end product can end up looking fragmented or inauthentic (Suzuki & Kanno, 2022); (Prandelli et al., 2024); (Caponigri, 2023); (Shen et al., 2017). These inconsistencies can chip away at consumer confidence and devalue perceived brand integrity. An error in co-branding—a poorly combined product experience or ad campaign that is unable to harmonize both brand stories—can alter consumer opinion, repositioning the luxury brand in an undesirable manner(Quamina et al., 2023); (Suzuki & Kanno, 2022); (Prandelli et al., 2024).. This is particularly important in luxury markets where positioning is so intricately connected with emotional appeal and aspirational worth. Therefore, intelligent luxury co-branding involves a very fine balancing act: blending innovation without sacrificing heritage, and escalating value without sacrificing the essence brand story (Quamina et al., 2023); (Caponigri, 2023); (Shen et al., 2017).

6.2 Consumer Suspicion of Tech-Enriched Luxury Items

Regardless of increased infusion of technology within luxury products, a large portion of luxury consumers is still dubious of the value proposition and authenticity of tech-enriched items (Ho et al., 2021); (S. Han & Kim, 2020); (Holmqvist et al., 2019). This skepticism stems from the conventional understanding of luxury, which is based on timeless design, hand craftsmanship, and long-term emotional value—qualities that tend to be at odds with the perceived disposability and obsolescence of technology (Ho et al., 2021); (S. Han & Kim, 2020). To the discerning consumer, smart features can appear to be a gimmick that devalues the very qualities that make a product "luxurious" (S. Han & Kim, 2020); (Holmqvist et al., 2019). For instance, a luxury watch has traditionally represented status, heritage, and careful engineering. When such a product incorporates digital functionalities, some consumers may perceive it as losing its timeless character or becoming more aligned with mass-market gadgets (Joy et al., 2022); (Xu & Mehta, 2022). The rapid advancement and short life cycles of technology can further reinforce this sentiment. As compared to a high-end handbag or mechanical watch that can be treasured and handed down the generations, technology-infused luxury products risk being outdated in just a few years through software changes or hardware constraints. Such volatility undermines the values of longevity and timeless value associated with the luxury sector (Joy et al., 2022); (Xu & Mehta, 2022); (Sanz-Lopez et al., 2024).

Consumers also frequently doubt the authenticity of such collaborations, assuming they are cynical marketing exercises instead of synergies with substance. If the technological integration comes across as forced or not having a functional value, it can elicit backlash or consumer indifference (Ruiz & Cruz, 2023); (Ting, 2023). Data privacy and the intrusiveness of linked products are also a concern since luxury should provide a sense of personalization, discretion, and control. For example, a luxury fashion product integrated with tracking sensors or AI analytics can raise concerns regarding surveillance, use of data, and intrusiveness, and therefore collide with the privacy expectations that go with luxury ownership (Ruiz & Cruz, 2023); (Ting, 2023). To transcend this cynicism, luxury brands need to prove that advanced features add to—not take away from—the product's exclusivity, craftsmanship, and emotional connection. Transparency, effortless design, and value addition are the key. The challenge is providing tech-luxury hybrids that appear natural, elegant, and irreplaceable rather than frivolous or ephemeral (Xu & Mehta, 2022); (S. Han & Kim, 2020).

6.3 Brand Equity Management and Dilution of Exclusivity

In luxury brand space, exclusivity is not merely a cost strategy—it's a central column of brand equity that signals prestige, heritage, and singularity(Suzuki & Kanno, 2022); (Chapman & Dilmperi, 2022). The marriage of technology into luxury goods, especially through co-branding associations, presents a precarious balancing act. While the partnership can stretch market presence and add richness to product functionality, it also inherently risks dilution of the brand if handled with strategic inexactness (Quamina et al., 2023); (Suzuki & Kanno, 2022); (Chapman & Dilmperi, 2022); (S et al., 2021). The major challenge that faces luxury brands when co-branding with technology businesses is the undermining of their symbolic value undermined (Quamina et al., 2023);(Choi et al., 2016). Their perceived exclusivity is frequently tied to scarcity, elite craftsmanship, and restricted accessibility. In collaborations between luxury brands and technology firms that are largely accessible, the aura of inaccessibility by the brand risks being undermined (Quamina et al., 2023); (Suzuki & Kanno, 2022); (Choi et al., 2016). For example, if a prestige brand that historically caters to a niche customer base collaborates with a technology company with mass-market products, it can create mixed messages among consumers and threaten its upscale positioning(Quamina et al., 2023); (Suzuki & Kanno, 2022);.

Moreover, co-branding initiatives can result in brand identity confusion, particularly when the partnering brands' values, aesthetics, or target customers are not congruent. A failed partnership will break the linear narrative luxury brands have constructed over years, stripping consumers of confidence and loyalty towards the brand (Suzuki & Kanno, 2022); (Choi et al., 2016). (Karangi & Lowe, 2021). For instance, if the partner's brand involved in the technological partnership goes through a reputational scandal—be it privacy invasions, work place issues, or product flaws—the negative contagion will mar the image of the luxury brand as well. Such a scenario is called “brand contagion” and its implications on the brand equity last for years (Quamina et al., 2023); (Suzuki & Kanno, 2022); (Karangi & Lowe, 2021).

Additionally, co-branded product overexposure can jeopardize exclusivity. If intelligent luxury products become commonly available or kept in frequent circulation, they might become less desirable to high-end consumers who place a premium on scarcity and eternal beauty(Quamina et al., 2023); (Suzuki & Kanno, 2022); (Y. Yu et al., 2020). Launching too many iterations or co-branded releases within a condensed timeframe also tends to tire the market and diffuse the product's point of distinction. Limited runs, custom features, and bespoke experiences are among the weapons luxury brands have to keep their products and services feeling rare even where they incorporate technology(Quamina et al., 2023); (Suzuki & Kanno, 2022); (Y. Yu et al., 2020); (Shen et al., 2017) . In order to protect brand equity in co-branding initiatives, luxury brands have to keep strict control over design, communication, and distribution. The collaboration should look like an organic extension of the luxury brand's DNA, not a digression (Quamina et al., 2023); (Suzuki & Kanno, 2022); (Shen et al., 2017). Strategic congruence, long-term perspective, and narrative consistency are crucial in making sure that the product enriches both brands without betraying their fundamental identities. The ultimate objective is to reinforce the value proposition—not weaken it—by blending innovation with eternal allure (Quamina et al., 2023); (Suzuki & Kanno, 2022); (Y. Yu et al., 2020); (Shen et al., 2017) .

7. FUTURE TRENDS AND STRATEGIC RECOMMENDATIONS

As intelligent luxury co-branding keeps developing, the environment is being driven by influential technological, social, and environmental forces(Quamina et al., 2023); (Suzuki & Kanno, 2022); (Abedsoltan et al., 2022). The intersection of innovation and exclusivity is forcing brands to reimagine the concept of luxury in the digital age. As consumer expectations change and the competitive landscape gets fiercer, luxury brands need to forecast trends and employ cutting-edge strategies in order to stay relevant and engage their audience(Quamina et al., 2023); (Suzuki & Kanno, 2022).

7.1 The Emergence of Immersive Experiences (AR/VR) in Luxury Co-Branding

The use of immersive technologies like Augmented Reality (AR) and Virtual Reality (VR) will revolutionize customer interaction in luxury co-branding (Nawres et al., 2023); (Lin, 2022). With these technologies, experiential storytelling is facilitated that amplifies emotional connection, enabling customers to "experience" the luxury prior to a purchase. For instance, AR apps enabling users to virtually try on high-end watches or handbags, or VR platforms mimicking vip fashion shows and showroom visits, provide a deeper emotional and sensory affinity with the brand (Nawres et al., 2023); (Lin, 2022); (Scholz & Duffy, 2018). For co-branded items, immersive technology opens doors for each brand story to be shared alongside the other (Tran et al., 2022); (Chakraborty & Biswal, 2024). Picture a virtual reality experience wherein consumers get to see the innovation of a luxury smartwatch as seen through both the fashion company and the technological partner—creating one cohesive brand narrative. These experiences meet the increasing need for personalization, exclusivity, and interaction among younger luxury customers who prefer technology-based interactions (Tran et al., 2022); (Chakraborty & Biswal, 2024); (Romero et al., 2024). To properly harness this trend, luxury brands will need to make investments in excellent digital craftsmanship that reflects the quality of their physical products. The experience needs to be as high-end as the product, both brands seamlessly integrating into an emotionally engaging trip (Pozharliev et al., 2015); (Bazi et al., 2020).

7.2 The Role of Sustainability and Ethical Branding in Smart Luxury Partnerships

In a world characterized by climate awareness and ethical consumption, the incorporation of sustainability into luxury branding is no longer a trend but a requirement (Athwal et al., 2019); (Mahmoud et al., 2023); (Di Leo et al., 2023). Contemporary luxury consumers, particularly Gen Z and Millennials, increasingly value purpose-driven brands that share their values. They demand transparency, sustainable sourcing, ethical labor, and environmentally friendly innovation. In this regard, smart luxury co-branding collaborations need to transcend luxury and aesthetics to incorporate a common vision for sustainability and ethics (Athwal et al., 2019); (Mahmoud et al., 2023).

Co-branding presents the special chance to combine technological innovation with sustainable practices (Barile et al., 2020); (Anadon et al., 2016). As an illustration, a partnership between a luxury fashion brand and a green-tech firm can yield smart products which are technologically sophisticated but also sustainable. Wearable luxury electronics constructed from reclaimed materials, fuelled by green energy solutions, or composed of modular, repairable parts can resonate strongly with the value-conscious consumer. These developments can reposition luxury as not merely a status symbol, but an expression of values (Barile et al., 2020); (Anadon et al., 2016); (Elkhwesky et al., 2022). In addition, ethical branding in co-branded smart products strengthens trust and authenticity. The customers examine not only the final product but the whole supply chain (Alwi et al., 2017); (Sénéchal et al., 2013). Thus, transparency in sourcing, sustainable production practices, and equitable labor practices must form the core of both partners' values. Luxury brands that clearly spell out these initiatives—supported by certifications, carbon-neutral commitments, or ESG (Environmental, Social, Governance) ratings—can build greater emotional affinity with their audience (Alwi et al., 2017); (Sénéchal et al., 2013); (Hunt, 2018).

Strategically, sustainability is also an influence on long-term brand value. As regulation grows and consumers keep demanding ethical responsibility, first-movers toward sustainable co-branding models will benefit from increased competitive advantage (Diallo et al., 2020); (Zein et al., 2019). Brands need thus to incorporate sustainability into the genetic makeup of co-branding plans—selecting partners that share comparable environmental and ethical values and innovating collectively to create products that embody luxury as well as responsibility (Diallo et al., 2020); (Zein et al., 2019); (Loh & Tan, 2020).

7.3 Strategic Implications for Brands Contemplating Penetration of Smart Luxury Co-Branding

Penetrating the world of smart luxury co-branding entails more than a great product concept—it takes sophisticated, research-grounded strategy in deference to the history of luxury but alignment with the pace of technological growth(Quamina et al., 2023); (Suzuki & Kanno, 2022); (Y. Yu et al., 2020);. For companies looking to dominate this segment, a number of strategic implications can act as a guide to map out effective collaboration, long-term customer retention, and brand distinctiveness (Quamina et al., 2023); (Suzuki & Kanno, 2022); (Y. Yu et al., 2020); (Shen et al., 2017) . Above all, consistency of brand values is important. Co-branding efforts in the luxury technology arena need to transcend transactional partnerships and adopt relational synergy. Both companies must have a shared vision and maintain a uniform level of quality, narrative, and consumer interaction (Ho et al., 2019); (Ho et al., 2017). A clash of brand personalities—e.g., a minimalist technology brand partnering with a theatrical luxury brand—can cause consumers to get confused and dilute brand value. Strategic due diligence must entail a thorough cultural audit and value compatibility analysis prior to engaging in any co-branding agreement (Ho et al., 2019); (Ho et al., 2017); (Suzuki & Kanno, 2022); (Y. Yu et al., 2020); (Shen et al., 2017).

Lastly, consumer-driven innovation must be the core of the collaboration. Savvy luxury consumers want more than functionality; they want smooth integration of technology that improves their lifestyle, represents their identity, and addresses their aspirational desires(Quamina et al., 2023); (Suzuki & Kanno, 2022). Co-branded intelligent products should thus emphasize intuitive design, personalization options, and ease of use while maintaining the sensory, emotional essence of luxury. AI- and machine-learning-powered data-driven personalization should be used ethically to provide customized experiences, from bespoke app content to responsive product functionality(Quamina et al., 2023); (Suzuki & Kanno, 2022); (Y. Yu et al., 2020); (Shen et al., 2017).

Another key takeaway is to leverage narrative branding as a strategic point of difference. Storytelling is amplified in the luxury category. Brands need to come together to create an engaging story that weaves the history and innovation of both partners together (Kim et al., 2015); (Rachbini et al., 2023). Whether it's an intelligent watch that writes the story of ageless craftsmanship and digital accuracy or a connected handbag that represents fashion-tech empowerment, emotional storylines enrich consumer attachment and fuels desirability (Ryu et al., 2018); (Rachbini et al., 2023); (Huang et al., 2018). Both scarcity and timing become strategic. Many co-branded luxury smart product successes rely on generating scarcity from limited editions, soft releases to high-end customers, or invitation-only retail environments(Shen et al., 2017); (Quamina et al., 2023). This retains the exclusivity aura that exists in luxury brands, while engendering hype and a sense of urgency by early adopters and loyal followers (Shen et al., 2017); (Quamina et al., 2023); (Suzuki & Kanno, 2022).

Finally, there is a need to adopt an iterative approach based on consumer feedback cycles and rapid adaptation. Co-branded offerings should not be isolated experiments but integrated part of a broader system of innovation (Bhandari et al., 2020); (Dhaoui & Webster, 2020). Performance analytics on post-launch performance, customer sentiment analysis, and co-creation websites can assist brands in optimizing subsequent collaborations and building community-driven brand devotion. Integrating real-time feedback systems and open innovation programs also allows the brand to stay culturally contemporary and technologically abreast (Bhandari et al., 2020); (Dhaoui & Webster, 2020); (J. Yang & Mundel, 2021). Essentially, smart luxury co-branding is more than a merger of brand identities but an orchestration of heritage, innovation, and consumer psychology. With a clear path of value alignment, user-centric design, narrative depth, exclusivity, and agile evolution, brands can move forward confidently in this space and create iconic, forward-looking partnerships(Shen et al., 2017); (Quamina et al., 2023); (Suzuki & Kanno, 2022).

8. CONCLUSION

The advent of smart luxury products at the convergence of technology and exclusivity marks a paradigm change in the manner in which value is communicated, consumers are interacted with, and meaning is co-created within the contemporary marketplace (J. Lin, 2023); (Zhu, 2024). From a consumer psychology and branding strategy perspective, this chapter has explored the distinctive dynamics underlying co-branding success within the smart luxury category—where innovation needs to find harmony with tradition, and consumer expectations are formulated by both emotional connection and functional utility (J. Lin, 2023); (Zhu, 2024); (Ma, 2023); (Pino et al., 2017).

The examination started with how technology is remodeling luxury, and this has created a new line of intelligent consumer goods that have elegance, wit, and bespoke utility. Co-branding is now a very effective strategy during this revolution as it helps established luxury brands remain relevant while also helping technology companies acquire aspirational value (Al-Issa & Thanasi, 2024); (Creevey et al., 2021). Yet effective co-branding within this niche is not so much a matter of logo or product feature combination—it is about matching brand DNA, having consistency of storytelling, and providing substance to consumer experiences that maintain the dignity and cachet of both partners (Al-Issa & Thanasi, 2024); (Creevey et al., 2021); (Shin & Jeong, 2022).

Throughout this chapter, the central position of consumer psychology has been highlighted. Emotional drivers like trust, exclusivity, and status signaling have a powerful impact on how intelligent luxury products are viewed and embraced (Oğuz, 2024); (Yadav, 2024). The strategic application of neuromarketing and behavioral economics—based on cognitive heuristics such as scarcity, social proof, and anchoring—has been effective in influencing buying behavior. Furthermore, the convergence of AI and data-driven personalization provides unprecedented opportunity to craft hyper-relevant brand experiences, at the same time that it also poses ethical issues related to data privacy and brand manipulation (Oğuz, 2024); (Yadav, 2024); (Xu & Mehta, 2022).

Concurrently, the path to smart luxury co-branding is also fraught with obstacles. From the danger of brand dilution to consumer cynicism regarding hybrid products, brands have to navigate cautiously to protect their equity and legitimacy (Quamina et al., 2023); (Suzuki & Kanno, 2022). The last parts of the chapter offered future-oriented insights, emphasizing immersive technologies, sustainability, and cultural authenticity as essential pillars for the future generation of co-branded smart luxury experiences(Quamina et al., 2023); (Suzuki & Kanno, 2022); (Y. Yu et al., 2020); (Shen et al., 2017). For companies and marketers, the stakes are high: appreciating the psychological subtleties of their customer base and making strategic use of that knowledge in designing co-branded stories and experiences will be the secret to standing out in an oversaturated market(Pangarkar & Shukla, 2023); (Leban et al., 2019). For researchers, this new field of study offers rich soil for deeper investigation into the ways digital disruption is reconfiguring the world of luxury consumption—especially through such prisms as neuropsychology, AI ethics, and cross-cultural conduct (Pangarkar & Shukla, 2023); (Leban et al., 2019); (Sanz-Lopez et al., 2024).

As consumer needs remain to be developed further toward experiential, personalized, and meaningful luxury, the blending of technology and psychology will characterize the competitive advantage (Okonkwo, 2009); (Quamina et al., 2023);

(Suzuki & Kanno, 2022); (Y. Yu et al., 2020); (Shen et al., 2017). Intelligent luxury co-branding, when practiced with strategic purpose and psychological acumen, is not only capable of fulfilling these aspirations but also for creating new ones—reconfiguring the meaning of luxury in the digital age (Chapman & Dilmperi, 2022); (Wu & Kim, 2023); (Okonkwo, 2009); (Quamina et al., 2023); (Suzuki & Kanno, 2022); (Y. Yu et al., 2020); (Shen et al., 2017).

REFERENCES

Abedsoltan, H., Taleizadeh, A. A., & Sarker, R. B. (2022). Optimal production of remanufactured products with collaboration and co-branding of luxury brands. *Computers & Industrial Engineering*, *172*, 108533. DOI: 10.1016/j.cie.2022.108533

Aguirre, E., Roggeveen, A. L., Grewal, D., & Wetzels, M. (2016). The personalization-privacy paradox: Implications for new media. *Journal of Consumer Marketing*, *33*(2), 98–110. DOI: 10.1108/JCM-06-2015-1458

Ajitha, S., & Sivakumar, V. (2017). Understanding the effect of personal and social value on attitude and usage behavior of luxury cosmetic brands. *Journal of Retailing and Consumer Services*, *39*, 103–113. DOI: 10.1016/j.jretconser.2017.07.009

Akram, U., & Lavuri, R. (2023). Understanding the consumer's luxury webrooming intention: Moderating role of perceived risk and review. *Journal of Consumer Behaviour*, *23*(3), 1602–1619. DOI: 10.1002/cb.2295

Al-Issa, N., & Thanasi, M. (2024). Mapping the Future of Tech-Infused Luxury: A Roadmap and research directions. *International Journal of Consumer Studies*, *48*(6), e13103. Advance online publication. DOI: 10.1111/ijcs.13103

Alexander, B., & Contreras, L. O. (2016). Inter-industry creative collaborations incorporating luxury fashion brands. *Journal of Fashion Marketing and Management*, *20*(3), 254–275. DOI: 10.1108/JFMM-09-2015-0075

Alsharif, A. H., Salleh, N. Z. M., Abdullah, M., Khraiwish, A., & Ashaari, A. (2023). Neuromarketing Tools used in the Marketing Mix: A Systematic literature and Future Research agenda. *SAGE Open*, *13*(1), 21582440231156563. Advance online publication. DOI: 10.1177/21582440231156563

Alwi, S. F. S., Ali, S. M., & Nguyen, B. (2017). The importance of Ethics in branding: Mediating effects of ethical branding on company reputation and brand loyalty. *Business Ethics Quarterly*, *27*(3), 393–422. DOI: 10.1017/beq.2017.20

Anadon, L. D., Chan, G., Harley, A. G., Matus, K., Moon, S., Murthy, S. L., & Clark, W. C. (2016). Making technological innovation work for sustainable development. *Proceedings of the National Academy of Sciences of the United States of America*, *113*(35), 9682–9690. DOI: 10.1073/pnas.1525004113 PMID: 27519800

Anisimova, T. A. (2007). The effects of corporate brand attributes on attitudinal and behavioural consumer loyalty. *Journal of Consumer Marketing*, *24*(7), 395–405. DOI: 10.1108/07363760710834816

Arrigo, E. (2018). Social media marketing in luxury brands. *Management Research Review, 41*(6), 657–679. DOI: 10.1108/MRR-04-2017-0134

Athwal, N., Wells, V. K., Carrigan, M., & Henninger, C. E. (2019). Sustainable Luxury Marketing: A synthesis and research agenda. *International Journal of Management Reviews, 21*(4), 405–426. DOI: 10.1111/ijmr.12195

Atwal, G., & Williams, A. (2009). Luxury brand marketing – The experience is everything! *Journal of Brand Management, 16*(5–6), 338–346. DOI: 10.1057/bm.2008.48

Babatunde, N. S. O., Odejide, N. O. A., Edunjobi, N. T. E., & Ogundipe, N. D. O. (2024). THE ROLE OF AI IN MARKETING PERSONALIZATION: A THEORETICAL EXPLORATION OF CONSUMER ENGAGEMENT STRATEGIES. *International Journal of Management & Entrepreneurship Research, 6*(3), 936–949. DOI: 10.51594/ijmer.v6i3.964

Baghi, I., & Gabrielli, V. (2018). Brand prominence in cause-related marketing: Luxury versus non-luxury. *Journal of Product and Brand Management, 27*(6), 716–731. DOI: 10.1108/JPBM-07-2017-1512

Bakir, V., Laffer, A., McStay, A., Miranda, D., & Urquhart, L. (2024). On manipulation by emotional AI: UK adults' views and governance implications. *Frontiers in Sociology, 9*, 1339834. Advance online publication. DOI: 10.3389/fsoc.2024.1339834 PMID: 38912311

Bakpayev, M., Baek, T. H., Van Esch, P., & Yoon, S. (2020). Programmatic creative: AI can think but it cannot feel. [AMJ]. *Australasian Marketing Journal, 30*(1), 90–95. DOI: 10.1016/j.ausmj.2020.04.002

Bandyopadhyay, A., Septianto, F., & Nallaperuma, K. (2021). Mixed feelings enhance the effectiveness of luxury advertising. [AMJ]. *Australasian Marketing Journal, 30*(1), 28–34. DOI: 10.1177/1839334921998848

Barile, S., Grimaldi, M., Loia, F., & Sirianni, C. A. (2020). Technology, Value Co-Creation and Innovation in service ecosystems: Toward Sustainable Co-Innovation. *Sustainability (Basel), 12*(7), 2759. DOI: 10.3390/su12072759

Barnes, F., & Higgins, D. M. (2017). Brand image, cultural association and marketing: 'New Zealand' butter and lamb exports to Britain, c. 1920–1938. *Business History, 62*(1), 70–97. DOI: 10.1080/00076791.2017.1344223

Baudier, P., De Boissieu, E., & Duchemin, M. (2022). Source Credibility and Emotions generated by Robot and Human Influencers: The perception of luxury brand representatives. *Technological Forecasting and Social Change*, *187*, 122255. DOI: 10.1016/j.techfore.2022.122255

Baumgarth, C., & Wieker, J. B. (2020). From the classical art to the urban art infusion effect: The effect of street art and graffiti on the consumer evaluation of products. *Creativity and Innovation Management*, *29*(S1), 116–127. DOI: 10.1111/caim.12362

Bazi, S., Filieri, R., & Gorton, M. (2020). Customers' motivation to engage with luxury brands on social media. *Journal of Business Research*, *112*, 223–235. DOI: 10.1016/j.jbusres.2020.02.032

Bhandari, M., Rodgers, S., & Pan, P. (2020). Brand feedback to negative eWOM messages: Effects of stability and controllability of problem causes on brand attitudes and purchase intentions. *Telematics and Informatics*, *58*, 101522. DOI: 10.1016/j.tele.2020.101522

Bi, M. (2024). Narrative brand design in the field of emotional consumption. *Highlights in Art and Design*, *6*(3), 22–25. DOI: 10.54097/yy2n8f93

Bishnoi, S. K., & Singh, S. (2021). A study on consumer buying behaviour for fashion and luxury brands under emotional influence. *Research Journal of Textile and Apparel*, *26*(4), 405–418. DOI: 10.1108/RJTA-03-2021-0026

Blazquez, M., Alexander, B., & Fung, K. (2020). Exploring Millennial's perceptions towards luxury fashion wearable technology. *Journal of Fashion Marketing and Management*, *24*(3), 343–359. DOI: 10.1108/JFMM-09-2019-0200

Bleier, A., Goldfarb, A., & Tucker, C. E. (2020). Consumer privacy and the future of Data-Based Innovation and marketing. SSRN *Electronic Journal*. https://doi.org/ DOI: 10.2139/ssrn.3570156

Boka, T. F., Niu, Z., & Neupane, R. B. (2024). A survey of sequential recommendation systems: Techniques, evaluation, and future directions. *Information Systems*, *125*, 102427. DOI: 10.1016/j.is.2024.102427

Bonifield, C., & Cole, C. (2006). Affective responses to service failure: Anger, regret, and retaliatory versus conciliatory responses. *Marketing Letters*, *18*(1–2), 85–99. DOI: 10.1007/s11002-006-9006-6

Budnitsky, S., & Jia, L. (2018). Branding Internet sovereignty: Digital media and the Chinese–Russian cyberalliance. *European Journal of Cultural Studies*, *21*(5), 594–613. DOI: 10.1177/1367549417751151

Burger, J. M. (1986). Increasing compliance by improving the deal: The that's-not-all technique. *Journal of Personality and Social Psychology*, *51*(2), 277–283. DOI: 10.1037/0022-3514.51.2.277

Campbell, C., Plangger, K., Sands, S., & Kietzmann, J. (2021). Preparing for an era of Deepfakes and AI-Generated Ads: A framework for understanding responses to manipulated advertising. *Journal of Advertising*, *51*(1), 22–38. DOI: 10.1080/00913367.2021.1909515

Caponigri, F. (2023). Splitting the atom of heritage to respond to luxury's challenges and crises. *Luxury Studies the in Pursuit of Luxury Journal*, *2*(1), 37–57. DOI: 10.1386/ipol_00019_1

Carnevale, M., Yucel-Aybat, O., & Kachersky, L. (2017). Meaningful stories and attitudes toward the brand: The moderating role of consumers' implicit mindsets. *Journal of Consumer Behaviour*, *17*(1). Advance online publication. DOI: 10.1002/cb.1687

Castillo-Villar, F. R., & Villasante-Arellano, A. J. (2020). Applying the multisensory sculpture technique to explore the role of brand usage on multisensory brand experiences. *Journal of Retailing and Consumer Services*, *57*, 102185. DOI: 10.1016/j.jretconser.2020.102185

Chakraborty, D., Polisetty, A. G. S., Rana, N. P., & Khorana, S. (2024). Unlocking the potential of AI: Enhancing consumer engagement in the beauty and cosmetic product purchases. *Journal of Retailing and Consumer Services*, *79*, 103842. DOI: 10.1016/j.jretconser.2024.103842

Chakraborty, U., & Biswal, S. K. (2024). Diffusion of innovation in direct-to-avatar (D2A): A study of immersive communication and branding in the metaverse. *Computers in Human Behavior*, *158*, 108318. DOI: 10.1016/j.chb.2024.108318

Chan, E. Y., & Northey, G. (2021). Luxury goods in online retail: How high/low positioning influences consumer processing fluency and preference. *Journal of Business Research*, *132*, 136–145. DOI: 10.1016/j.jbusres.2021.04.031

Chapman, A., & Dilmperi, A. (2022). Luxury brand value co-creation with online brand communities in the service encounter. *Journal of Business Research*, *144*, 902–921. DOI: 10.1016/j.jbusres.2022.01.068

Chauhan, S. (2024). The growing energy demand of data centers: Impacts of AI and cloud computing. *International Journal for Multidisciplinary Research*, *6*(4), 26591. Advance online publication. DOI: 10.36948/ijfmr.2024.v06i04.26591

Cheah, I., Shimul, A. S., & Parker, J. (2021). Guest editorial: Luxury Branding – Strategy, Innovation and Sustainability. [AMJ]. *Australasian Marketing Journal, 29*(4), 275–276. DOI: 10.1177/18393349211052617

Chen, C., Guo, L., Zhou, M., Tsaur, W., Sun, H., Zhan, W., Deng, Y., & Li, C. (2022). Blockchain-Based Anti-Counterfeiting Management System for traceable luxury products. *Sustainability (Basel), 14*(19), 12814. DOI: 10.3390/su141912814

Chen, X., & Li, Z. (2022). Investigating the consumer evaluation of the co-branding of luxury brands. *International Journal of Research in Business and Social Science (2147-4478), 11*(4), 01–15. https://doi.org/DOI: 10.20525/ijrbs.v11i4.1808

Chernev, A., Hamilton, R., & Gal, D. (2011). Competing for consumer identity: Limits to Self-Expression and the perils of lifestyle branding. *Journal of Marketing, 75*(3), 66–82. DOI: 10.1509/jmkg.75.3.66

Cherubino, P., Martinez-Levy, A. C., Caratù, M., Cartocci, G., Di Flumeri, G., Modica, E., Rossi, D., Mancini, M., & Trettel, A. (2019). Consumer Behaviour through the Eyes of Neurophysiological Measures: State-of-the-Art and Future Trends. *Computational Intelligence and Neuroscience, 2019*, 1–41. DOI: 10.1155/2019/1976847 PMID: 31641346

Cheshire, P., & Sheppard, S. (1989). British planning policy and access to housing: Some empirical estimates. *Urban Studies (Edinburgh, Scotland), 26*(5), 469–485. DOI: 10.1080/00420988920080541

Chiang, C. (2023). A systematic literature network analysis of green information technology for sustainability: Toward smart and sustainable livelihoods. *Technological Forecasting and Social Change, 199*, 123053. DOI: 10.1016/j.techfore.2023.123053

Chiang, C., & Jang, S. S. (2007). The effects of perceived price and brand image on value and purchase intention: Leisure travelers' attitudes toward online hotel booking. *Journal of Hospitality & Leisure Marketing, 15*(3), 49–69. DOI: 10.1300/J150v15n03_04

Childs, M., & Jin, B. E. (2020). Brand and retailer co-branding. *Journal of Fashion Marketing and Management, 24*(1), 49–65. DOI: 10.1108/JFMM-03-2019-0061

Choi, C. (2024). The impact of perceived brand prestige on brand engagement: the mediating roles of brand trust and brand commitment. *Korea International Trade Research Institute, 20*(2), 351–368. https://doi.org/DOI: 10.16980/jitc.20.2.202404.351

Choi, E., Ko, E., & Kim, A. J. (2016). Explaining and predicting purchase intentions following luxury-fashion brand value co-creation encounters. *Journal of Business Research, 69*(12), 5827–5832. DOI: 10.1016/j.jbusres.2016.04.180

Choi, J., & Kim, S. (2016). Is the smartwatch an IT product or a fashion product? A study on factors affecting the intention to use smartwatches. *Computers in Human Behavior*, *63*, 777–786. DOI: 10.1016/j.chb.2016.06.007

Christodoulides, G. (2009). Branding in the post-internet era. *Marketing Theory*, *9*(1), 141–144. DOI: 10.1177/1470593108100071

Creevey, D., Coughlan, J., & O'Connor, C. (2021). Social media and luxury: A systematic literature review. *International Journal of Management Reviews*, *24*(1), 99–129. DOI: 10.1111/ijmr.12271

Cunha, M.Jr, Forehand, M. R., & Angle, J. W. (2014). Riding Coattails: When Co-Branding Helps versus Hurts Less-Known Brands. *The Journal of Consumer Research*, *41*(5), 1284–1300. DOI: 10.1086/679119

Cutrona, V., De Paoli, F., Košmerlj, A., Nikolov, N., Palmonari, M., Perales, F., & Roman, D. (2019). Semantically-Enabled optimization of digital marketing campaigns. In *Lecture notes in computer science* (pp. 345–362). https://doi.org/DOI: 10.1007/978-3-030-30796-7_22

Dai, D., An, Z., Pan, Q., & Yang, L. (2024). Harnessing NFC to generate standard optical barcodes for NFC-Missing smartphones. *IEEE Transactions on Mobile Computing*, *23*(12), 12952–12968. DOI: 10.1109/TMC.2024.3420410

Das, M., Habib, M., Saha, V., & Jebarajakirthy, C. (2021). Bandwagon vs snob luxuries: Targeting consumers based on uniqueness dominance. *Journal of Retailing and Consumer Services*, *61*, 102582. DOI: 10.1016/j.jretconser.2021.102582

Davenport, T. H. (2023). Hyper-Personalization for Customer Engagement with Artificial Intelligence. *Management and Business Review.*, *3*(1–2), 29–36. DOI: 10.1177/26941058202303010006

De Boissieu, E., Kondrateva, G., Baudier, P., & Ammi, C. (2021). The use of blockchain in the luxury industry: Supply chains and the traceability of goods. *Journal of Enterprise Information Management*, *34*(5), 1318–1338. DOI: 10.1108/JEIM-11-2020-0471

De Kerviler, G., & Rodriguez, C. M. (2019). Luxury brand experiences and relationship quality for Millennials: The role of self-expansion. *Journal of Business Research*, *102*, 250–262. DOI: 10.1016/j.jbusres.2019.01.046

Desai, D. (2021). Hyper-Personalization. In *Advances in marketing, customer relationship management, and e-services book series* (pp. 40–53). https://doi.org/DOI: 10.4018/978-1-7998-7959-6.ch003

Dhaoui, C., & Webster, C. M. (2020). Brand and consumer engagement behaviors on Facebook brand pages: Let's have a (positive) conversation. *International Journal of Research in Marketing*, *38*(1), 155–175. DOI: 10.1016/j.ijresmar.2020.06.005

Di Leo, A., Sfodera, F., Cucari, N., Mattia, G., & Dezi, L. (2023). Sustainability reporting practices: An explorative analysis of luxury fashion brands. *Management Decision*, *61*(5), 1274–1297. DOI: 10.1108/MD-02-2022-0142

Diallo, M. F., Mouelhi, N. B. D., Gadekar, M., & Schill, M. (2020). CSR actions, brand value, and willingness to pay a premium price for luxury brands: Does Long-Term orientation matter? *Journal of Business Ethics*, *169*(2), 241–260. DOI: 10.1007/s10551-020-04486-5

Diawara, M. (1995). Dial Ouaga for AFRICAN Cinema. *Nka Journal of Contemporary African Art*, *1995*(3), 46–49. DOI: 10.1215/10757163-3-1-46

Dogaru, I., Furnham, A., & McClelland, A. (2024). Understanding how the presence of music in advertisements influences consumer behaviour. *Acta Psychologica*, *248*, 104333. DOI: 10.1016/j.actpsy.2024.104333 PMID: 39032271

Dubois, D., Jung, S., & Ordabayeva, N. (2020). The psychology of luxury consumption. *Current Opinion in Psychology*, *39*, 82–87. DOI: 10.1016/j.copsyc.2020.07.011 PMID: 32861090

Elkhwesky, Z., Manzani, Y. E., & Salem, I. E. (2022). Driving hospitality and tourism to foster sustainable innovation: A systematic review of COVID-19-related studies and practical implications in the digital era. *Tourism and Hospitality Research*, *24*(1), 115–133. DOI: 10.1177/14673584221126792 PMID: 40479339

Endres, T., Eitel, A., Renninger, K. A., Vössing, C., & Renkl, A. (2024). Why narrative frames matter for instructional videos: A value-evoking narrative frame is essential to foster sustained learning with emotional design videos. *Learning and Instruction*, *94*, 101962. DOI: 10.1016/j.learninstruc.2024.101962

Escobedo, P., Bhattacharjee, M., Nikbakhtnasrabadi, F., & Dahiya, R. (2020). Smart bandage with wireless strain and temperature sensors and batteryless NFC tag. *IEEE Internet of Things Journal*, *8*(6), 5093–5100. DOI: 10.1109/JIOT.2020.3048282

Fionda, A. M., & Moore, C. M. (2009). The anatomy of the luxury fashion brand. *Journal of Brand Management*, *16*(5–6), 347–363. DOI: 10.1057/bm.2008.45

Frew, B., Brinkman, G., Denholm, P., Narwade, V., Stephen, G., Bloom, A., & Lau, J. (2021). Impact of operating reserve rules on electricity prices with high penetrations of renewable energy. *Energy Policy*, *156*, 112443. DOI: 10.1016/j.enpol.2021.112443

Fuchs, C., & Diamantopoulos, A. (2010). Evaluating the effectiveness of brand-positioning strategies from a consumer perspective. *European Journal of Marketing, 44*(11/12), 1763–1786. DOI: 10.1108/03090561011079873

Gao, E., & Yang, X. (2022). The Effectiveness of High-tech Product on Customer Satisfaction towards Luxury Hotel. *BCP Business & Management, 20*, 858–864. DOI: 10.54691/bcpbm.v20i.1073

Giglio, S., Pantano, E., Bilotta, E., & Melewar, T. (2019). Branding luxury hotels: Evidence from the analysis of consumers' "big" visual data on TripAdvisor. *Journal of Business Research, 119*, 495–501. DOI: 10.1016/j.jbusres.2019.10.053

Gil, M. Á. C. (2019). Standard and precordial leads obtained with an Apple watch. *Annals of Internal Medicine, 172*(6), 436–437. DOI: 10.7326/M19-2018 PMID: 31766051

Godey, B., Manthiou, A., Pederzoli, D., Rokka, J., Aiello, G., Donvito, R., & Singh, R. (2016). Social media marketing efforts of luxury brands: Influence on brand equity and consumer behavior. *Journal of Business Research, 69*(12), 5833–5841. DOI: 10.1016/j.jbusres.2016.04.181

Gonçalves, A. R., Pinto, D. C., Rita, P., & Pires, T. (2023). Artificial intelligence and its ethical implications for marketing. *Emerging Science Journal, 7*(2), 313–327. DOI: 10.28991/ESJ-2023-07-02-01

Grandhi, B., Patwa, N., & Saleem, K. (2020). Data-driven marketing for growth and profitability. *EuroMed Journal of Business, 16*(4), 381–398. DOI: 10.1108/EMJB-09-2018-0054

Grewal, D., Krishnan, R., Baker, J., & Borin, N. (1998). The effect of store name, brand name and price discounts on consumers' evaluations and purchase intentions. *Journal of Retailing, 74*(3), 331–352. DOI: 10.1016/S0022-4359(99)80099-2

Guan, Y. (2023). Emotional branding Strategies and effects in the U.S market. *Communications in Humanities Research, 4*(1), 524–530. DOI: 10.54254/2753-7064/4/20220838

Gungunawat, A., Khandelwal, N., & Gupta, N. (2024). AI-Powered Personalization in Digital Marketing: Transforming consumer engagement and strategy. *RESEARCH REVIEW International Journal of Multidisciplinary, 9*(11), 183–191. DOI: 10.31305/rrijm.2024.v09.n11.026

Hamzah, M. I., & Pontes, N. (2022). What drives car buyers to accept a rejuvenated brand? the mediating effects of value and pricing in a consumer-brand relationship. *Journal of Strategic Marketing*, *32*(2), 114–136. DOI: 10.1080/0965254X.2022.2129749

Han, S., & Kim, K. (2020). Role of consumption values in the luxury brand experience: Moderating effects of category and the generation gap. *Journal of Retailing and Consumer Services*, *57*, 102249. DOI: 10.1016/j.jretconser.2020.102249

Han, Y. J., Nunes, J. C., & Drèze, X. (2010). Signaling Status with Luxury Goods: The Role of Brand Prominence. *Journal of Marketing*, *74*(4), 15–30. DOI: 10.1509/jmkg.74.4.015

Harmon-Kizer, T. R., Kumar, A., Ortinau, D., & Stock, J. (2013). When multiple identities compete: The role of centrality in self-brand connections. *Journal of Consumer Behaviour*, *12*(6), 483–495. DOI: 10.1002/cb.1449

Hassan, S., Husić-Mehmedović, M., & Duverger, P. (2015). Retaining the allure of luxury brands during an economic downturn. *Journal of Fashion Marketing and Management*, *19*(4), 416–429. DOI: 10.1108/JFMM-03-2015-0030

Hassan, S. Q. U., Ilyas, M., & Rehman, C. A. (2016). Corporate branding, emotional attachment and brand loyalty : The case of luxury fashion branding. *Deleted Journal*, *6*(2), 43–57. DOI: 10.12816/0031561

Hayes, J. L., Brinson, N. H., Bott, G. J., & Moeller, C. M. (2021). The influence of Consumer–Brand Relationship on the personalized Advertising privacy calculus in social media. *Journal of Interactive Marketing*, *55*(1), 16–30. DOI: 10.1016/j.intmar.2021.01.001

Heath, T. B., & Chatterjee, S. (1995). Asymmetric Decoy Effects on Lower-Quality versus Higher-Quality Brands: Meta-Analytic and Experimental Evidence. *The Journal of Consumer Research*, *22*(3), 268. DOI: 10.1086/209449

Hermann, E. (2021). Leveraging Artificial Intelligence in Marketing for Social Good—An Ethical Perspective. *Journal of Business Ethics*, *179*(1), 43–61. DOI: 10.1007/s10551-021-04843-y PMID: 34054170

Ho, H., Chiu, C. L., Liu, S., Lee, L. L., Lado, N., & Cesaronic, F. (2019). Understanding the role of attitude components in co-branding: A comparison of Spanish and Taiwanese consumers. *Journal of International Consumer Marketing*, *31*(3), 203–224. DOI: 10.1080/08961530.2018.1531363

Ho, H., Lado, N., & Rivera-Torres, P. (2017). Detangling consumer attitudes to better explain co-branding success. *Journal of Product and Brand Management, 26*(7), 704–721. DOI: 10.1108/JPBM-11-2015-1039

Ho, S. C., Yeh, Y. Y., & Lin, T. M. Y. (2021). Effects of luxury perceptions on purchase intention of High-Tech products. *Journal of International Consumer Marketing, 34*(2), 168–183. DOI: 10.1080/08961530.2021.1937769

Hoang, D., Kousi, S., & Martinez, L. F. (2022). Online customer engagement in the post-pandemic scenario: A hybrid thematic analysis of the luxury fashion industry. *Electronic Commerce Research, 23*(3), 1401–1428. DOI: 10.1007/s10660-022-09635-8

Hoffmann, J., Ramirez, R., & Lecamp, L. (2018). Right on time - Socioecological strategy and implications of turbulence in the Swiss watchmaking field. *Technological Forecasting and Social Change, 137*, 101–117. DOI: 10.1016/j.techfore.2018.06.047

Holmqvist, J., Ruiz, C. D., & Peñaloza, L. (2019). Moments of luxury: Hedonic escapism as a luxury experience. *Journal of Business Research, 116*, 503–513. DOI: 10.1016/j.jbusres.2019.10.015

Hoyer, W. D., Kroschke, M., Schmitt, B., Kraume, K., & Shankar, V. (2020). Transforming the customer experience through new technologies. *Journal of Interactive Marketing, 51*(1), 57–71. DOI: 10.1016/j.intmar.2020.04.001

Hsiao, K. (2017). What drives smartwatch adoption intention? Comparing Apple and non-Apple watches. *Library Hi Tech, 35*(1), 186–206. DOI: 10.1108/LHT-09-2016-0105

Hsieh, S. H., & Chang, A. (2016). The psychological mechanism of brand co-creation engagement. *Journal of Interactive Marketing, 33*(1), 13–26. DOI: 10.1016/j.intmar.2015.10.001

Huang, R., Ha, S., & Kim, S. (2018). Narrative persuasion in social media: An empirical study of luxury brand advertising. *Journal of Research in Interactive Marketing, 12*(3), 274–292. DOI: 10.1108/JRIM-07-2017-0059

Huang, X., Yang, F., Zheng, J., Feng, C., & Zhang, L. (2023). Personalized human resource management via HR analytics and artificial intelligence: Theory and implications. *Asia Pacific Management Review, 28*(4), 598–610. DOI: 10.1016/j.apmrv.2023.04.004

Huh, J., Kim, H., & Lee, G. (2023). "Oh, happy day!" Examining the role of AI-powered voice assistants as a positive technology in the formation of brand loyalty. *Journal of Research in Interactive Marketing*, *17*(5), 794–812. DOI: 10.1108/JRIM-10-2022-0328

Hunt, S. D. (2018). The ethics of branding, customer-brand relationships, brand-equity strategy, and branding as a societal institution. *Journal of Business Research*, *95*, 408–416. DOI: 10.1016/j.jbusres.2018.07.044

Husain, R., Paul, J., & Koles, B. (2022). The role of brand experience, brand resonance and brand trust in luxury consumption. *Journal of Retailing and Consumer Services*, *66*, 102895. DOI: 10.1016/j.jretconser.2021.102895

Jain, N. C. (2024). Predictive analytics for personalized health interventions using wearable data. *Deleted Journal*, *12*(3), 188–198. DOI: 10.36676/dira.v12.i3.66

Jiang, X., Guangkuo, G., & Xuezheng, Y. (2022). Evolutionary game analysis on live streaming commerce considering brand awareness and anchor influence. *Kybernetes*, *52*(12), 6467–6493. DOI: 10.1108/K-04-2022-0593

Jihua, S., & Lin, S. (2024). Research on Data-driven Marketing Strategy Optimization. *Journal of Business and Marketing.*, *1*(2), 157–161. DOI: 10.62517/jbm.202409224

Johnson, D. S., Muzellec, L., Sihi, D., & Zahay, D. (2019). The marketing organization's journey to become data-driven. *Journal of Research in Interactive Marketing*, *13*(2), 162–178. DOI: 10.1108/JRIM-12-2018-0157

Joshi, I. (2024). Neuromarketing in Fashion Advertising: Enhancing consumer engagement through sensory and neuroscientific insights. *International Journal of Social Science and Economic Research*, *09*(11), 5637–5642. DOI: 10.46609/IJSSER.2024.v09i11.043

Joy, A., Zhu, Y., Peña, C., & Brouard, M. (2022). Digital future of luxury brands: Metaverse, digital fashion, and non-fungible tokens. *Strategic Change*, *31*(3), 337–343. DOI: 10.1002/jsc.2502

Jun, S., & Park, D. (2017). Visualization of brand positioning based on consumer web search information. *Internet Research*, *27*(2), 381–407. DOI: 10.1108/IntR-02-2016-0037

Jun, Y., & Lee, H. (2021). A sound brand identity design: The interplay between sound symbolism and typography on brand attitude and memory. *Journal of Retailing and Consumer Services*, *64*, 102724. DOI: 10.1016/j.jretconser.2021.102724

Kajla, T., Raj, S., Kansra, P., Gupta, S. L., & Singh, N. (2023). Neuromarketing and consumer behavior: A bibliometric analysis. *Journal of Consumer Behaviour, 23*(2), 959–975. DOI: 10.1002/cb.2256

Kapferer, J., & Valette-Florence, P. (2017). The impact of brand penetration and awareness on luxury brand desirability. *Journal of Business Research, 83*, 38–50. DOI: 10.1016/j.jbusres.2017.09.025

Karangi, S. W., & Lowe, B. (2021). Haptics and brands: The effect of touch on product evaluation of branded products. *Journal of Consumer Behaviour, 20*(6), 1480–1496. DOI: 10.1002/cb.1959

Kauppinen-Räisänen, H., Mühlbacher, H., & Taishoff, M. (2020). Exploring consumers' subjective shopping experiences in directly operated luxury brand stores. *Journal of Retailing and Consumer Services, 57*, 102251. DOI: 10.1016/j.jretconser.2020.102251

Kaur, K., & Soch, H. (2025). From prominence to preference: Social and emotional value perception as mediator between the brand prominence and purchase intentions in the luxury market. *Journal of Fashion Marketing and Management, 29*(4), 639–661. Advance online publication. DOI: 10.1108/JFMM-05-2024-0164

Keller, K. L. (2009). Managing the growth tradeoff: Challenges and opportunities in luxury branding. *Journal of Brand Management, 16*(5–6), 290–301. DOI: 10.1057/bm.2008.47

Kim, J., Lloyd, S., & Cervellon, M. (2015). Narrative-transportation storylines in luxury brand advertising: Motivating consumer engagement. *Journal of Business Research, 69*(1), 304–313. DOI: 10.1016/j.jbusres.2015.08.002

Kim, J. C., Park, B., & Dubois, D. (2018). How consumers' political ideology and Status-Maintenance goals interact to shape their desire for luxury goods. *Journal of Marketing, 82*(6), 132–149. DOI: 10.1177/0022242918799699

Kim, K., Ko, E., Lee, M., Mattila, P., & Kim, K. H. (2014). Fashion collaboration effects on consumer response and customer equity in global luxury and SPA brand marketing. *Journal of Global Scholars of Marketing Science, 24*(3), 350–364. DOI: 10.1080/21639159.2014.913376

Kim, Y. (2018). Power moderates the impact of desire for exclusivity on luxury experiential consumption. *Psychology and Marketing, 35*(4), 283–293. DOI: 10.1002/mar.21086

Kim, Y., & Sullivan, P. (2019). Emotional branding speaks to consumers' heart: The case of fashion brands. *Fashion and Textiles*, *6*(1), 2. Advance online publication. DOI: 10.1186/s40691-018-0164-y

Knoeferle, K. M., Knoeferle, P., Velasco, C., & Spence, C. (2016). Multisensory brand search: How the meaning of sounds guides consumers' visual attention. *Journal of Experimental Psychology. Applied*, *22*(2), 196–210. DOI: 10.1037/xap0000084 PMID: 27295466

Ko, E., Phau, I., & Aiello, G. (2016). Luxury brand strategies and customer experiences: Contributions to theory and practice. *Journal of Business Research*, *69*(12), 5749–5752. DOI: 10.1016/j.jbusres.2016.04.170

Kongsri, N., & Jaroenwanit, P. (2024). Key aspects of personal brand identity in social media commerce: Impact on successful personal branding. *International Journal of Data and Network Science*, *8*(4), 2663–2676. DOI: 10.5267/j.ijdns.2024.4.012

Kordzadeh, N., & Ghasemaghaei, M. (2021). Algorithmic bias: Review, synthesis, and future research directions. *European Journal of Information Systems*, *31*(3), 388–409. DOI: 10.1080/0960085X.2021.1927212

Kowalczyk, C. M., & Mitchell, N. A. (2021). Understanding the antecedents to luxury brand consumer behavior. *Journal of Product and Brand Management*, *31*(3), 438–453. DOI: 10.1108/JPBM-09-2020-3126

Kressmann, F., Sirgy, M. J., Herrmann, A., Huber, F., Huber, S., & Lee, D. (2006). Direct and indirect effects of self-image congruence on brand loyalty. *Journal of Business Research*, *59*(9), 955–964. DOI: 10.1016/j.jbusres.2006.06.001

Kumagai, K., & Nagasawa, S. (2019). Psychological switching mechanism of consumers' luxury and non-luxury brand attitude formation: The effect of store location prestige and self-congruity. *Heliyon*, *5*(5), e01581. DOI: 10.1016/j.heliyon.2019.e01581 PMID: 31193173

Kumar, D., & Suthar, N. (2024). Ethical and legal challenges of AI in marketing: An exploration of solutions. *Journal of Information Communication and Ethics in Society*, *22*(1), 124–144. DOI: 10.1108/JICES-05-2023-0068

Kumar, J. (2024). Psychological mechanisms behind access-based luxury brand consumption: Empirical investigation from the lens of new ownership paradigm. *Journal of Business Research*, *175*, 114539. DOI: 10.1016/j.jbusres.2024.114539

Kumar, J., & Nayak, J. K. (2019). Consumer psychological motivations to customer brand engagement: A case of brand community. *Journal of Consumer Marketing*, *36*(1), 168–177. DOI: 10.1108/JCM-01-2018-2519

Kumar, V., & Kaushal, V. (2021). Perceived brand authenticity and social exclusion as drivers of psychological brand ownership. *Journal of Retailing and Consumer Services*, *61*, 102579. DOI: 10.1016/j.jretconser.2021.102579

Kumar, V., Rajan, B., Venkatesan, R., & Lecinski, J. (2019). Understanding the role of artificial intelligence in personalized engagement marketing. *California Management Review*, *61*(4), 135–155. DOI: 10.1177/0008125619859317

Kunz, J., May, S., & Schmidt, H. J. (2020). Sustainable luxury: Current status and perspectives for future research. *BuR -. Business Research*, *13*(2), 541–601. DOI: 10.1007/s40685-020-00111-3

Kur, A. (2016). Brand symbols, the consumer, and the internet. SSRN *Electronic Journal*. https://doi.org/DOI: 10.2139/ssrn.2721740

Kurtz, J. (2020). On foreign investor 'Privilege' and the limits of the law: A reply to Ivar Alvik. *European Journal of International Law*, *31*(1), 313–319. DOI: 10.1093/ejil/chaa020

Kusumasondjaja, S. (2019). Exploring the role of visual aesthetics and presentation modality in luxury fashion brand communication on Instagram. *Journal of Fashion Marketing and Management*, *24*(1), 15–31. DOI: 10.1108/JFMM-02-2019-0019

Lankton, N., McKnight, D. H., & Tripp, J. (2015). Technology, Humanness, and Trust: Rethinking trust in technology. *Journal of the Association for Information Systems*, *16*(10), 880–918. DOI: 10.17705/1jais.00411

Leban, M., Seo, Y., & Voyer, B. G. (2019). Transformational effects of social media lurking practices on luxury consumption. *Journal of Business Research*, *116*, 514–521. DOI: 10.1016/j.jbusres.2019.09.010

Lee, J. E., Hur, S., & Watkins, B. (2018). Visual communication of luxury fashion brands on social media: Effects of visual complexity and brand familiarity. *Journal of Brand Management*, *25*(5), 449–462. DOI: 10.1057/s41262-018-0092-6

Li, A., Sun, J., Li, P., Pu, Y., Li, H., & Chen, Y. (2021). Hermes. *Proceedings of the 28th Annual International Conference on Mobile Computing and Networking*, 420–437. https://doi.org/DOI: 10.1145/3447993.3483278

Li, Y., Liu, J., & Ren, J. (2019). Social recommendation model based on user interaction in complex social networks. *PLoS One*, *14*(7), e0218957. DOI: 10.1371/journal.pone.0218957 PMID: 31291288

Liang, Y., Xu, Q., & Jin, L. (2021). The effect of smart and connected products on consumer brand choice concentration. *Journal of Business Research, 135*, 163–172. DOI: 10.1016/j.jbusres.2021.06.039

Lin, J. (2023). Analysis of luxury brand marketing strategies based on consumer psychology. *Advances in Economics Management and Political Sciences, 24*(1), 240–245. DOI: 10.54254/2754-1169/24/20230443

Lin, Y. (2022). Augmented Reality in customer interaction for branding of high-end vehicles. *BCP Business & Management, 21*, 330–339. DOI: 10.54691/bcpbm.v21i.1258

Liu, X., Shin, H., & Burns, A. C. (2019). Examining the impact of luxury brand's social media marketing on customer engagement: Using big data analytics and natural language processing. *Journal of Business Research, 125*, 815–826. DOI: 10.1016/j.jbusres.2019.04.042

Loh, L., & Tan, S. (2020). Impact of sustainability Reporting on brand Value: An examination of 100 leading brands in Singapore. *Sustainability (Basel), 12*(18), 7392. DOI: 10.3390/su12187392

Loureiro, S. M. C., Bilro, R. G., & Japutra, A. (2019). The effect of consumer-generated media stimuli on emotions and consumer brand engagement. *Journal of Product and Brand Management, 29*(3), 387–408. DOI: 10.1108/JPBM-11-2018-2120

Lowe, M. L., & Haws, K. L. (2016). Sounds big: The effects of acoustic pitch on product perceptions. *JMR, Journal of Marketing Research, 54*(2), 331–346. DOI: 10.1509/jmr.14.0300

Ma, Y. (2023). Analysis of the Hybridity Translation Strategy of Moment in Peking from the Perspective of Post-colonialism. *Communications in Humanities Research, 6*(1), 304–310. DOI: 10.54254/2753-7064/6/20230284

Machado, P. M., Lawson-Tovey, S., Strangfeld, A., Mateus, E., Hyrich, K., Gossec, L., Carmona, L., Rodrigues, A., Raffeiner, B., Duarte, C., Hachulla, E., Veillard, E., Strakova, E., Burmester, G. R., Yardimci, G. K., Gómez-Puerta, J. A., Zepa, J., Kearsley-Fleet, L., Trefond, L., & Mariette, X. (2022). OA01 Safety of vaccination against SARS-CoV-2 in people with rheumatic and musculoskeletal diseases: Results from the EULAR Coronavirus Vaccine (COVAX) physician-reported registry. *Lara D. Rheumatology (Oxford, England), 61*(Supplement_1), keac132. Advance online publication. DOI: 10.1093/rheumatology/keac132

Mahmoud, A. B., Kumar, V., Fuxman, L., & Mohr, I. (2023). Climate change risks, sustainability and luxury branding: Friend or a foe. *Industrial Marketing Management*, *115*, 57–68. DOI: 10.1016/j.indmarman.2023.09.008

Majeed, M. U., Aftab, H., Arslan, A., & Shakeel, Z. (2024). Determining online consumer's luxury purchase intention: The influence of antecedent factors and the moderating role of brand awareness, perceived risk, and web atmospherics. *PLoS One*, *19*(2), e0295514. DOI: 10.1371/journal.pone.0295514 PMID: 38394125

Maldonado-Canca, L., Casado-Molina, A., Cabrera-Sánchez, J., & Bermúdez-González, G. (2024). Beyond the post: An SLR of enterprise artificial intelligence in social media. *Social Network Analysis and Mining*, *14*(1), 219. Advance online publication. DOI: 10.1007/s13278-024-01382-y

Mallipeddi, R., Kumar, S., Sriskandarajah, C., & Zhu, Y. (2018). A framework for Analyzing influencer marketing in Social Networks: Selection and Scheduling of Influencers. SSRN *Electronic Journal*. https://doi.org/DOI: 10.2139/ssrn.3255198

Mandler, T., Johnen, M., & Gräve, J. (2019). Can't help falling in love? How brand luxury generates positive consumer affect in social media. *Journal of Business Research*, *120*, 330–342. DOI: 10.1016/j.jbusres.2019.10.010

Margariti, K., Boutsouki, C., & Hatzithomas, L. (2019). The dilemma of luxury brand extensions. *Journal of Global Fashion Marketing*, *10*(4), 305–323. DOI: 10.1080/20932685.2019.1649167

Meyerding, S. G., & Mehlhose, C. M. (2018). Can neuromarketing add value to the traditional marketing research? An exemplary experiment with functional near-infrared spectroscopy (fNIRS). *Journal of Business Research*, *107*, 172–185. DOI: 10.1016/j.jbusres.2018.10.052

Migicovsky, Z., Gardner, K. M., Richards, C., Chao, C. T., Schwaninger, H. R., Fazio, G., Zhong, G., & Myles, S. (2021). Genomic consequences of apple improvement. *Horticulture Research*, *8*(1), 9. Advance online publication. DOI: 10.1038/s41438-020-00441-7 PMID: 33384408

Mirzajani, H., Istif, E., Abbasiasl, T., Mirlou, F., Özkahraman, E. E., Hasanreisoglu, M., & Beker, L. (2022). Femtosecond laser ablation assisted NFC antenna fabrication for smart contact lenses. *Advanced Materials Technologies*, *7*(10), 2101629. Advance online publication. DOI: 10.1002/admt.202101629

Modhoriye, P., Yadav, P., & Jadhav, S. (2023). AI Transformation in Business: Unveiling the dual effects of advancement and challenges. *INTERANTIONAL JOURNAL OF SCIENTIFIC RESEARCH IN ENGINEERING AND MANAGEMENT, 07*(11), 1–11. DOI: 10.55041/IJSREM27359

Moon, H., & Sprott, D. E. (2015). SELECTING THE RIGHT PARTNERS FOR THE RIGHT REASONS: THE IMPORTANCE OF FIT IN INGREDIENT BRANDING STRATEGY IN LUXURY BRAND. *Global Fashion Management Conference, 2*(4), 199–200. https://doi.org/DOI: 10.15444/GFMC2015.02.04.04

Moreau, C. P., Prandelli, E., Schreier, M., & Hieke, S. (2020). Customization in luxury brands: Can Valentino get personal? *JMR, Journal of Marketing Research, 57*(5), 937–947. DOI: 10.1177/0022243720943191

Mrad, M., Farah, M. F., & Haddad, S. (2018). From Karl Lagerfeld to Erdem: A series of collaborations between designer luxury brands and fast-fashion brands. *Journal of Brand Management, 26*(5), 567–582. DOI: 10.1057/s41262-018-00146-2

Nadeem, W., Tan, T. M., Tajvidi, M., & Hajli, N. (2021). How do experiences enhance brand relationship performance and value co-creation in social commerce? The role of consumer engagement and self brand-connection. *Technological Forecasting and Social Change, 171*, 120952. DOI: 10.1016/j.techfore.2021.120952

Nawres, D., Nedra, B., Yousaf, A., & Mishra, A. (2023). The role of augmented reality in shaping purchase intentions and WOM for luxury products. *Journal of Business Research, 171*, 114368. DOI: 10.1016/j.jbusres.2023.114368

Nieroda, M. E., Mrad, M., & Solomon, M. R. (2018). How do consumers think about hybrid products? Computer wearables have an identity problem. *Journal of Business Research, 89*, 159–170. DOI: 10.1016/j.jbusres.2018.04.024

Norori, N., Hu, Q., Aellen, F. M., Faraci, F. D., & Tzovara, A. (2021). Addressing bias in big data and AI for health care: A call for open science. *Patterns (New York, N.Y.), 2*(10), 100347. DOI: 10.1016/j.patter.2021.100347 PMID: 34693373

Ochkovskaya, M. S., Mkhitaryan, S. V., & Indichekova, E. P. (2021). Influence of collaborations with visual art on luxury brand identity: Pilot study. *Vestnik of Saint Petersburg University Management, 20*(1), 3–29. DOI: 10.21638/11701/spbu08.2021.101

Oğuz, A. (2024). Consumer behavior in the era of AI-Driven marketing. *Human-Computer Interaction, 8*(1), 147. DOI: 10.62802/h9frxh42

Okonkwo, U. (2009). Sustaining the luxury brand on the Internet. *Journal of Brand Management, 16*(5–6), 302–310. DOI: 10.1057/bm.2009.2

Pangarkar, A., & Shukla, P. (2023). Conspicuous and inconspicuous consumption of luxury goods in a digital world: Insights, implications, and future research directions. *International Journal of Advertising, 42*(7), 1226–1238. DOI: 10.1080/02650487.2023.2246260

Pantano, E., Passavanti, R., Priporas, C., & Verteramo, S. (2018). To what extent luxury retailing can be smart? *Journal of Retailing and Consumer Services, 43*, 94–100. DOI: 10.1016/j.jretconser.2018.03.012

Parashar, J., Jain, A., & Ali, S. (2023). Artificial intelligence impact on human fitness: Exploring emerging trends. *International Journal for Research in Applied Science and Engineering Technology, 11*(10), 287–293. DOI: 10.22214/ijraset.2023.55971

Parguel, B., Delécolle, T., & Valette-Florence, P. (2015). How price display influences consumer luxury perceptions. *Journal of Business Research, 69*(1), 341–348. DOI: 10.1016/j.jbusres.2015.08.006

Park, J., & Ahn, S. (2024). Traditional vs. AI-generated brand personalities: Impact on brand preference and purchase intention. *Journal of Retailing and Consumer Services, 81*, 104009. DOI: 10.1016/j.jretconser.2024.104009

Paska, M., & Budnik, K. (2023). Trust as a participant in the adaptation of technological changes. *Scientific Papers of Silesian University of Technology Organization and Management Series, 2023*(170), 383–391. DOI: 10.29119/1641-3466.2023.170.24

Pavashe, A. S., Kadam, P. D., Zirange, V. B., & Katkar, R. D. (2023). The impact of artificial intelligence on employment and workforce trends in the Post-Pandemic era. *International Journal for Research in Applied Science and Engineering Technology, 11*(11), 154–157. DOI: 10.22214/ijraset.2023.56418

Peck, J., & Shu, S. B. (2009). The effect of mere touch on perceived ownership. *The Journal of Consumer Research, 36*(3), 434–447. DOI: 10.1086/598614

Pennington, C. R., Monk, R. L., Qureshi, A. W., Kulkarni, R., Li, W., Li, J., & Heim, D. (2022). Dark Nudges: Branding magnifies the decoy effect in alcohol purchasing decisions. *Journal of Studies on Alcohol and Drugs, 83*(6), 857–866. DOI: 10.15288/jsad.21-00398 PMID: 36484583

Pentina, I., Guilloux, V., & Micu, A. C. (2018). Exploring social media engagement behaviors in the context of luxury brands. *Journal of Advertising, 47*(1), 55–69. DOI: 10.1080/00913367.2017.1405756

Petrova, S., & Pruzhansky, V. (2011). The Economics of Luxury Goods: Utility based on exclusivity. SSRN *Electronic Journal*. https://doi.org/DOI: 10.2139/ssrn.1930361

Phau, I., & Prendergast, G. (2000). Consuming luxury brands: The relevance of the 'Rarity Principle.'. *Journal of Brand Management*, *8*(2), 122–138. DOI: 10.1057/palgrave.bm.2540013

Philiastides, M. G., & Ratcliff, R. (2013). Influence of branding on Preference-Based Decision making. *Psychological Science*, *24*(7), 1208–1215. DOI: 10.1177/0956797612470701 PMID: 23696199

Pinello, C., Picone, P. M., & Destri, A. M. L. (2022). Co-branding research: Where we are and where we could go from here. *European Journal of Marketing*, *56*(2), 584–621. DOI: 10.1108/EJM-02-2021-0073

Pino, G., Amatulli, C., Peluso, A. M., Nataraajan, R., & Guido, G. (2017). Brand prominence and social status in luxury consumption: A comparison of emerging and mature markets. *Journal of Retailing and Consumer Services*, *46*, 163–172. DOI: 10.1016/j.jretconser.2017.11.006

Porbundarwala, T., Lalji, S., Patil, H., & Dongardive, J. (2023). Impact of AI in personalised health monitoring using wearable devices. *International Journal for Multidisciplinary Research*, *5*(5), 7144. Advance online publication. DOI: 10.36948/ijfmr.2023.v05i05.7144

Pozharliev, R., Verbeke, W. J., Van Strien, J. W., & Bagozzi, R. P. (2015). Merely Being with you Increases My Attention to Luxury Products: Using EEG to Understand Consumers' Emotional Experience with Luxury Branded Products. *JMR, Journal of Marketing Research*, *52*(4), 546–558. DOI: 10.1509/jmr.13.0560

Prandelli, E., Wang, Y., & Weijo, H. (2024). Luxury branding and the creator Economy: Emerging challenges and future avenues. *International Journal of Research in Marketing*, *41*(3), 455–467. DOI: 10.1016/j.ijresmar.2024.07.004

Prentice, C., & Loureiro, S. M. C. (2018). Consumer-based approach to customer engagement – The case of luxury brands. *Journal of Retailing and Consumer Services*, *43*, 325–332. DOI: 10.1016/j.jretconser.2018.05.003

Quamina, L. T., Xue, M. T., & Chawdhary, R. (2023). 'Co-branding as a masstige strategy for luxury brands: Desirable or not? *Journal of Business Research*, *158*, 113704. DOI: 10.1016/j.jbusres.2023.113704

Rachbini, W., Soeharso, S. Y., Nugroho, A. S., Evi, T., Rahmawati, E., & Suyanto, N. (2023). The Impact of Brand Society and Narrative Branding on Brand Image and its Implications on Consumer Purchase Behavior: A Case Study of Luxury Fashion Brands in Jakarta. *Journal of Law and Sustainable Development*, *11*(7), e1387. DOI: 10.55908/sdgs.v11i7.1387

Raffaelli, R. (2018). Technology Reemergence: Creating new value for old technologies in Swiss mechanical watchmaking, 1970–2008. *Administrative Science Quarterly*, *64*(3), 576–618. DOI: 10.1177/0001839218778505

Rahmadani, D., Fauzi, H., Lubis, R. A., & Ariyanti, M. (2022). Study of Neuromarketing in Consumer Behavior due to Product Logos Color Changes Effect. [CEPAT]. *Journal of Computer Engineering Progress Application and Technology*, *1*(02), 41. DOI: 10.25124/cepat.v1i02.5245

Rahman, M. S., Bag, S., Hossain, M. A., Fattah, F. M. A., Gani, M. O., & Rana, N. P. (2023). The new wave of AI-powered luxury brands online shopping experience: The role of digital multisensory cues and customers' engagement. *Journal of Retailing and Consumer Services*, *72*, 103273. DOI: 10.1016/j.jretconser.2023.103273

Ramaswamy, V., & Ozcan, K. (2018). Offerings as Digitalized Interactive Platforms: A Conceptual framework and implications. *Journal of Marketing*, *82*(4), 19–31. DOI: 10.1509/jm.15.0365

Ranfagni, S., Faraoni, M., Zollo, L., & Vannucci, V. (2021). Combining online market research methods for investigating brand alignment: The case of Nespresso. *British Food Journal*, *123*(13), 37–58. DOI: 10.1108/BFJ-06-2020-0462

Reimann, M., Castaño, R., Zaichkowsky, J., & Bechara, A. (2011). How we relate to brands: Psychological and neurophysiological insights into consumer–brand relationships. *Journal of Consumer Psychology*, *22*(1), 128–142. DOI: 10.1016/j.jcps.2011.11.003

Reis, R. (2022). Losing the inflation anchor. *Brookings Papers on Economic Activity*, *2021*(2), 307–379. DOI: 10.1353/eca.2022.0004

Reyes-Menendez, A., Palos-Sanchez, P., Saura, J. R., & Santos, C. R. (2021). Revisiting the impact of perceived social value on consumer behavior toward luxury brands. *European Management Journal*, *40*(2), 224–233. DOI: 10.1016/j.emj.2021.06.006

Rodrigues, C. (2019). Multisensory brand experiences and brand love. In *IGI Global eBooks* (pp. 615–635). https://doi.org/DOI: 10.4018/978-1-5225-9282-2.ch029

Romero, P. M., Lim, R. E., & Park, H. (2024). Branded virtual worlds in fashion gaming: Examining the role of brand narratives and social presence in consumer engagement. *Computers in Human Behavior*, *108505*. Advance online publication. DOI: 10.1016/j.chb.2024.108505

Rosendo-Rios, V., & Shukla, P. (2022). The effects of masstige on loss of scarcity and behavioral intentions for traditional luxury consumers. *Journal of Business Research*, *156*, 113490. DOI: 10.1016/j.jbusres.2022.113490

Roy, S. K., Singh, G., Hope, M., Nguyen, B., & Harrigan, P. (2019). The rise of smart consumers: Role of smart servicescape and smart consumer experience co-creation. *Journal of Marketing Management, 35*(15–16), 1480–1513. DOI: 10.1080/0267257X.2019.1680569

Royo-Vela, M., & Sánchez, M. P. (2022). Downward price-based luxury brand line extension: Effects on premium luxury buyer's perception and consequences on buying intention and brand loyalty. *European Research on Management and Business Economics, 28*(3), 100198. DOI: 10.1016/j.iedeen.2022.100198

Ruiz, C. D., & Cruz, A. G. B. (2023). Unconventional luxury brand collaborations: A new form of luxury consumption among young adults in China. *International Marketing Review, 40*(7), 1–21. DOI: 10.1108/IMR-04-2022-0099

Ryu, K., Lehto, X. Y., Gordon, S. E., & Fu, X. (2018). Effect of a brand story structure on narrative transportation and perceived brand image of luxury hotels. *Tourism Management, 71*, 348–363. DOI: 10.1016/j.tourman.2018.10.021

Sahne, S. S. K., & Daronkola, H. K. (2025). Enhancing loyalty in luxury fashion with AI: The mediation role of customer relationships. *Asia-Pacific Journal of Business Administration*. Advance online publication. DOI: 10.1108/APJBA-08-2024-0456

Santos, J., & Boote, J. (2003). A theoretical exploration and model of consumer expectations, post-purchase affective states and affective behaviour. *Journal of Consumer Behaviour, 3*(2), 142–156. DOI: 10.1002/cb.129

Sanz-Lopez, F., Gallego-Losada, R., Montero-Navarro, A., & García-Abajo, E. (2024). IS THE DIGITALISATION THE FUTURE OF THE LUXURY INDUSTRY? *Heliyon, 10*(21), e40029. DOI: 10.1016/j.heliyon.2024.e40029 PMID: 39553661

Sarkar, S., & Banerjee, S. (2019). Brand co-creation through triadic stakeholder participation. *European Business Review, 31*(5), 585–609. DOI: 10.1108/EBR-04-2018-0079

Saura, J. R., Palacios-Marqués, D., & Ribeiro-Soriano, D. (2021). Digital marketing in SMEs via data-driven strategies: Reviewing the current state of research. *Journal of Small Business Management, 61*(3), 1278–1313. DOI: 10.1080/00472778.2021.1955127

Schaefer, M., & Rotte, M. (2007). Favorite brands as cultural objects modulate reward circuit. *Neuroreport, 18*(2), 141–145. DOI: 10.1097/WNR.0b013e328010ac84 PMID: 17301679

Schmitt, B. (2011). The consumer psychology of brands. *Journal of Consumer Psychology, 22*(1), 7–17. DOI: 10.1016/j.jcps.2011.09.005

Scholz, J., & Duffy, K. (2018). We ARe at home: How augmented reality reshapes mobile marketing and consumer-brand relationships. *Journal of Retailing and Consumer Services*, *44*, 11–23. DOI: 10.1016/j.jretconser.2018.05.004

Schuetz, S., Kuai, L., Lacity, M. C., & Steelman, Z. (2024). A qualitative systematic review of trust in technology. *Journal of Information Technology*. Advance online publication. DOI: 10.1177/02683962241254392

Sénéchal, S., Georges, L., & Pernin, J. L. (2013). Alliances between corporate and fair trade brands: Examining the antecedents of overall evaluation of the co-branded product. *Journal of Business Ethics*, *124*(3), 365–381. DOI: 10.1007/s10551-013-1875-z

Septianto, F., Seo, Y., Sung, B., & Zhao, F. (2020). Authenticity and exclusivity appeals in luxury advertising: The role of promotion and prevention pride. *European Journal of Marketing*, *54*(6), 1305–1323. DOI: 10.1108/EJM-10-2018-0690

Shah, A. M., & Ali, M. (2023). Social media marketing activities and luxury fashion brands in the post-pandemic world. *Asia Pacific Journal of Marketing and Logistics*, *36*(9), 2104–2120. DOI: 10.1108/APJML-10-2022-0863

Shah, D., & Murthi, B. (2020). Marketing in a data-driven digital world: Implications for the role and scope of marketing. *Journal of Business Research*, *125*, 772–779. DOI: 10.1016/j.jbusres.2020.06.062

Shahid, S., Iqbal, M., Saeed, H., Hira, S., Batool, A., & Goyal, A. (2024). Abstract 4136358: Diagnostic accuracy of Apple Watch Electrocardiogram for Atrial Fibrillation: A Systematic Review and Meta-Analysis. *Circulation*, *150*(Suppl_1). Advance online publication. DOI: 10.1161/circ.150.suppl_1.4136358

Shahid, S., & Paul, J. (2021). Intrinsic motivation of luxury consumers in an emerging market. *Journal of Retailing and Consumer Services*, *61*, 102531. DOI: 10.1016/j.jretconser.2021.102531

Shahzad, M. F., Xu, S., An, X., & Javed, I. (2024). Assessing the impact of AI-chatbot service quality on user e-brand loyalty through chatbot user trust, experience and electronic word of mouth. *Journal of Retailing and Consumer Services*, *79*, 103867. DOI: 10.1016/j.jretconser.2024.103867

Shaikh, S., & Gummerum, M. (2025). A luxury consumption perspective on motivations for brand value co-creation in emerging markets. *International Journal of Emerging Markets*. Advance online publication. DOI: 10.1108/IJOEM-10-2023-1637

Shamim, M. A., Dwivedi, P., Aggarwal, P., Dwivedi, R., Anil, A., Saravanan, A., Tiwari, K., Dodiya, R. P., Yadav, I., Sukhija, S., Varthya, S. B., Shukla, R., Singh, S., & Singh, K. (2024). 802-P: A New Dawn beyond Basal Insulin? Dose-Response Meta-analysis for Efficacy and Safety of Tirzepatide Compared with Basal Insulin in Uncontrolled Diabetes Mellitus. *Diabetes*, *73*(Supplement_1), 802-P. Advance online publication. DOI: 10.2337/db24-802-P

Shankar, A., & Jain, S. (2020). Factors affecting luxury consumers' webrooming intention: A moderated-mediation approach. *Journal of Retailing and Consumer Services*, *58*, 102306. DOI: 10.1016/j.jretconser.2020.102306

Shankar, S. (2024). Enhancing energy efficiency in AI-Powered Data Centers: Challenges and solutions. *International Journal of High School Research*, *6*(11), 93–98. DOI: 10.36838/v6i11.15

Shankar, V. (2024). Managing the Twin Faces of AI: A commentary on "Is AI changing the world for better or worse?". *Journal of Macromarketing*, *44*(4), 892–899. Advance online publication. DOI: 10.1177/02761467241286483

Sharma, P., Pradhan, S., & Srivastava, A. (2021). Understanding the luxury purchase intentions of young consumers: A qualitative analysis. *Asia-Pacific Journal of Business Administration*, *13*(4), 452–476. DOI: 10.1108/APJBA-01-2021-0019

Shen, B., Choi, T., & Chow, P. (2017). Brand loyalties in designer luxury and fast fashion co-branding alliances. *Journal of Business Research*, *81*, 173–180. DOI: 10.1016/j.jbusres.2017.06.017

Shi, X., Li, F., & Chumnumpan, P. (2020). The use of product scarcity in marketing. *European Journal of Marketing*, *54*(2), 380–418. DOI: 10.1108/EJM-04-2018-0285

Shin, H. H., & Jeong, M. (2022). Redefining luxury service with technology implementation: The impact of technology on guest satisfaction and loyalty in a luxury hotel. *International Journal of Contemporary Hospitality Management*, *34*(4), 1491–1514. DOI: 10.1108/IJCHM-06-2021-0798

Singh, P. (2024). UNVEILING ZARA'S LUXURY PERSONA: A COMPREHENSIVE ANALYSIS OF SOCIAL MEDIA STRATEGIES AND CONSUMER BEHAVIOUR. *Deleted Journal*, *147–151*, 147–151. Advance online publication. DOI: 10.62823/7.1(II).6462

Solem, B. A. (2016). Influences of customer participation and customer brand engagement on brand loyalty. *Journal of Consumer Marketing*, *33*(5), 332–342. DOI: 10.1108/JCM-04-2015-1390

Song, G., Gazi, M. I., Waaje, A., Roshid, M. M., Karim, R., Rahaman, M. A., Min, Z., & Senathirajah, A. R. B. S. (2025). The Neuromarketing: Bridging neuroscience and marketing for enhanced consumer engagement. *IEEE Access : Practical Innovations, Open Solutions*, *1*, 40331–40353. Advance online publication. DOI: 10.1109/ACCESS.2025.3545742

Song, X., & Bonanni, C. (2024). AI-Driven Business model: How AI-Powered Try-On technology is refining the luxury shopping experience and customer satisfaction. *Journal of Theoretical and Applied Electronic Commerce Research*, *19*(4), 3067–3087. DOI: 10.3390/jtaer19040148

Soni, N., & Kumar, S. (2024). What drives new luxury consumption? Application of schema congruity theory and heuristic systematic framework. *Asia Pacific Journal of Marketing and Logistics*, *36*(9), 2213–2233. DOI: 10.1108/APJML-04-2023-0319

Sprott, D. E., & Liu, R. L. (2016). Research trends on branding in consumer psychology. *Current Opinion in Psychology*, *10*, 124–128. DOI: 10.1016/j.copsyc.2016.01.002

Sung, B., Hatton-Jones, S., Teah, M., Cheah, I., & Phau, I. (2020). Shelf-based scarcity as a cue of luxuriousness: An application of psychophysiology. *European Journal of Marketing*, *55*(2), 497–516. DOI: 10.1108/EJM-06-2018-0418

Suzuki, S., & Kanno, S. (2022). The role of brand coolness in the masstige co-branding of luxury and mass brands. *Journal of Business Research*, *149*, 240–249. DOI: 10.1016/j.jbusres.2022.04.061

Tajeddini, K., & Mueller, S. L. (2011). Corporate entrepreneurship in Switzerland: Evidence from a case study of Swiss watch manufacturers. *The International Entrepreneurship and Management Journal*, *8*(3), 355–372. DOI: 10.1007/s11365-011-0179-y

Tasci, A. D., & Guillet, B. D. (2011). It affects, it affects not: A quasi-experiment on the transfer effect of co-branding on consumer-based brand equity of hospitality products. *International Journal of Hospitality Management*, *30*(4), 774–782. DOI: 10.1016/j.ijhm.2010.12.009

Terenteva, N., & Vovchenko, D. (2023). FEATURES OF THE SPACE OF NEUROMARKETING SENSORY TECHNOLOGIES. *Aktual'ni Problemi Ekonomìki*, *1*(268), 23–30. DOI: 10.32752/1993-6788-2023-1-268-23-30

Ting, L. (2023). Factors that Enhance Consumer Purchase Intention of Luxury Brand Collaboration Products. *Korea Industrial Technology Convergence Society, 28*(2), 115–129. https://doi.org/DOI: 10.29279/jitr.2023.28.2.115

Tran, T., Taylor, D. G., & Wen, C. (2022). Value co-creation through branded apps: Enhancing perceived quality and brand loyalty. *Journal of Research in Interactive Marketing, 17*(4), 562–580. DOI: 10.1108/JRIM-04-2022-0128

Turan, C. P. (2021). Success drivers of co-branding: A meta-analysis. *International Journal of Consumer Studies, 45*(4), 911–936. DOI: 10.1111/ijcs.12682

Tynan, C., McKechnie, S., & Chhuon, C. (2009). Co-creating value for luxury brands. *Journal of Business Research, 63*(11), 1156–1163. DOI: 10.1016/j.jbusres.2009.10.012

Ukanwa, K. (2024). Algorithmic bias: Social science research integration through the 3-D Dependable AI Framework. *Current Opinion in Psychology, 58*, 101836. DOI: 10.1016/j.copsyc.2024.101836 PMID: 38981371

Umarova, N., & Jurayeva, M. (2024). NEURO-LINGUISTIC PROGRAMMING ELEMENTS IN ADVERTISING: CRAFTING PERSUASIVE DISCOURSE. *CURRENT RESEARCH JOURNAL OF PHILOLOGICAL SCIENCES, 5*(10), 42–49. DOI: 10.37547/philological-crjps-05-10-08

Wang, R. W. Y., Chang, Y., & Chuang, S. (2016). EEG Spectral Dynamics of Video Commercials: Impact of the narrative on the branding product preference. *Scientific Reports, 6*(1), 36487. Advance online publication. DOI: 10.1038/srep36487 PMID: 27819348

Wang, X., Sung, B., & Phau, I. (2021). Examining the influences of perceived exclusivity and perceived rarity on consumers' perception of luxury. *Journal of Fashion Marketing and Management, 26*(2), 365–382. DOI: 10.1108/JFMM-12-2020-0254

Wang, X., Yin, H., & Ma, D. (2023). Understanding consumer reference price behavior in co-branding: From the perspective of a dynamic supply chain. *Electronic Commerce Research*. Advance online publication. DOI: 10.1007/s10660-023-09759-5

Wang, Y., & Griskevicius, V. (2013). Conspicuous consumption, relationships, and rivals: Women's luxury products as signals to other women. *The Journal of Consumer Research, 40*(5), 834–854. DOI: 10.1086/673256

Weatherill, S. (2017). The several internal markets. *Yearbook of European Law, 36*, 125–178. DOI: 10.1093/yel/yex007

Wei, Y. (2022). *A study of the luxury brand marketing strategy and its influence on consumers*. Advances in Economics, Business and Management Research/Advances in Economics, Business and Management Research., DOI: 10.2991/aebmr.k.220603.046

Wiedmann, K., Labenz, F., Haase, J., & Hennigs, N. (2017). The power of experiential marketing: Exploring the causal relationships among multisensory marketing, brand experience, customer perceived value and brand strength. *Journal of Brand Management, 25*(2), 101–118. DOI: 10.1057/s41262-017-0061-5

Wu, S., & Kim, S. Y. (2023). The Co-creation of Luxury Brand Hedonic Experiences in Social Media: Luxury Brand Strategies from a Design Perspective. *Archives of Design Research, 36*(3), 27–47. DOI: 10.15187/adr.2023.08.36.3.27

Xiang, D., & Cho, J. (2024). Impact of Yayoi Kusama's art on Commercial Design: A case study of brand Collaborations. *Asia-pacific Journal of Convergent Research Interchange, 10*(9), 469–485. DOI: 10.47116/apjcri.2024.09.38

Xie, Q., & Lou, C. (2020). Curating luxe experiences online? Explicating the mechanisms of luxury content marketing in cultivating brand loyalty. *Journal of Interactive Advertising, 20*(3), 209–224. DOI: 10.1080/15252019.2020.1811177

Xu, L., & Mehta, R. (2022). Technology devalues luxury? Exploring consumer responses to AI-designed luxury products. *Journal of the Academy of Marketing Science, 50*(6), 1135–1152. DOI: 10.1007/s11747-022-00854-x

Xue, J. (2024). The influence of brand culture and effects on consumer behavior. *SHS Web of Conferences, 199*, 03010. https://doi.org/DOI: 10.1051/shsconf/202419903010

Yadav, T. (2024). AI in Neuromarketing: Understanding Consumer Emotions and Behavior through Machine Learning. *International Journal for Multidisciplinary Research, 6*(5), 28154. Advance online publication. DOI: 10.36948/ijfmr.2024.v06i05.28154

Yang, J., & Mundel, J. (2021). Effects of brand feedback to negative eWOM on brand love/hate: An expectancy violation approach. *Journal of Product and Brand Management, 31*(2), 279–292. DOI: 10.1108/JPBM-05-2020-2900

Yassin, C. A., & Soares, A. M. (2023). Understanding impulse buying: A two-faceted tale. *Journal of Consumer Marketing, 40*(7), 884–896. DOI: 10.1108/JCM-09-2020-4116

Yoganathan, V., Osburg, V., & Bartikowski, B. (2021). Building better employer brands through employee social media competence and online social capital. *Psychology and Marketing, 38*(3), 524–536. DOI: 10.1002/mar.21451

Yu, B. (2021). How consumer opinions are affected by marketers: An empirical examination by deep learning approach. *Journal of Research in Interactive Marketing, 16*(4), 601–614. DOI: 10.1108/JRIM-04-2021-0106

Yu, H., Lee, D., Cottingham, M., & Maeng, L. (2019). Consumer perceptions to luxury co-branding partnership in sport wearable market. *INTERNATIONAL JOURNAL OF HUMAN MOVEMENT SCIENCE, 13*(03), 17–29. DOI: 10.23949/ijhms.2019.12.13.3.2

Yu, J., & Wang, J. (2024). Analysis of the Impact of luxury marketing methods on customer Consumption Psychology. *Advances in Economics Management and Political Sciences, 81*(1), 76–83. DOI: 10.54254/2754-1169/81/20241325

Yu, Y., Rothenberg, L., & Moore, M. (2020). Exploring young consumer's decision-making for luxury co-branding combinations. *International Journal of Retail & Distribution Management, 49*(3), 341–358. DOI: 10.1108/IJRDM-12-2019-0399

Yuan, Y., & Xiao, T. (2022). Retailer's decoy strategy versus consumers' reference price effect in a retailer-Stackelberg supply chain. *Journal of Retailing and Consumer Services, 68*, 103081. DOI: 10.1016/j.jretconser.2022.103081

Zakiyah, N. T. A. (2023). The impact of AI-Driven approaches on product branding in ethical business practices. *SAMMAJIVA Jurnal Penelitian Bisnis Dan Manajemen, 1*(2), 299–306. DOI: 10.47861/sammajiva.v1i2.949

Zein, S. E., Consolacion-Segura, C., & Huertas-Garcia, R. (2019). The role of sustainability in brand equity value in the financial sector. *Sustainability (Basel), 12*(1), 254. DOI: 10.3390/su12010254

Zhang, Q., Chen, J., & Lin, J. (2021). Market targeting with social influences and risk aversion in a co-branding alliance. *European Journal of Operational Research, 297*(1), 301–318. DOI: 10.1016/j.ejor.2021.05.022

Zhang, Y. (2024). Attention-based multilayer linear diffusion model. *Academic Journal of Computing & Information Science, 7*(4). Advance online publication. DOI: 10.25236/AJCIS.2024.070408

Zhao, Y. (2024). The study of the impact of the crossover co-branding marketing strategy in luxury firms. *Journal of Economics Business and Management*, 383–391. https://doi.org/DOI: 10.18178/joebm.2024.12.4.826

Zhou, Y., & Zhao, J. (2024). Leveraging computational systems for lifecycle management and enhancement of circular economy in fashion: A study on tracking, recycling, and reuse technologies. *Applied and Computational Engineering, 87*(1), 148–153. DOI: 10.54254/2755-2721/87/20241531

Zhu, D. (2022). The strategic and financial analysis of Hermès. *BCP Business & Management, 26*, 733–740. DOI: 10.54691/bcpbm.v26i.2033

Zhu, L. (2024). Marketing Strategy Analysis based on Consumer Psychology-Taking Tiffany as an example. *Advances in Economics Management and Political Sciences*, *69*(1), 271–276. DOI: 10.54254/2754-1169/69/20231448

Zhu, Y., & Kanjanamekanant, K. (2020). No trespassing: Exploring privacy boundaries in personalized advertisement and its effects on ad attitude and purchase intentions on social media. *Information & Management*, *58*(2), 103314. DOI: 10.1016/j.im.2020.103314

Zhu, Z. (2023). The impact of digital transformation on the fashion industry in the Post-Pandemic era. *BCP Business & Management*, *44*, 696–705. DOI: 10.54691/bcpbm.v44i.4920

Zuo, Y. (2024). Exploring the Synergy: AI Enhancing Blockchain, Blockchain Empowering AI, and their Convergence across IoT Applications and Beyond. *IEEE Internet of Things Journal*, *1*. Advance online publication. DOI: 10.1109/JIOT.2024.3507746

Chapter 5
Ethical and Sustainable Co-Branding in the Luxury Industry:
Strategies for a New Era of Symbolic Value and Corporate Responsibility

Noelia Araújo Vila
https://orcid.org/0000-0002-3395-8536
University of Vigo, Spain

Lucía Rubio-Escuderos
University of Valencia, Spain

ABSTRACT

The luxury industry's transformation in recent decades reflects growing integration of ethical and sustainable values due to societal pressure and evolving consumer expectations. This chapter examines ethical and sustainable co-branding as an emerging strategy in luxury, addressing demands for transparency, responsibility, and environmental commitment. Using a qualitative, multiple case study approach, it analyzes four collaborations: Burberry x The RealReal, Chantecaille x Space for Giants, Stella McCartney x Parley for the Oceans, and IWC x Laureus Sport for Good. Findings indicate that co-branding can effectively integrate sustainable practices while preserving luxury's symbolic value, expanding audiences, reinforcing brand authenticity, and supporting the Sustainable Development Goals (SDGs). Challenges include the risk of greenwashing, value alignment between partners, and equity in collaboration.

DOI: 10.4018/979-8-3373-3010-5.ch005

1. INTRODUCTION

The luxury industry has undergone a significant transformation in recent decades, driven by a growing environmental and social awareness among consumers. This evolution has led luxury brands to integrate sustainable and ethical practices into their operations and marketing strategies. The luxury sector has experienced steady growth, and consequently, luxury products have attracted the interest of numerous academics. However, only recently has research begun to explore the potential relationship between luxury and corporate social responsibility (CSR) and sustainability activities (Kunz et al., 2020).

Over the past two decades, the luxury industry has begun to redefine its traditional principles in response to a global cultural transformation. While historically, luxury has been associated with exclusivity, excess, and material opulence, it is now increasingly linked to values such as authenticity, transparency, and sustainability. This shift responds not only to internal industry developments but especially to a change in consumer expectations: new generations—particularly Millennials and Generation Z—demand that the brands they support demonstrate a genuine commitment to the environment, human rights, and business ethics (Kapferer & Michaut-Denizeau, 2017). In this context, sustainability is no longer seen as a secondary attribute but rather as a core value in the value proposition of many luxury brands (Becker-Leifhold & Iran, 2018).

As the concept of the luxury industry has evolved, so too have co-branding strategies, which have gained increasing relevance as mechanisms for expanding competencies, reaching new audiences, and legitimizing values that a single brand could not convey on its own. In the specific case of sustainable luxury, co-branding offers a strategic pathway for brands historically associated with aspiration and exclusivity to incorporate environmental or ethical credentials through partnerships with innovative companies, emerging brands, or other organizations. These types of alliances not only facilitate the transition toward more sustainable business models but also allow brands to respond more agilely to an increasingly informed and demanding consumer (Osburg et al., 2021).

Therefore, the main objective of this chapter is to analyze how co-branding strategies can be used by luxury brands to integrate sustainable and ethical practices without losing their symbolic essence or their prestigious market position. The text will focus on understanding the strategic potential of these collaborations, evaluating their opportunities, limitations, and conditions for success through a combination of theoretical review and case study analysis. The aim is not only to enrich the conceptual framework surrounding ethical co-branding in luxury but also to provide applicable insights for professionals in marketing, innovation, and corporate sustainability.

First, a theoretical review of key concepts will be conducted to help readers understand the intersection between luxury, sustainability, and co-branding. This includes exploring the new paradigm of conscious luxury, sustainability in fashion and high-end products, brand ethics, and co-branding as a strategy for shared symbolic value.

As the practical component of the chapter, a series of case studies on ethical and sustainable co-branding in luxury brands will be presented. This section will focus on detailed analysis of real-life co-branding initiatives that combine the prestige of luxury with sustainable innovation. The cases will be selected based on criteria such as geographical diversity, sector, type of alliance, and profile of the collaborating brand.

Each case will be analyzed from strategic, communicational, and ethical perspectives. Their impacts will be assessed in terms of consumer perception, brand reputation, scalability of the alliance, and contributions to the Sustainable Development Goals (SDGs). Furthermore, the extent to which these collaborations have managed to avoid the risks of greenwashing and drive structural transformation within the involved brands will be discussed.

The chapter will conclude with a critical reflection on the possibilities and limitations of co-branding as a tool for transformation within the luxury industry. Key questions will be posed, such as whether co-branding can genuinely reconcile the logic of exclusivity with the urgency of collective and sustainable action, in an attempt to redress the impact of consumerism, and what conditions must be met for such alliances to be not only effective from a marketing standpoint but also ethically responsible.

2. BACKGROUND

2.1. New Paradigm of Conscious Luxury

Until recently, the luxury industry has been associated with exclusivity, opulence, and conspicuous consumption—an image that persists in the minds of many individuals regarding the sector and its products. In fact, luxury brands remain strongly associated with exclusivity, quality, and popularity (Tarnanidis et al., 2025). However, we can affirm that a transformation is underway toward a more conscious and sustainable model. This shift is a direct response to the broader social movement toward sustainability and the growing consumer demand for responsible business practices and products that align with ethical and environmental values (Kapferer & Bastien, 2012). Over the past decades, the concept of luxury has evolved from a synonym of ostentation into a more complex notion encompassing sustainability, social responsibility, and environmental awareness (Silverstein & Fiske, 2003). This new approach, often referred to as "conscious luxury," entails a profound reassessment of the values that underpin the industry. This is a global phenomenon, as luxury brands successfully navigate diverse cultural landscapes to appeal to a broad range of international consumers (Vimalnath et al., 2025).

It is crucial here to consider younger generations—namely Millennials and Generation Z—who increasingly demand authenticity, transparency, and engagement with social causes. Within this context, luxury is no longer solely defined by material exclusivity. Rather than seeking ostentation, today's consumers value brands' efforts to align their practices with ethical and environmental principles (Csaba & Skjold, 2018). According to a KPMG report (Muret, 2024), 21% of professionals in the luxury sector believe that sustainable development and ethical criteria are key factors for consumers when purchasing luxury goods. Similarly, a McKinsey & Company study (2020) reveals that sustainability is gaining ground as one of the top priorities among luxury consumers. This phenomenon is also driven by the rise of movements such as fair trade and ethical production, which have begun to significantly influence luxury purchasing decisions.

But is this shift reflected in the business strategies of luxury brands? Clearly, yes. Companies such as LVMH (best known for Louis Vuitton) have integrated sustainability into their business models through initiatives like LIFE 360, which focuses on circular economy practices, biodiversity, and transparency. This initiative aims to mitigate environmental impact and improve product traceability (LVMH, 2020). Additionally, brands like Nona Source and Bee Respect promote the reuse of materials and supply chain traceability, aligning themselves with the United Nations' Sustainable Development Goals (SDGs). The integration of these values into corporate strategies reflects an effort not only to enhance brand reputation but also to adapt to a market that increasingly rewards sustainability (Heath & Potter, 2005).

The concept of the circular economy is also becoming central to the luxury industry, representing a model that seeks to reduce waste and maximize resource reuse. Brands like Hermès and Prada are at the forefront of this trend. Hermès, for example, has launched repair and maintenance initiatives to extend the lifespan of its products, and developed the "Petit h" line, which uses leftover materials to create new items (Mohtar & Taha, 2024). Meanwhile, Prada has introduced the Re-Nylon initiative, which uses regenerated nylon made from recycled plastics collected from the ocean (Prada Group, 2020). These initiatives not only respond to the demands of conscious consumers but also represent a strategic opportunity for differentiation in an increasingly competitive market.

Another concept closely associated with this industry is digitalization and the adoption of new technologies, which are essential in the transformation of luxury towards more sustainable models. Artificial intelligence, blockchain, and digital traceability enable deeper personalization and greater transparency in the production of luxury goods (Brynjolfsson & McAfee, 2014). Blockchain in particular has revolutionized the way luxury brands ensure the authenticity and traceability of their products. This technology not only strengthens consumer trust (Manda et al., 2024), but also allows for detailed tracking of the supply chain, helping to ensure ethical and sustainable practices (Comert, 2020). The ability to verify a product's origin and production process has become a crucial tool for brands seeking to align with the expectations of increasingly informed and demanding consumers.

Current data only further supports the notion that the sustainable luxury market is experiencing remarkable growth. It is expected to grow by 9.1% annually between 2024 and 2034, outpacing the growth of the overall luxury goods market (McKinsey & Company, 2020). This growth clearly reflects a shift in consumer mindset, as buyers increasingly seek not only high-quality products but also those that respect the environment and promote human well-being. Brands that have embraced sustainability and social responsibility principles hold a competitive advantage, as they connect more deeply with consumers' concerns and values (Kapferer & Michaut-Denizeau, 2019).

For all these reasons, we are not witnessing a passing trend, but rather a paradigm shift responding to new socio-economic and environmental realities. Luxury brands that manage to integrate these principles authentically and effectively will be well-positioned as industry leaders, as sustainability and social responsibility have become crucial factors for long-term success. According to Bain & Company (2020), luxury brands that implement sustainable and ethical practices not only experience greater growth but also enjoy increased consumer loyalty and satisfaction.

2.2. Sustainability in Fashion and High-End Products

As discussed in the previous section, sustainability has become a strategic imperative for fashion brands, including those operating in the luxury segment, the focus of this chapter. This paradigm shift responds not only to regulatory and environmental pressures, but also to profound changes in consumer expectations and the need to preserve the cultural legitimacy of brands. In this sense, the high-end fashion industry is undergoing a transformation, in which traditional values of exclusivity and craftsmanship coexist with new principles of sustainability, traceability, and social responsibility (Joy et al., 2012).

This is an industry that promotes durability, quality materials, and craftsmanship, which in some ways naturally positions it toward structural sustainability, the opposite of fast or low-cost fashion. Practices linked to overproduction, the use of exotic skins, and opacity in supply chains have generated criticism, forcing many fashion houses to review their business models (Henninger et al., 2016).

In an effort to achieve this new trend, companies in the luxury industry have implemented measures such as traceability and transparency. Traceability allows brands to guarantee the ethical origin of their raw materials, improve working conditions in their supply chains, and ensure compliance with environmental standards. Technologies such as blockchain have been adopted by brands like LVMH and Kering to provide consumers with verifiable information about the origin of their products (Tapscott & Tapscott, 2017). This transparency not only strengthens trust but also promotes a more conscious and responsible consumer culture.

Another key variable in this new sustainability landscape is the role that sustainable materials play in luxury fashion. The use of recycled, organic, or low-environmental-impact fibers has been progressively adopted by numerous brands. For example, Stella McCartney, a leading British designer in this industry, has been one of the pioneers in this field, using organic cotton, sustainable viscose, and recycled polyester in her collections (Quiles-Soler et al., 2023). Similarly, Gucci, part of the Kering group, has introduced the Off the Grid line, based on recycled materials and low-carbon processes (Kering, 2021).

We must once again allude to the concept of the circular economy, which has gained an important place in luxury fashion as a strategy for closing the life cycle of products. This logic involves rethinking the entire production and consumption system, from design to the recycling or reuse of garments. In fact, recent studies indicate that the circular economy can be especially effective in luxury due to the longevity of products and the high perceived value that facilitates their resale or restoration (Niinimäki, 2018).

We can't continue talking about sustainability without mentioning the more social aspect of a sustainable model. Although many luxury brands pride themselves on their artisanal and localized production, globalization has led to an increasing outsourcing of parts of production. In this context, ensuring decent working conditions has become a key challenge. The integration of external audits, codes of conduct, and continuous improvement programs is increasingly common, and many brands are publishing sustainability reports as part of their corporate social responsibility strategy (Caniato et al., 2012).

In addition to the factors mentioned so far, the institutional and regulatory environment has played a significant role in pushing the sector toward more sustainable practices. The European Union, for example, has launched plans such as the Green Deal and the EU Strategy for Sustainable and Circular Textiles, which seek to accelerate the transition toward a more environmentally friendly economy. These regulations have pressured luxury sector players to adapt their processes and report their environmental impact more rigorously (European Commission, 2022).

At the consumer level, sustainability is also consolidating as a key criterion in the purchase decision for luxury goods. According to a Deloitte study (2023), 37% of luxury product consumers indicate they are willing to pay more for brands that demonstrate a clear commitment to sustainability. This behavior is particularly evident among younger generations, such as millennials and Generation Z mentioned in the previous section, who value authenticity, ethics, and the positive impact of their consumer decisions (Varan, 2020).

Faced with this situation, communication and marketing have also had to transform. Luxury brands not only communicate aesthetic or historical attributes, but also their commitment to the environment and society. Campaigns highlighting the use of recycled materials, collaborations with NGOs, or sustainability certifications are increasingly common. However, there is a risk of greenwashing, that is, the exaggeration or falsification of sustainable commitments for advertising purposes. In this sense, transparency and consistency become essential to maintaining credibility (Delmas & Burbano, 2011).

A complementary aspect that has not been considered until now is the growth of the secondhand luxury market, driven by platforms such as Vestiaire Collective or The RealReal, similar to better-known platforms like Vinted, but in the luxury sector. These platforms not only extend the lifespan of products but also allow consumers to access luxury in a more accessible way that is aligned with the principles of responsible consumption. According to Bain & Company (2023), the luxury resale market is growing at twice the rate of the primary market, a sign of a structural shift in the way consumers relate to high-end products.

Therefore, sustainability in luxury fashion cannot be approached as an isolated or superficial strategy, but rather as a comprehensive process that spans the entire value chain. From conscious design and material selection to ethical production, reverse logistics, and transparent communication, every element must reflect a genuine commitment to the principles of sustainable development. Far from representing a threat, this transformation represents an opportunity to renew the symbolic value of luxury in the 21st century, aligning it with the ideals of responsibility, authenticity, and lasting beauty.

2.3. Brand Ethics

In light of the preceding sections, it is abundantly clear that the current social orientation toward transparency and sustainability has positioned brand ethics as a fundamental strategic axis within the luxury sector. But what exactly do we mean by brand ethics? It refers to the set of moral and normative principles that guide a brand's behavior in its interactions with stakeholders. These principles are reflected in decisions ranging from corporate governance to labor practices, marketing, production, and environmental management (Melé, 2003). In the context of luxury, brand ethics acquires a particularly complex dimension, as it must reconcile the traditional symbolic value of the sector—rooted in scarcity, craftsmanship, and exclusivity—with growing societal demands for greater integrity, equity, and responsibility.

Importantly, referring to an ethical brand goes beyond mere compliance with current laws and regulations. It implies responsible behavior across all operations—for instance, respecting human rights, promoting fair labor conditions, minimizing environmental impact, and engaging in honest, non-manipulative communication with consumers (Carrington et al., 2010), all of which have surfaced throughout this chapter. In this sense, ethics becomes an inextricable dimension of brand value, influencing reputation, consumer loyalty, and long-term resilience (Kapferer & Michaut-Denizeau, 2019).

The luxury industry, historically distant from debates on corporate responsibility, has begun to reconstruct itself as consumers increasingly demand consistency between a brand's stated values and its actual practices. As previously mentioned, today's consumers—particularly Millennials and Gen Z—expect brands to demonstrate tangible ethical commitment and are inclined to penalize those that display inconsistent or misleading behavior (Achabou & Dekhili, 2013). This heightened ethical sensitivity has led traditional values of luxury—such as exclusivity and status—to be reinterpreted through the lens of authenticity and social responsibility.

One of the core areas of brand ethics is supply chain management. Many high-end fashion houses have faced criticism for outsourcing production to countries with weak labor regulations, often resulting in worker exploitation, unfair wages, or unsafe working conditions—precisely the issues this ethical turn seeks to address. In response, some brands have adopted auditing and certification policies to ensure their suppliers comply with international labor and working condition standards. Transparency in this area has become essential for building trust and reinforcing a brand's ethical legitimacy (Perry & Towers, 2013).

Material and production choices also reflect a brand's ethical commitment. The use of exotic skins, for instance, has sparked controversy due to its impact on biodiversity and animal welfare. Brands such as Chanel and Burberry have ceased using natural fur in response to public pressure and internal value reassessment (PETA, 2020). Similarly, the use of sustainable, low-impact raw materials has become a key indicator of brand ethics. This focus, however, is not limited to environmental issues but encompasses a holistic vision of sustainability that includes social, cultural, and economic dimensions (Kozlowski et al., 2015).

Beyond supply chain and materials, a third pillar of brand ethics is ethical communication. In a context where storytelling has become central to brand value construction, the temptation to overstate sustainable or ethical commitments—known as greenwashing or ethics-washing—presents serious reputational risks. According to Delmas and Burbano (2011), greenwashing fosters consumer distrust and can severely undermine a brand's credibility, especially when discrepancies between rhetoric and practice come to light. Brand ethics, therefore, demands truthful, verifiable, and consistent communication that respects consumers' intelligence and moral sensibilities.

In this regard, various self-assessment tools and external certifications have emerged as mechanisms for ensuring ethical compliance. Labels such as B Corp, Fair Trade, or the Global Organic Textile Standard (GOTS) offer objective and verifiable criteria that can both guide internal corporate decisions and transparently communicate ethical commitments. While not yet widespread in the luxury industry, these certifications are gaining ground as markers of corporate integrity and competitive differentiation.

Nonetheless, brand ethics cannot be understood merely as a collection of isolated practices or reactive responses to public pressure. It requires an integrated strategic vision and an organizational culture that embraces responsibility as a core principle. In this sense, authors such as Crane et al. (2019) emphasize the need to embed ethics in decision-making at all levels of the company—from product design to advertising, including supplier and employee relations. In luxury brands, where every detail is meticulously curated to communicate a symbolic universe, ethics must permeate every component of the business model.

A paradigmatic example is the brand Chloé, which in October 2021 became the first luxury maison to be certified as a B Corp. This achievement represents not only a declaration of intent but also a structural transformation of its governance model, which now includes social and environmental objectives in its corporate statutes (Chloé, 2021). Chloé's experience demonstrates that it is possible to integrate ethics at the core of luxury without relinquishing its symbolic value, but rather by re-signifying it in accordance with new contemporary ideals.

2.4. Co-branding as a Shared Symbolic Value Strategy

In today's landscape of conscious luxury, co-branding has emerged as a key strategy for generating shared value. Through this approach, various brands join forces to align their identities with ethical and sustainable principles. Co-branding, which involves collaboration between two or more brands to develop joint products or services, allows them to combine resources, capabilities, and symbolic values—offering richer and more resonant value propositions to increasingly discerning consumers.

More precisely, co-branding—or brand alliance—is a strategic partnership between two or more brands with the aim of creating synergies that enhance perceived value for consumers (Helmig, et al, 2008). This strategy has evolved from its initial use as a tactical marketing tool into a more sophisticated practice that seeks to integrate shared values and meanings among participating brands.

In the luxury sector, co-branding has taken on a significant symbolic dimension, enabling brands to collaborate with others that share their ethical and sustainable values. These partnerships not only expand market reach but also reinforce legitimacy and authenticity in a market where social and environmental responsibility is increasingly valued by consumers.

The symbolic value generated through co-branding refers to the creation of shared meanings and values that resonate with the identities and aspirations of consumers. As Kapferer and Bastien (2012) note, luxury brands operate on a symbolic level, where perceived value extends far beyond functional attributes.

Co-branding can take various forms. Already in the 1990s, Rao and Ruekert (1994) identified *complementary co-branding*—the collaboration between brands with different but complementary competencies. For instance, a luxury fashion brand might partner with a company specializing in sustainable materials to develop an eco-friendly product line. Such collaborations allow each brand to leverage its strengths and jointly offer a product that reflects their shared values.

More recently, Becker-Leifhold and Iran (2018) introduced the concept of *innovation co-branding*, which focuses on the joint creation of innovative products or services that embed ethical and sustainable principles. The most recent trend is *cause-related co-branding*, which involves partnerships between brands and nonprofit organizations or social initiatives to support specific causes. These collaborations enable luxury brands to demonstrate their commitment to social responsibility and connect with consumers who value corporate activism (Varan, 2020).

It is clear that co-branding offers a range of strategic advantages for luxury brands seeking to integrate ethical and sustainable values into their value proposition. Among the most notable benefits:

- Market reach expansion: By partnering with brands that have a different but value-aligned consumer base, luxury brands can tap into new market segments and boost visibility.
- Authenticity reinforcement: Collaborations with brands that have a strong reputation for sustainability and ethics can enhance the perceived authenticity of luxury brands—an increasingly important factor in a market wary of greenwashing (Delmas & Burbano, 2011).
- Innovation and differentiation: Co-branding facilitates the exploration of new ideas and sustainable technologies, often resulting in innovative products that stand out in the market.

- Risk management: By sharing resources and expertise, brands can mitigate risks associated with implementing ethical and sustainable practices, such as supply chain complexity or investments in new technologies.

However, not all aspects of co-branding are positive. Successful and coherent partnerships require careful attention to a number of ethical and strategic considerations:

- Value alignment: It is essential that partnering brands share similar ethical values and commitments. Discrepancies can lead to negative perceptions and reputational damage for both parties (Helmig et al., 2008).
- Transparency and communication: Brands must clearly and transparently communicate the goals and outcomes of the partnership, avoiding exaggerations or misleading claims that may be perceived as greenwashing.
- Equity in the collaboration: Fair and respectful relationships are vital, with due recognition given to each party's contributions.
- Impact assessment: Brands should implement mechanisms to evaluate the social and environmental impact of the collaboration, ensuring that ethical and sustainability goals are effectively met.

As we will explore in the practical section of this chapter, there are already several successful examples of co-branding in the luxury sector that have generated shared symbolic value.

3. OBJECTIVE AND METHODOLOGY

3.1. General Objective

The main objective of this chapter is to analyze how co-branding strategies can be used by luxury brands as mechanisms to integrate sustainable and ethical practices without compromising their symbolic value or their prestigious positioning in the market.

3.2. Specific Objectives

1. Identify the key elements that define ethical and sustainable co-branding in the luxury sector.
2. Evaluate the strategic motivations behind alliances between luxury brands and partners with sustainability credentials.
3. Analyze the benefits of these collaborations.

4. Examine the reputational risks and associated ethical challenges, such as greenwashing or the trivialization of sustainable commitment.
5. Establish success criteria and enabling conditions for these co-branding strategies to generate real and lasting shared value.

3.3. Methodological Justification

To achieve these objectives, a qualitative methodology based on multiple case studies was chosen. This choice responds to the need to understand a complex, contemporary, and contextual phenomenon such as ethical co-branding in the luxury sector, which involves multiple symbolic, strategic, and ethical dimensions.

Case studies are an appropriate methodological strategy for exploring phenomena in depth, especially when the object of study is little known or requires a holistic approach (Yin, 2018). Through this methodology, it is possible to reconstruct the processes, decisions, and outcomes of strategic collaborations between brands, capturing not only objective data but also the meanings, perceptions, and discourses that shape their public narrative.

As early as the 1980s and 1990s, Eisenhardt (1989), and later Stake (1995), asserted that case studies allow for the formulation of theories based on rich and contextualized empirical data, which is especially valuable in fields where interactions between variables are difficult to isolate. In this sense, the multiple-case study does not aim to achieve statistical representativeness, but rather analytical one: to offer a deep understanding of strategically selected cases to illustrate different modalities, outcomes, and approaches within the same phenomenon.

The use of the multiple-case approach, rather than a single case, also responds to the recommendation of authors such as Baxter and Jack (2008), who argue that this strategy allows for the identification of common and divergent patterns, reinforcing the external validity of the findings. In this chapter, the analysis covers four cases of sustainable co-branding within the luxury sector, selected according to the following criteria:

- Geographic diversity: Include brands from different countries to capture cultural and market differences.
- Alliance sector: Understand the type of ethical alliance according to the specific commercial sector.
- Alliance modality: Consider collaborations between traditional luxury brands and sustainable startups, as well as alliances with NGOs, ethical suppliers, or independent designers.
- Collaborating brand profile: Examine cases where the partner brand provides environmental, social, or technological credentials of high strategic value.

3.4. Analysis Process

The methodology will be developed in three main phases:

a) Documentary Collection and Preliminary Analysis

In the first stage, secondary information on the selected alliances will be collected through academic sources, corporate reports, press releases, official websites, and specialized media outlets. This preliminary analysis will help establish the context, stated objectives, communicative messages, and institutional narrative of each case.

This process aligns with what Yin (2018) refers to as "source triangulation," which allows information to be validated through multiple types of documents and perspectives, thereby increasing the credibility of the findings.

b) Thematic and Comparative Analysis

The second phase will involve developing a thematic analysis to categorize common and divergent elements among the cases. Open coding will be applied based on axes such as:

- Purpose of the alliance
- Communication actions (storytelling, campaigns, projected values)
- Reception by the press and consumers
- Impact on brand image
- Associated risks and mitigation strategies
- Contribution to the SDGs

This approach will enable the comparison and contrast of cases around key dimensions, thus identifying patterns that support theorizing about the conditions that favor the success of ethical co-branding in the luxury industry.

c) Critical Reflection and Theoretical Integration

Finally, the empirical findings will be integrated with the previous conceptual review in a feedback exercise between theory and practice. This step is essential to strengthen the explanatory nature of the study, allowing for the development of interpretative models that connect case analysis with current theoretical frameworks on branding, sustainability, and responsible innovation.

4. RESULTS

4.1. The Fashion Industry and the Circular Economy: Burberry x TheRealReal

As previously stated, fashion industry consumers are showing increasing concern over environmental issues and the waste it generates. In response, some brands have begun to incorporate circular economy principles, either through specific actions or by directly integrating them into their production processes and surplus management.

The first case analyzed is the co-branding initiative between British fashion brand Burberry and U.S.-based luxury resale digital platform The RealReal, who partnered in 2019 to promote circular economy practices through the resale of luxury goods.

a) Documentary Collection and Preliminary Analysis

The collaboration between both brands aims to extend the lifecycle of garments, reduce textile waste, and appeal to a younger, sustainability-minded audience (Burberry, 2025; Vogue Business, 2019; Forbes, 2019), within a context in which digital transformation increasingly shapes the retail value chain, particularly through the rise of e-commerce (Kaur & Siddhey, 2024).

On the one hand, Burberry encourages customers to consign their products on The RealReal. Customers whose Burberry items are accepted for consignment receive a personalized shopping experience, including a tea service in select Burberry stores in the U.S. (Vogue Business, 2019). Additionally, both brands have supported Materials for the Arts, an organization dedicated to the creative reuse of materials, and have launched joint campaigns that frame recycling as part of modern luxury: a stylish, conscious choice (Burberry, 2025; Forbes, 2019).

Prior to Burberry, Stella McCartney was the first luxury brand to partner with The RealReal, in 2017. As part of that initiative, the brand offered a $100 store credit to customers consigning Stella McCartney items on the platform. This resulted in a 65% increase in the number of consignors selling Stella McCartney products within the first year (Vogue Business, 2019).

Currently, luxury brands do not earn direct revenue from the resale of their items on The RealReal. However, the strategy aims to influence purchasing behavior: by recovering some of the value of their investment through consignment, consumers may be more inclined to buy new luxury goods (Vogue Business, 2019).

Burberry, which had previously faced heavy criticism for destroying unsold inventory (BBC, 2018), has used this partnership to improve its brand image and align with more sustainable practices. By collaborating with a platform focused on waste reduction and extending the life of garments, Burberry connects with a younger, more digital and eco-conscious audience (Forbes, 2019).

At the same time, The RealReal gains legitimacy by partnering with established luxury brands, strengthening its position as a respected channel for high-end resale. The visibility it gains through the benefits offered in Burberry's physical stores also enhances its market presence.

According to data from both companies, demand for secondhand Burberry products has increased by 64% year-over-year, with Millennial and Gen Z users leading the growth in search interest on the platform (Burberry, 2025). This demonstrates the positive impact of co-branding in terms of sustainability, brand image, and market reach.

b) Thematic and Comparative Analysis

Based on the categories proposed in the methodology, the following elements can be highlighted from this co-branding initiative:

- Purpose of the alliance: To promote circularity and legitimize resale as part of the luxury consumption cycle, while also encouraging sustainability as an aspirational value within the luxury market.
- Communication actions:
    -Joint campaign centered on the theme "giving luxury a second life."
    -Messaging focused on responsibility, design longevity, and conscious fashion.
    -Incentives for reselling Burberry items on The RealReal, such as VIP experiences or exclusive product access.
- Reception by press and consumers: Generally positive, as consumers gain access to luxury at more affordable prices while participating in the circular economy. The initiative is appreciated for its innovation and practical approach.
- Impact on brand image: For Burberry, the partnership signaled a repositioning toward a more responsible and youthful image, without abandoning traditional luxury. For The RealReal, it consolidated its role as a bridge between sustainability and luxury, enhancing institutional legitimacy.
- Associated risks and mitigation strategies:
    -Greenwashing: Although public reception was mostly positive, Burberry did face criticism for the contradiction between promoting circularity and its past practices.
    -Loss of exclusivity: The risk of diluting luxury through resale was countered by branding aligned with the brand's core values.

- Contribution to the SDGs:
  -SDG 12 (Responsible Consumption and Production): Directly related through the promotion of reuse.
  -SDG 13 (Climate Action): Waste reduction and lower carbon footprint through extended garment lifecycle.

c) Critical Reflection and Theoretical Integration

This first case study offers a tangible and relatable example of how a luxury brand can engage in sustainable practices through co-branding, thus contributing to the growing demand for circular economy models and sustainability—while maintaining brand prestige and identity.

This cooperation reaffirms several theoretical elements discussed earlier in this chapter. On one hand, it shows the integration of tradition—concepts such as exclusivity and craftsmanship—with ethical values and the use of technology (in this case, The RealReal's digital platform), resulting in a brand repositioning aligned with contemporary values. It also illustrates the preservation of luxury's value—not devalued but rather enhanced by the adoption of ethical and current values. Finally, it demonstrates the use of transparency by the brand towards consumers, ensuring they perceive coherence between brand, sustainability, and ethics.

From a theoretical perspective, this alliance can be analyzed as a form of symbiotic co-branding, where both brands benefit not only commercially but also in terms of reputational capital and institutional legitimacy. This type of collaboration reflects a shift toward relational branding, rooted more in shared values than in traditional product logic.

4.2. The Cosmetics Industry and Animal Welfare: The Case of Chantecaille x Space for Giants

Animal welfare has become a decisive factor for consumers in the food and cosmetics industries, with labels such as "vegan" and "cruelty-free" gaining particular relevance. The latter is especially valued in the cosmetics field, as it indicates that the product formulation has not been tested on animals—a practice widely considered "cruel" (Grappe et al., 2021).

Although many countries, including the European Union, have banned animal testing for cosmetics, exceptions still exist. For instance, China continues to require mandatory testing for imported products. As a result, consumers concerned with animal welfare often research whether brands export to that country or not (Grappe et al., 2021).

Organizations advocating for animal welfare, such as PETA (People for the Ethical Treatment of Animals), offer certifications like "PETA Approved" or "Leaping Bunny" (the latter issued by the Coalition for Consumer Information on Cosmetics - CCIC) to companies that meet strict standards. These certifications ensure that neither the company nor its suppliers conduct, commission, finance, or allow animal testing for ingredients, formulations, or finished products anywhere in the world—and that they commit to not doing so in the future (PETA, 2025).

Both Leaping Bunny and PETA offer independent databases of certified vegan brands (those using no animal-derived ingredients) and/or cruelty-free brands. These databases are continuously expanding, reflecting growing social and corporate awareness of animal welfare issues.

a) Documentary Collection and Preliminary Analysis

In recent years, some luxury brands have joined the animal welfare movement, although they still remain a minority. A prominent example of co-branding aimed at this positioning is the French luxury brand Chantecaille Beauté, which specializes in high-end skincare and makeup, using botanical ingredients and an ecological approach.

Chantecaille is not only certified by PETA as a vegan and cruelty-free brand, but it also holds the "PETA Business Friend" distinction. Since its foundation, the brand has stood out for its social and environmental commitment, encapsulated in its slogan "Beauty with impact." The company has engaged in multiple collaborations with environmental organizations, creating a form of cause-related co-branding. Although these partnerships involve companies and NGOs, both parties benefit from each other's reach, image, and values, generating mutual organizational reinforcement: the brand gains ethical legitimacy, while NGOs receive visibility and funding.

Chantecaille develops beauty products linked to environmental or animal welfare causes, promoting a specific NGO associated with that cause, to which a portion of the profits is donated. For example, it collaborated with Space for Giants (Kenya) to launch a lipstick collection aimed at supporting elephant conservation in African ecosystems. The product packaging features an engraved elephant, the formula includes responsibly sourced natural ingredients from Africa—such as Baobab Oil certified by FairWild—and the point of sale incorporates storytelling that emotionally connects the product with the organization and the cause.

The proceeds from the lipstick sales have been used, for instance, to build electrified smart fences in Kenya, which help keep elephant migration routes safe and prevent conflicts with local communities. Additionally, tens of thousands of trees have been planted in protected areas, contributing to ecosystem restoration and improving the well-being of both wildlife and nearby human populations (Chantecaille,

2025). The initiative has been notably successful. In Laikipia, Kenya, this approach, combined with the conservation efforts of local partners, helped reduce the number of elephants killed by poaching by up to 97% from its peak in 2012. As a result, Space for Giants has continued to expand, doubling the number of countries it operates in to nine last year—home to 85% of the remaining elephants in Africa, from Gabon to Botswana and Uganda. In terms of sales, the Lip Veil x Space for Giants line generated over $150,000 USD for the nonprofit in 2017 (Chantecaille, 2025).

Kenyan media outlets confirm Chantecaille's conservation efforts. For example, the Kenya Wildlife Service (KWS), a public institution under the Kenyan state, received a donation valued at $1.8 million USD, including 20 upgraded Land Cruiser vehicles, a truck, and 50 tents (Kenyans.co.ke, 2025). These resources were officially delivered by the brand's co-founder and CEO and received by KWS Director General Erustus Kanga. According to KWS, "the donations will enhance mobility, strengthen communication, support rangers, and promote conservation collaboration" (Kenyans.co.ke, 2025). This is not an isolated action—Chantecaille has previously made significant contributions, including the protection of approximately 1.2 million acres of land in the Amboseli region and the monthly support of 67 schools near wildlife habitats in Kenya (Kenyans.co.ke, 2025).

Following a comprehensive search, no significant public controversies regarding Chantecaille's business practices have been found to date. On the contrary, the company is widely recognized for its sustainability and wildlife conservation commitments. Founded in 1997 by Sylvie Chantecaille, the brand has been known for its philanthropic approach and active participation in wildlife protection initiatives. For example, it was the first company to support the "Stop the Illegal Wildlife Trade" campaign by The Independent newspaper, contributing with donations and partnerships to fight the illegal trafficking of endangered species like rhinos and tigers. In 2021, Chantecaille was acquired by German company Beiersdorf, owner of brands such as Nivea and La Prairie. Despite this acquisition, Chantecaille continues to operate independently, maintaining its focus on natural beauty and sustainability.

b) Thematic and Comparative Analysis

Highlights of the Chantecaille x Space for Giants co-branding include:

- Purpose of the alliance: Support the conservation of African elephants and their natural habitats. Through the sale of its Lip Veil line, Chantecaille donates funds to projects that aim to protect these animals' migration routes and reduce human-wildlife conflict in Africa.
- Communication action:

-Information on Space for Giants and the elephant conservation cause is shared at Lip Veil product points of sale (both online and physical).
-Chantecaille produced a documentary about Space for Giants' work, published on its official website.
-Information is disseminated via the brand's social media platforms, such as Instagram.
-Space for Giants publicly recognizes Chantecaille as an official partner.
-Updates about the projects funded by Chantecaille are shared on Space for Giants' social media.

- Reception by the press and consumers: Official acts in Kenya had a positive media impact, especially in local press outlets.
- Impact on brand image: This co-branding effort is aligned with Chantecaille's established identity as a supporter of global wildlife conservation efforts. It also amplifies the visibility of the Kenyan organization Space for Giants worldwide.
- Associated risks and mitigation strategies:

To date, no reputational risks have been identified. On the contrary, Chantecaille is positively acknowledged by media and animal welfare organizations (such as PETA and Space for Giants), reinforcing the idea that luxury brands can have a positive impact—or at least mitigate the negative effects of capitalist consumerism.

- Contribution to the SDGs:
  -SDG 12 (Responsible Consumption and Production): through the use of responsibly sourced natural ingredients.
  -SDG 11 (Sustainable Cities and Communities): by supporting local communities in preserving their natural heritage and wildlife.
  -SDG 15 (Life on Land): through initiatives aimed at protecting biodiversity and ecosystems.

c) Critical Reflection and Theoretical Integration

This case clearly illustrates the strong connection between the cosmetics industry and animal welfare. Chantecaille goes beyond the current "vegan" and "cruelty-free" trend, building its brand identity around the conservation of endangered wildlife and ecosystems.

The collaboration between Chantecaille and Space for Giants demonstrates how the beauty industry can become a powerful driver of positive change beyond aesthetics. Rather than simply selling cosmetic products, the brand chooses to link its launches—such as the Lip Veil lipstick—to concrete and urgent causes like elephant conservation in Africa.

This alliance not only raises awareness about biodiversity loss but also redefines the role of a company in today's society: Chantecaille proves that it is possible to generate meaningful impact through everyday consumer choices. The fact that something as simple as buying a lipstick can help preserve migration corridors or reduce conflicts between wildlife and human populations reveals the transformative potential of informed consumption.

Moreover, by supporting initiatives such as reforestation or the installation of smart fences, the brand does not act symbolically—it actively contributes to long-term sustainable solutions. This symbiosis between beauty and nature conservation not only enriches the brand's value proposition but also inspires a new way of understanding luxury: as an opportunity to care for the world, not just for oneself.

4.3. Fashion for Ocean Conservation: Stella McCartney x Parley for the Oceans

Returning to the fashion sector, garment manufacturing represents one of the greatest threats to ocean health due to the release of microplastics, the intensive use of chemicals, and the generation of textile waste (Brooks et al., 2017: Karypidis et al., 2024). In response to this issue, some luxury fashion brands have begun to adopt a more committed stance on sustainability. A prominent example is Stella McCartney, known for her environmental activism and for embedding ethical principles at the core of her business model (Forbes, 2017).

The brand has developed numerous initiatives with social and environmental impact. For instance, as previously mentioned, it has partnered with the luxury resale platform TheRealReal, promoting the circular economy and extending the life cycle of its products. On its official website, Stella McCartney features a dedicated sustainability section, where it outlines its efforts under headings such as "Measuring Our Impact." This section details the brand's strategies to mitigate environmental impacts related to material use, waste management, and energy consumption (Stella McCartney, 2025).

In addition, the brand applies labor standards inspired by the conventions of the International Labour Organization (ILO), which translates into clear guidelines for its network of suppliers. Documents such as the Supplier Code of Conduct, the Responsible Sourcing Guide, the Modern Slavery Policy, and the Subcontracting Policy establish control and transparency mechanisms to ensure that all individuals involved in its supply chain are treated with dignity and fairness (Stella McCartney, 2025).

a) Documentary Collection and Preliminary Analysis

In line with her commitment to environmental sustainability, designer Stella McCartney has established a strategic partnership with the international organization Parley for the Oceans, aiming to address the issue of marine pollution through innovation in the luxury fashion industry. This collaboration seeks to transform plastic waste recovered from the oceans into functional materials that enable the production of high-end garments, thereby contributing to the circular economy and reducing the use of virgin materials.

Through this initiative, collections have been developed that incorporate innovative materials such as recycled polyester and fibers recovered by Parley, which are combined with wool and other sustainable textiles for the production of coats, footwear, clothing, and accessories. This strategy not only decreases the demand for non-renewable natural resources but also positions fashion as an influential platform for raising awareness about the oceanic environmental crisis. In the words of McCartney herself: "Fashion should always have an ethical edge, and to be at the forefront, we must first raise awareness and improve the planet" (Forbes, 2017).

Parley for the Oceans is a nonprofit environmental organization that promotes awareness projects on the health of marine ecosystems through collaboration with creative sectors, government entities, private companies, and civil society organizations (Parley, 2025). Among its key programs are the Parley Ocean School, aimed at environmental education for local communities and business groups, as well as artistic initiatives, scientific research, and the development of a material known as Ocean Plastic, created through the upcycling of plastic waste recovered from marine environments (Parley, 2025).

The collaboration between Stella McCartney and Parley initially took shape through a joint partnership with sportswear brand Adidas, under the "adidas by Stella McCartney" line, which focused on producing footwear made from ocean plastic. In later stages, the designer deepened her commitment by replacing traditional polyester—both virgin and recycled—in footwear, handbags, and outerwear with upcycled materials developed by Parley. This shift represents a significant step toward eliminating ocean plastic waste and is a tangible action in the transition toward more sustainable production models (Parley, 2025).

Parley itself recognizes the Stella McCartney brand as "deeply rooted in its commitment as a modern and responsible company" (Parley, 2025), reflecting the growing convergence between environmental ethics and innovation within contemporary fashion design.

b) Thematic and Comparative Analysis

The following information can be highlighted about the co-branding between Stella McCartney and Parley for the Oceans:

- Purpose of the alliance: To transform plastic waste extracted from the oceans into sustainable materials for the production of luxury fashion items. This collaboration aims to reduce the use of virgin materials, raise awareness of marine pollution, and promote a circular economy within the textile industry.
- Communication action: The campaign has been promoted through global platforms such as fashion shows, specialized media (Forbes, Vogue), social media, and collaborations with well-known brands like Adidas. The ethical and sustainable focus of the project has been emphasized, using direct testimonies from Stella McCartney and visual materials showcasing the recycling process.
- Reception by the press and consumers: The initiative has been well received by both the media and the public, highlighting the brand's genuine commitment to sustainability. Specialized media have praised McCartney's consistency with her environmental values, and environmentally conscious consumers appreciate the effort to offer ethical products without compromising design.
- Impact on brand image: The collaboration has strengthened Stella McCartney's identity as a pioneer in sustainable fashion. The partnership with Parley has reinforced her positioning as an ethical and environmentally responsible brand, enhancing her prestige in the conscious luxury sector. Parley, in turn, has benefited by being recognized as an environmental organization supported by luxury brands like Stella McCartney and Adidas, gaining significant visibility for their work.
- Associated risks and mitigation strategies: Risks include perceived greenwashing and dependency on recycled materials in high-demand contexts. Transparency in production processes, material traceability, and collaboration with accredited organizations like Parley have been key strategies to mitigate these risks.
- Contribution to the SDGs:

  -SDG 12 (Responsible consumption and production): through the reuse of polluting materials recovered from the oceans.

  -SDG 13 (Climate action): Caring for the oceans significantly contributes to the fight against climate change.

  -SDG 14 (Life below water): Through waste reduction, innovation in materials, and marine environmental awareness.

c) Critical Reflection and Theoretical Integration

The collaboration between Stella McCartney and Parley for the Oceans represents an exemplary case of how the creative industries—particularly luxury fashion—can become active agents of environmental change. In a global context marked by climate urgency and the ocean crisis, this alliance goes beyond the commercial logic of traditional co-branding to stand as a concrete expression of corporate responsibility and ethical commitment.

Far from being a superficial green marketing strategy, this partnership has demonstrated consistency between the values and practices of both entities. Stella McCartney, known for her cruelty-free approach and refusal to use fur or leather, finds in Parley a like-minded partner to strengthen her vision of sustainable fashion. For its part, Parley gains access to a high-impact global platform from which it can amplify its message and expand its upcycled recycling practices.

The value of this alliance lies not only in the products it creates—which combine aesthetics, functionality, and sustainability—but also in its ability to reshape social imaginaries around luxury and responsible consumption. By repurposing marine plastic waste into desirable objects, this initiative turns an environmental problem into a creative solution, fostering greater public awareness of ocean pollution and the possibilities of regenerative design.

Ultimately, the alliance between Stella McCartney and Parley shows that it is possible to combine innovation, profitability, and sustainability without compromising ethical integrity. It stands as a replicable model that encourages other brands to take an active role in facing global environmental challenges, and underscores that sustainability is no longer an option, but an urgent and inescapable necessity.

4.4. Watches with a Social Cause: IWC x Laureus Sport for Good Foundation

Lastly, a representative example of initiatives aimed at social sustainability within the luxury sector is presented. While corporate social responsibility (CSR) has gained relevance in recent years, within the realm of fashion and luxury it is often subordinated to environmental sustainability strategies, despite its significance (McCormick & Ram, 2022). As previously discussed, social sustainability encompasses multiple dimensions. Firstly, it involves respect for labor rights and the dignity of workers, which requires rigorous corporate policies and effective auditing systems, especially in contexts of outsourced production. This aspect, however, remains a topic of debate, as outsourcing is often driven by cost-cutting strategies, leading to operations in countries where labor conditions and workers' rights are poorly regulated or weakly enforced. It also entails transparency regarding the origin of products, working conditions, and manufacturing processes for consumers (Tai & Chuang, 2014). Moreover, companies act as agents of development in the communities where they operate, and thus must contribute to local development (Tai & Chuang, 2014).

As an example of co-branding for social sustainability, we present the case of the luxury watch brand IWC and the Laureus Sport for Good Foundation.

a) Documentary Collection and Preliminary Analysis

The Swiss brand IWC Schaffhausen collaborates with Laureus Sport for Good, an organization that supports vulnerable youth through sport. This collaboration is based on the design of limited-edition watches bearing the joint insignia, aiming to raise awareness of various social causes. A portion of the profits from these limited editions is allocated to social programs (IWC, 2025).

Founded in 1868, IWC Schaffhausen is one of Switzerland's most prestigious watchmaking houses, known for combining precision engineering with classic design. In recent decades, the brand has evolved toward a stronger commitment to environmental and social sustainability, becoming a benchmark for responsible luxury in the watchmaking sector (Laureus, 2025). Laureus Sport for Good Foundation, on the other hand, is a non-profit organization that uses sport as a tool for integration and social development. The initiative focuses on disadvantaged children and adolescents, who often receive visits from internationally renowned athletes at the foundation's academy (Laureus, 2025).

Since the creation of Laureus Sport for Good, more than six million children and young people have participated in over 250 projects across 40 countries—programs that use sport to promote social change and fight violence and discrimination. All programs supported by Laureus address at least one of the following six social issues, aligned with the United Nations Sustainable Development Goals: health, education, women and girls, employability, inclusive society, and peaceful society (IWC, 2025).

Since 2006, IWC has partnered with the Laureus Sport for Good Foundation, reflecting a clear commitment to social sustainability. Each year, IWC releases an exclusive watch with a special feature: in addition to a distinctive blue dial, the back of the case is engraved with the winning drawing from an international children's art contest organized by Laureus. Part of the sales revenue is used to fund projects in over 50 countries focused on violence prevention, gender equality, social inclusion, and education. The campaigns highlight real stories of children who have benefited from the programs, connecting luxury with purpose (IWC, 2025; Laureus, 2025).

In addition to sales, the brand also organizes auctions of special watches, with all proceeds donated to the foundation. In one such auction, it donated as much as €234,671 (El Periódico, 2024).

b) Thematic and Comparative Analysis

- Purpose of the alliance: To promote social sustainability through the use of sport as a tool for transformation and inclusion for vulnerable children and youth. Through co-branding, the aim is to raise awareness of social causes and generate funds for community impact programs.
- Communication actions:

  -Annual watch launch campaigns featuring storytelling centered on real stories of beneficiaries.

  -Use of sports ambassadors and celebrities as spokespeople.

  -Children's drawing contests that directly involve kids in the brand narrative.

  -Features in media outlets such as Vogue, Forbes, and social media with high visual and media coverage.

  -Charity auctions with broad promotion and transparent reporting.
- Reception by the press and consumers: The initiative has generally been well received. Luxury and CSR-focused media highlight the authenticity and coherence of the project. According to Business of Fashion and Forbes articles, the commitment of a luxury brand to tangible social sustainability efforts is positively viewed. Consumers appreciate the opportunity to purchase high-end products with a clear social purpose, which enhances the emotional value of the product.
- Impact on brand image: The impact has been notably positive. IWC has managed to position itself not only as a technical and aesthetic benchmark in watchmaking but also as a conscious and committed brand. This alliance has strengthened its identity as an active player in sustainability, attracting a younger, socially aware audience and increasing loyalty among existing customers.
- Associated risks and mitigation strategies:

  -Greenwashing/perceived opportunism: A mitigation strategy includes transparency regarding the allocation of funds, partnership with a reputable social such as Laureus and involving the beneficiary community in the brand narrative.

  -Lack of tangible or measurable outcomes for the brand: The main benefit remains a positive brand image.
- Contribution to the SDGs:

  -SDG 3: Good health and well-being – by promoting healthy activities like sports.

  -SDG 4: Quality education – through the establishment of schools in disadvantaged areas.

  -SDG 5: Gender equality – many of the joint initiatives aim to promote the inclusion and empowerment of women and girls.

-SDG 8: Decent work and economic growth – contributing to local development.

-SDG 10: Reduced inequalities

c) Critical Reflection and Theoretical Integration

The alliance between IWC Schaffhausen and Laureus Sport for Good can be understood as a form of purpose-driven co-branding, where the prestige and symbolic capital of a luxury brand are combined with the moral legitimacy and social mission of a non-profit organization.

According to the classical perspective of CSR (Carroll, 1991), companies have economic, legal, ethical, and philanthropic responsibilities. IWC's collaboration clearly aligns with this last dimension: its actions are neither mandated by law nor necessarily profitable in the short term, but they contribute to social well-being. The brand takes on an active role in global causes that go beyond its immediate commercial interests, thereby strengthening its legitimacy as a responsible company.

IWC incorporates elements of ethical consumption as a differentiation strategy. By partnering with Laureus, the brand transforms a luxury object into a carrier of social meaning, appealing to consumers who seek not only exclusivity but also values. This trend reflects the growing consumer pressure for brands with purpose.

5. CONCLUSIONS

The research presented in this chapter offers meaningful insight into ethical and sustainable co-branding within the luxury sector, demonstrating that such strategies are not only possible but also desirable from both a reputational and strategic perspective. The alliances analyzed—Burberry x The RealReal, Chantecaille x Space for Giants, Stella McCartney x Parley for the Oceans, and IWC x Laureus—help identify a set of common patterns that point toward a new paradigm in which luxury is redefined through a commitment to social and environmental responsibility.

It becomes evident that co-branding in the context of luxury acts as a facilitator for integrating ethical competencies that a single brand, due to its trajectory or structural limitations, may not credibly adopt on its own. In all four cases studied, these collaborations have enabled brands to expand their audiences, reinforce the authenticity of their brand narrative, and access symbolic resources from NGOs or specialized organizations. Partnering with entities that possess strong ethical legitimacy—as in the case of Chantecaille and Space for Giants—enhances the brand's symbolic value and fosters a deeper emotional connection with consumers, aligning consumption with activism.

It can be concluded that the key variables for achieving ethical co-branding in the luxury sector are alignment and sustainable innovation, as illustrated in the model proposed in Figure 1. Moreover, variables such as greenwashing represent a significant risk to maintaining a positive image and reputation within this industry.

Figure 1. Ethical co-branding model in luxury industry

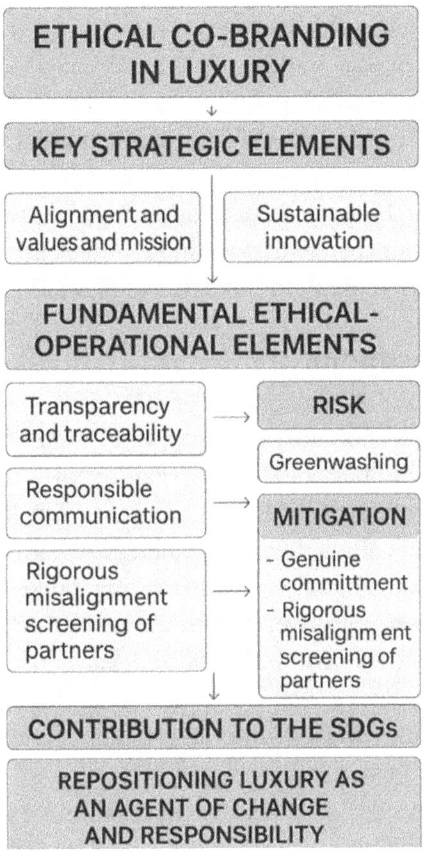

Source. Own elaboration

One of the most significant findings of this study is co-branding's potential to address—or at least mitigate—long-standing tensions in the luxury industry: the contradiction between exclusivity and sustainability, between image and authenticity, between opulence and responsibility. Far from being an obstacle, these dilemmas become opportunities to redefine the value of luxury as a cultural good that reflects the contemporary values of new generations.

However, the picture is not entirely positive. While the effectiveness of these alliances is evident, the chapter also identifies important ethical and strategic challenges that brands must consider. Consistency between a brand's stated values and its actual practices is a prerequisite for the success of any sustainable co-branding initiative. The strategies analyzed show that when this consistency is perceived, the reputational and commercial benefits are substantial. However, when signs of inconsistency emerge—as in the case of Burberry, which had previously been criticized for destroying unsold inventory—the risk of greenwashing can erode credibility and provoke backlash.

Another key challenge is ensuring equity in the relationship between luxury brands and their sustainable or social partners. There is an underlying risk of asymmetry in these collaborations, where the dominant brand may appropriate the symbolic value of its partner without establishing a fair relationship. To prevent this, it is essential to foster horizontal and transparent partnerships, with clear commitments regarding the distribution of benefits, mutual visibility, and recognition of contributions.

5.1. Practical Implications

This study draws implications for marketing, innovation, and sustainability professionals, suggesting that co-branding strategies can be powerful tools for accelerating ethical transformation processes in luxury companies. The key is to select strategic partners whose mission is aligned with the values they wish to project, and to design collaborations that are not only visually striking but also structurally coherent.

From a management perspective, it is recommended to incorporate social and environmental impact metrics into the evaluation of partnership performance, beyond traditional indicators of sales or awareness. This allows for building solid narratives for demanding stakeholders and avoiding superficial practices that compromise the project's legitimacy.

Likewise, it is suggested to strengthen internal and external communication systems, integrating principles of transparency and traceability at every stage of the collaborative process. Digital tools, such as blockchain or external verification platforms, can play a key role in this regard.

5.2. Limitations of the Study and Future Lines of Research

Like all qualitative exploratory research, this study does not aim to generalize its findings to the entire universe of luxury brands, but rather to contribute to the understanding of emerging practices in a changing field. The richness and diversity of the selected cases will allow us to extract lessons transferable to similar contexts.

Based on this theoretical and empirical analysis, several lines of future research emerge that could enrich the study of co-branding in the high-end fashion sector with a focus on sustainability. First, it would be pertinent to delve deeper into longitudinal studies that evaluate the real impact of these collaborations in environmental, social, and economic terms, beyond the brand narrative. Likewise, comparative research across different regions or consumer cultures could offer a more nuanced view of how sustainability is perceived in the context of co-created luxury.

Furthermore, the insights derived from these case studies could serve as a foundation for the development of quantitative instruments aimed at measuring consumer perceptions of ethical legitimacy, brand authenticity, and trust in sustainability-related co-branding. Surveys or experimental designs could be used to test whether associations with nonprofit organizations effectively enhance brand equity, or to validate the proposed theoretical connections between symbolic value and consumer loyalty.

Another promising avenue is to analyze the role of artificial intelligence and digital traceability in strengthening ethical practices within strategic alliances. Finally, incorporating qualitative methodologies focused on consumer experience—such as in-depth interviews—would allow for exploring how symbolic value and ethical legitimacy are constructed in these emerging forms of sustainable co-branding.

REFERENCES

Achabou, M. A., & Dekhili, S. (2013). Luxury and sustainable development: Is there a match? *Journal of Business Research*, *66*(10), 1896–1903. DOI: 10.1016/j.jbusres.2013.02.011

Bain & Company. (2020). *The luxury goods market: Outlook 2020 and beyond.* https://www.bain.com

Baxter, P., & Jack, S. (2008). Qualitative case study methodology: Study design and implementation for novice researchers. *The qualitative report*, *13*(4), 544–559.

BBC. (2018, July 19). Burberry burns bags, clothes and perfume worth millions. *BBC News*. https://www.bbc.com/news/business-44885983

Becker-Leifhold, C., & Iran, S. (2018). Collaborative fashion consumption–drivers, barriers and future pathways. *Journal of Fashion Marketing and Management*, *22*(2), 189–208. DOI: 10.1108/JFMM-10-2017-0109

Brooks, A., Fletcher, K., Francis, R. A., Rigby, E. D., & Roberts, T. (2017). Fashion, sustainability, and the anthropocene. *Utopian Studies*, *28*(3), 482–504. DOI: 10.5325/utopianstudies.28.3.0482

Brynjolfsson, E., & McAfee, A. (2014). *The second machine age: Work, progress, and prosperity in a time of brilliant technologies.* W. W. Norton & Company.

Burberry (2025, April 23). Burberry and The RealReal join forces to make fashion circular. https://www.burberryplc.com/news/sustainability/2019/burberry-and-the-realreal-join-forces-to-make-fashion-circular

Caniato, F., Caridi, M., Crippa, L., & Moretto, A. (2012). Environmental sustainability in fashion supply chains: An exploratory case based research. *International Journal of Production Economics*, *135*(2), 659–670. DOI: 10.1016/j.ijpe.2011.06.001

Carrington, M. J., Neville, B. A., & Whitwell, G. J. (2010). Why ethical consumers don't walk their talk: Towards a framework for understanding the gap between the ethical purchase intentions and actual buying behaviour of ethically minded consumers. *Journal of Business Ethics*, *97*(1), 139–158. DOI: 10.1007/s10551-010-0501-6

Carroll, A. B. (1991). The pyramid of corporate social responsibility: Toward the moral management of organizational stakeholders. *Business Horizons*, *34*(4), 39–48. DOI: 10.1016/0007-6813(91)90005-G

Chantecaille. (2025, April 15). *Space for Giants.* https://chantecaille.com/pages/space-for-giants

Chloé. (2021). Chloé becomes the first luxury fashion Maison to receive B Corp certification. https://www.chloe.com

Comert, O. (2020). *Blockchain revolution: how the technology behind bitcoin and other cryptocurrencies is changing the world.* Hein Online.

Crane, A., Matten, D., Glozer, S., & Spence, L. J. (2019). *Business ethics: Managing corporate citizenship and sustainability in the age of globalization.* Oxford university press. DOI: 10.1093/hebz/9780198810070.001.0001

Csaba, F. F., & Skjold, E. (2018). Fur and sustainability: Oxymoron or key to 'deeper' luxury? *Journal of Design. Business & Society, 4*(2), 131–149.

Delmas, M. A., & Burbano, V. C. (2011). The drivers of greenwashing. *California Management Review, 54*(1), 64–87. DOI: 10.1525/cmr.2011.54.1.64

Deloitte. (2023). *Global Powers of Luxury Goods 2023.* https://www.deloitte.com/global/en/Industries/consumer/analysis/gx-cb-global-powers-of-luxury-goods.html

Eisenhardt, K. M. (1989). Building theories from case study research. *Academy of Management Review, 14*(4), 532–550. DOI: 10.2307/258557

El Periódico. (2024, April 2). El cronógrafo IWC de Lewis Hamilton recauda más de 230.000 euros para la Fundación Laureus. https://gentleman.elperiodico.com/estilo/el-cronografo-iwc-de-lewis-hamilton-recauda-mas-de-230-0000-euros-para-la-fundacion-laureus/

European Commission. (2022). *EU Strategy for Sustainable and Circular Textiles.* https://ec.europa.eu

Forbes (2017). La moda lucha por los océanos: Stella McCartney se alía con Parley for the Oceans. https://forbes.com.mx/forbes-life/stella-mccartney-parley-for-the-oceans/

Forbes (2019). What Is Really Driving The RealReal's New Partnership With Burberry. https://www.forbes.com/sites/pamdanziger/2019/10/07/circular-fashion-is-not-whats-most-important-about-burberry-partnering-with-the-realreal/

Grappe, C. G., Lombart, C., Louis, D., & Durif, F. (2021). "Not tested on animals": How consumers react to cruelty-free cosmetics proposed by manufacturers and retailers? *International Journal of Retail & Distribution Management, 49*(11), 1532–1553. DOI: 10.1108/IJRDM-12-2020-0489

Heath, J., & Potter, A. (2005). *The rebel sell: Why the culture can't be jammed.* HarperCollins.

Helmig, B., Huber, J. A., & Leeflang, P. S. (2008). Co-branding: The state of the art. *Schmalenbach Business Review*, *60*(4), 359–377. DOI: 10.1007/BF03396775

Henninger, C. E., Alevizou, P. J., & Oates, C. J. (2016). What is sustainable fashion? *Journal of Fashion Marketing and Management*, *20*(4), 400–416. DOI: 10.1108/JFMM-07-2015-0052

IWC. (2025, April 20). *Laureus Sport for Good.* https://www.iwc.com/es/es/company/partnerships/laureus-sport.html

Joy, A., Sherry, J. F.Jr, Venkatesh, A., Wang, J., & Chan, R. (2012). Fast fashion, sustainability, and the ethical appeal of luxury brands. *Fashion Theory*, *16*(3), 273–295. DOI: 10.2752/175174112X13340749707123

Kapferer, J. N., & Bastien, V. (2012). *The luxury strategy: Break the rules of marketing to build luxury brands*. Kogan Page Publishers.

Kapferer, J. N., & Michaut-Denizeau, A. (2017). Is luxury compatible with sustainability? Luxury consumers' viewpoint. In *Advances in Luxury Brand Management* (pp. 123–156). Springer. DOI: 10.1007/978-3-319-51127-6_7

Kapferer, J. N., & Michaut-Denizeau, A. (2019). Are millennials really more sensitive to sustainable luxury? A cross-generational international comparison of sustainability consciousness when buying luxury. *Journal of Brand Management*, *26*, 123–138.

Karypidis, M., Tarnanidis, T., & Papachristou, E. (2024). Treating Textile Effluents for Sustainable Fashion and Green Marketing. In *Driving Green Marketing in Fashion and Retail* (pp. 166–179). IGI Global. DOI: 10.4018/979-8-3693-3049-4.ch010

Kaur, U., & Siddhey, P. K. (2024). The Influence of Digital Transformation on the Retail Value Chain. In *Reshaping Marketing Science in Wholesaling and Retailing* (pp. 334–353). IGI Global. DOI: 10.4018/979-8-3693-6145-0.ch014

Kenyans.co.ke. (2025, April 10). *Chantecaille Foundation: Details of make-up company that just donated Land Cruisers worth Ksh 200M to KWS*. https://www.kenyans.co.ke/news/108862-chantecaille-foundation-details-make-company-just-donated-landcruisers-worth-ksh-200m

Kering. (2021). *Sustainability progress report.* https://www.kering.com

Kozlowski, A., Searcy, C., & Bardecki, M. (2015). Corporate sustainability reporting in the apparel industry: An analysis of indicators disclosed. *International Journal of Productivity and Performance Management*, *64*(3), 377–397. DOI: 10.1108/IJPPM-10-2014-0152

Kunz, J., May, S., & Schmidt, H. J. (2020). Sustainable luxury: Current status and perspectives for future research. *Business Research, 13*(2), 541–601. DOI: 10.1007/s40685-020-00111-3

Laureus (2025). *Sporting legends launch IWC watch design to raise funds for Laureus.* https://www.laureus.com/news/sporting-legends-launch-iwc-watch-design-to-raise-funds-for-laureus

LVMH. (2020). [*Our sustainability program.* https://www.lvmh.com]. *Life (Chicago, Ill.), 360*, •••.

Manda, V. K., Sagi, S., & Yadav, A. (2024). Blockchain in advertising and marketing: Revolutionizing the industry through transparency and trust. In *New trends in marketing and consumer science* (pp. 89–112). IGI Global. DOI: 10.4018/979-8-3693-2754-8.ch005

McCormick, H., & Ram, P. (2022). 'Take a stand': The importance of social sustainability and its effect on Generation Z consumption of luxury fashion brands. In Gardetti, M., & Torres, A. J. (Eds.), *Sustainable luxury: An international perspective* (pp. 219–239). Springer International Publishing. DOI: 10.1007/978-3-031-06928-4_11

McKinsey & Company. (2020). *The state offashion 2020.* https://www.mckinsey.com

Melé, D. (2003). The challenge of humanistic management. *Journal of Business Ethics, 44*(1), 77–88. DOI: 10.1023/A:1023298710412

Mohtar, M., & Taha, A. Z. (2024). Case 12: From scraps to style—The art of Hermès' upcycling journey with "Petit H". In *Marketing case studies in emerging markets: Contemporary multi-industry issues and best-practices* (pp. 163–176). Springer Nature Switzerland.

Muret, D. (2024, March 21). *La percepción del lujo atraviesa un profundo cambio impulsada por la sostenibilidad y la tecnología.* FashionNetwork.com. https://es.fashionnetwork.com/news/La-percepcion-del-lujo-atraviesa-un-profundo-cambio-impulsada-por-la-sostenibilidad-y-la-tecnologia,1648338.html

Niinimäki, K. (2018). *Sustainable fashion in a circular economy.* Aalto University.

Osburg, V. S., Davies, I., Yoganathan, V., & McLeay, F. (2021). Perspectives, opportunities and tensions in ethical and sustainable luxury: Introduction to the thematic symposium. *Journal of Business Ethics, 169*(2), 201–210. DOI: 10.1007/s10551-020-04487-4

Parley. (2025). *Where creators, thinkers and leaders collaborate for the oceans.* https://parley.tv/

Perry, P., & Towers, N. (2013). Conceptual framework development: CSR implementation in fashion supply chains. *International Journal of Physical Distribution & Logistics Management, 43*(5/6), 478–501. DOI: 10.1108/IJPDLM-03-2012-0107

PETA. (2020). *Burberry and Chanel ban fur.* https://www.peta.org

PETA. (2025). *Beauty without bunnies.* https://crueltyfree.peta.org/

Prada Group. (2020). *Re-Nylon: A sustainable future.* https://www.pradagroup.com

Quiles-Soler, C., Martínez-Sala, A. M., & Monserrat-Gauchi, J. (2023). Fashion industry's environmental policy: Social media and corporate website as vehicles for communicating corporate social responsibility. *Corporate Social Responsibility and Environmental Management, 30*(1), 180–191. DOI: 10.1002/csr.2347

Rao, A. R., & Ruekert, R. W. (1994). Brand alliances as signals of product quality. *Sloan Management Review, 36*(1), 87–87.

Silverstein, M. J., & Fiske, N. (2003). Luxury for the masses. *Harvard Business Review, 81*(4), 48–57. PMID: 12687919

Stake, R. E. (1995). *The art of case study research.* SAGE Publications.

Stella McCartney. (2025). *Social responsibility & human rights.* https://www.stellamccartney.com/es/en/sustainability/social-sustainability.html

Tai, F. M., & Chuang, S. H. (2014). Corporate social responsibility. *Ibusiness, 6*(3), 117–130. DOI: 10.4236/ib.2014.63013

Tapscott, D., & Tapscott, A. (2017). *Blockchain revolution: How the technology behind bitcoin is changing money, business, and the world.* Penguin.

Tarnanidis, T., Papachristou, E., Karypidis, M., & Manda, V. K. (2025). Luxury fashion consumption and branding in the affluent market. In Tarnanidis, T., & Gardetti, M. (Eds.), *Cross-cultural perspectives on luxury marketing* (pp. 161–172). IGI Global.

Varan, N. (2020). *The effect of brand equity on intrinsic and extrinsic motivations of consumers for luxury consumption tendency* (Master's thesis).

Vimalnath, V., Karthick, K. K., Ganesan, M., & Veluchamy, R. (2025). Understanding cross-cultural marketing dynamics. In Tarnanidis, T., & Gardetti, M. (Eds.), *Cross-cultural perspectives on luxury marketing* (pp. 45–66). IGI Global.

Vogue Business. (2019, November 5). *Burberry's partnership with The RealReal signifies a real shift.* https://www.voguebusiness.com/companies/burberrys-partnership-realreal-secondhand

Yin, R. K. (2018). *Case study research and applications: Design and methods* (6th ed.). SAGE Publications.

KEY TERMS AND DEFINITIONS

Circular economy: An economic model that seeks to minimize waste and maximize the reuse of resources, promoting product durability, repair, recycling, and reuse. In the luxury sector, the circular economy manifests itself through initiatives such as the reuse of materials and extending product lifespans.

Co-branding: Marketing strategy in which two or more brands partner to create a joint offering, combining their competencies and values to broaden their reach and generate a shared value proposition. In the context of luxury, co-branding can be integrated with sustainable and ethical practices to strengthen the image of both brands.

Conscious luxury: An emerging concept in the luxury industry that links traditional exclusivity with sustainable and responsible practices. Conscious luxury seeks to integrate sustainability, ethics, and environmental respect without compromising the quality or status of the products.

Corporate Social Responsibility (CSR): A business approach that integrates actions into its strategy that promote social, ethical, and environmental well-being, beyond legal obligations. In the luxury sector, high-end brands are increasingly incorporating CSR as an integral part of their value proposition.

Greenwashing: A deceptive practice in which a company claims to have environmentally friendly or sustainable credentials to enhance its image, while in reality it does not implement genuine environmental practices. In the context of luxury, greenwashing can pose a significant reputational risk if brands are not transparent and sincere in their sustainability commitments.

Sustainable Development Goals (SDGs): A set of 17 global goals established by the UN to address issues such as poverty, inequality, climate change, environmental degradation, peace, and justice. Luxury brands that embrace ethical and sustainable principles align their strategies with these SDGs.

Chapter 6
Advanced Co-Branding Strategies for Smart Luxury Products:
A Multi-Criteria Decision-Making Approach Using Neutrosophic Sets

Ajoy Kanti Das
https://orcid.org/0000-0002-9326-1677
Tripura University, India

Carlos Granados
https://orcid.org/0000-0002-7754-1468
Universidad de Antioquia, Colombia

Nandini Gupta
https://orcid.org/0009-0000-5450-9783
Bir Bikram Memorial College, India

Bijoy Krishna Debnath
https://orcid.org/0000-0001-7215-9473
Tezpur University, India

Takaaki Fujita
Independent Researcher, Japan

Rakhal Das
ICFAI University, Tripura, India

Suman Das
National Institute of Technology, Agartala, India

ABSTRACT

The convergence of luxury branding and smart technology has led to a new paradigm in strategic co-branding. This chapter introduces a Multi-Criteria Decision-Making (MCDM) framework that integrates Neutrosophic Sets to address uncertainty, hesitation, and conflicting expert opinions in evaluating smart luxury co-branding partnerships. Traditional decision-making models often fail to capture the complexities

DOI: 10.4018/979-8-3373-3010-5.ch006

of consumer perception, technological integration, and market acceptance, which involve varying degrees of truth, indeterminacy, and falsity. Neutrosophic MCDM enhances decision robustness by incorporating Neutrosophic Weighted Aggregation Operators and Similarity Measures to quantify brand synergy and compatibility. Case studies, including collaborations between luxury brands and technology firms, illustrate the framework's application in selecting optimal co-branding partners. The findings demonstrate that the Neutrosophic approach provides a more nuanced and data-driven methodology for evaluating strategic alliances in the smart luxury sector, ensuring informed MCDM.

1. INTRODUCTION

The luxury industry has been greatly impacted by the quick development of smart technology, which has resulted in the creation of smart luxury goods that skillfully combine contemporary innovation with exclusivity, craftsmanship, and legacy (Liu, 2023; Rathi, 2022). Luxury companies are using co-branding tactics to work with tech companies as consumers' demands for both technological sophistication and brand prestige grow (Javornik, 2021; Huang, 2023; Tang, 2023). These collaborations seek to improve brand equity, differentiate products, and develop a distinctive market offering that appeals to both tech-savvy and traditional luxury consumers. In an increasingly digital and interconnected world, Smart luxury items, which combine cutting-edge technology with exquisite craftsmanship, have become a new frontier of innovation (Milanesi et al., 2022; Liu, 2023; Remy et al., 2023). To maintain their competitive edge and appeal to discriminating customers, brands in this market must carefully manage alliances (Halwani, 2020; Xu, 2020). Co-branding, which is a strategic partnership between two brands, has been shown to be a successful strategy for increasing technology capabilities, market reach, and brand equity (Anuar et al., 2021; Fakhravar & Tahami, 2022).

The most recent advancements in robotics are examined by Manda et al. (2025a), who highlight sophisticated robotic applications that improve automation and efficiency. Similar to this, Manda et al. (2025b) talk about the future of virtual reality (VR), emphasizing immersive experiences and how they can be used in practical situations. These studies highlight how VR and robotics have the potential to revolutionize a variety of industries. The function of multiple-criteria decision-making (MCDM) in managerial and marketing decision-making is examined by Tarnanidis et al. (2025c, 2025d). Their works apply statistical and decision-making models to enhance marketing strategies, particularly in digital and metaverse-based marketing. Furthermore, Tarnanidis (2024b) and Tarnanidis & Gkiouzepas (2024) examine consumer experiences and decision-making processes in the metaverse, providing insights into how immersive technologies shape purchasing behaviors. In their exploration of sustainable practices in the fashion and retail sectors, Tarnanidis et al. (2025a, 2025b) pinpoint slow fashion consumption trends in wealthy markets. Treating textile effluents for sustainable fashion is covered in more detail by Karypidis et al. (2024), which supports green marketing tactics. Drakou et al. (2024) draw attention to the effects of AI-driven 3D design on sustainability, creativity, and fashion education. In their analysis of blockchain's application to marketing and advertising, Manda et al. (2024) highlight how it can improve trust and transparency. Similar to this, Khaliq & Manda (2024) investigate customer experience in the Web 3.0 era, focusing on how blockchain and the metaverse interact to shape digital interactions. In his compilation of cross-cultural viewpoints on luxury marketing, Tarnanidis (2025) discusses branding tactics and consumer behavior in upscale markets. Tarnanidis et al. (2025e) examine luxury fashion consumption and its effects on international markets in other studies. These pieces shed light on how various cultural groups engage with luxury brands and make decisions about what to buy. In their discussion of the development of smart cities, Jitta et al. (2025) emphasize the significance of geospatial technologies for infrastructure management and urban planning. These technologies make it easier to make data-driven decisions that will enhance sustainability and city services. Tarnanidis (2024a) examines how mobile marketing affects consumer behavior, focusing on digital tactics that improve customer interaction. In addition, Tarnanidis et al. (2023b) investigate the impact of social media on consumer behavior in the fashion industry, and Tarnanidis et al. (2023c) expand on this topic to include the grocery and food industries. In their discussion of knowledge management procedures in the public sector, Ismyrlis et al. (2020) place a strong emphasis on how information systems can improve decision-making. For the analysis of milk supply chains, Tsaples & Tarnanidis (2020) offer a system dynamics model that provides important insights into supply chain efficiency and logistics.

Research has also focused on visual communication and advertising; Gkiouzepas and Tarnanidis (2024) conceptualize the extreme claims made in advertising and how they affect consumer perception. According to Tarnanidis and Sklavounos (2024), who edited a volume on new marketing trends and consumer science and provided insights into modern consumer behavior, this is in line with more general trends in marketing science. Tarnanidis et al. (2024a) have thoroughly examined the relationship between green marketing and retail fashion, highlighting sustainable marketing tactics. In a similar vein, Tarnanidis et al. (2024b) investigated cross-cultural consumer influences by looking at how young consumers react to both positive and negative framing in marketing messages. With an emphasis on the fashion industry, Tarnanidis et al. (2023a) have investigated the role of social media in consumer decision-making. Tarnanidis et al. (2023e) have also examined the entrepreneurial potential of young consumers, focusing on educational interventions to promote entrepreneurship. Kampakaki et al. (2022) examined the digital credentials needed in the fashion industry and evaluated key competencies for workers in the field. In parallel, Manda et al. (2022) have examined the function of blockchain in asset management, emphasizing its revolutionary potential in financial operations. Another important area of study has been sustainability in entrepreneurship; Tarnanidis (2022) created a framework for evaluating sustainable entrepreneurship in the food industry. Collectively, these studies contribute to a deeper understanding of contemporary marketing strategies, consumer behavior, and decision-making methodologies, offering valuable insights for both academia and industry practitioners. However, choosing the best co-branding partner is a difficult option that calls for weighing various frequently incompatible factors.

AHP, TOPSIS, and VIKOR are examples of traditional MCDM techniques that offer structured methods for ranking options (Wittstruck & Teuteberg, 2012; Büyüközkan & Görener, 2015; Dey et al., 2016; Kamalakannan et al., 2020). The field of marketing and managerial decision-making has been significantly influenced by MCDM methods, as highlighted by Tarnanidis et al. (2025f), who reviewed the applications of MCDM in marketing, emphasizing its practical implications and future research directions. Tarnanidis et al. (2023d) reviewed MCDM approaches in marketing applications, emphasizing the PROMETHEE method's usefulness in marketing for the food and grocery industries. Tarnanidis (2023) investigated reference points in purchasing decisions in order to further investigate consumer decision-making processes. Furthermore, the SERVQUAL approach has been thoroughly examined in the field of education. Tarnanidis and Tarnanidis (2024) examined service quality disparities in senior high schools and identified important variables affecting academic achievement. Tarnanidis (2024c) has studied consumer behavior, specifically in response to crises like the COVID-19 pandemic. She looked at the reactions of fashion consumers and their long-term effects. Additionally, Tarnanidis (2024d, 2024e) has thoroughly examined the changing wholesale and retailing landscape, offering insights into new marketing tactics and industry challenges.

To improve the robustness and dependability of co-branding strategy selection in the smart luxury sector, this chapter presents a novel approach that incorporates neutrosophic sets (Smarandache, 1998; Smarandache, 2005) into the MCDM framework. The ambiguity and uncertainty included in expert opinions, customer preferences, and market dynamics, however, are difficult for these traditional approaches to convey (Pramanik et al., 2018; Pramanik et al., 2023). A significant contribution by Das and Granados (2022a) involves the development of a fuzzy parameterized intuitionistic fuzzy soft multiset theory for MCDM. Their investigation of FP-intuitionistic multi-fuzzy N-soft sets, which increase the adaptability of MCDM models, complements this work (Das and Granados, 2022b). They presented IFP-intuitionistic multi-fuzzy N-soft sets, further expanding this field of study and proving their usefulness in MCDM situations (Das and Granados, 2022c). Furthermore, Das (2018) suggested weighted fuzzy soft multisets as a novel tool for MCDM, while Das (2016) and Das & Granados (2023) explored intuitionistic fuzzy rough relations and preference intuitionistic fuzzy rough relations, highlighting their partial inclusion properties. Their collaborative work in 2021 also presented a fuzzy soft group MCDM approach based on weighted average ratings, further refining the theoretical and practical aspects of decision analysis (Das and Granados, 2021). Applications-wise, Das et al. (2022) suggested new operations on fuzzy soft sets, demonstrating their usefulness in situations involving MCDM. Similarly, Das et al. (2023) introduced the concept of weighted hesitant bipolar-valued fuzzy soft sets, which provide a more nuanced framework for evaluating alternatives. Das et al.'s (2024) development of a fuzzy MCDM model for examining human influences on urban river water quality is one of their noteworthy contributions to environmental MCDM. This research is extended in 2025 with weighted hesitant fuzzy soft set and fuzzy soft set models for rating water pollution and a case study applying fuzzy soft MCDM model to assess pollution levels in the Haora and Manu Rivers (Das et al., 2025a; 2025b). Granados et al. (2022) provided continuous neutrosophic distributions with neutrosophic parameters, offering a probabilistic approach to uncertainty modeling. In 2023, Granados et al. (2023) proposed a weighted neutrosophic soft multiset approach, demonstrating its application in decision-making problems that require handling multiple levels of uncertainty. Fuzzy soft multi-set theories in decision-making have also benefited greatly from the contributions of Mukherjee and Das (2016a). Their work in 2015 applied interval-valued intuitionistic fuzzy soft multi-sets in real-world MCDM problems (Mukherjee & Das, 2015c), while their research explored the use of interval-valued intuitionistic fuzzy soft sets for investment decision-making (Mukherjee & Das, 2015a). In 2022, they invented Einstein operations for fuzzy soft multisets, further enhancing computing tools for decision analysis (Mukherjee & Das, 2022). Fuzzy soft multisets are thoroughly explored in their 2023 book, which covers both the theoretical underpinnings and real-world applications (Mukherjee & Das, 2023). The relational and topological structures of fuzzy and rough set models were further investigated by Mukherjee & Das (2015b; 2016b), and their 2025 collaboration presents an extended interval-valued neutrosophic rough soft set framework for water quality evaluation (Mukherjee et al. 2025). All things considered, these studies provide significant contributions to the development of fuzzy (Zadeh, 1965), soft, and neutrosophic set theories, which have applications in computational intelligence, environmental sustainability, and MCDM. Their work offers useful approaches for managing uncertainty, enhancing the precision of decisions, and tackling challenging real-world issues.

However, co-branding in smart luxury markets presents unique challenges. While technology brands place more emphasis on innovation, accessibility, and quick development, luxury brands function on the tenets of exclusivity, heritage, and craftsmanship. It takes strategic decision-making that takes into account various variables, such as market alignment, consumer perception, technological viability, and brand synergy, to combine these disparate identities. Because of uncertainty, divergent expert opinions, and changing market trends, traditional decision-making models frequently fall short of capturing the complexities of these collaborations. To address these challenges, this chapter presents a Neutrosophic MCDM framework, allowing decision-makers to assess co-branding opportunities in the face of ambiguity, hesitancy, and contradiction. Neutrosophic Similarity Measures, Neutrosophic Sets, and Neutrosophic Weighted Aggregation Operators are all included in this framework, which offers an organized method for evaluating the viability, dangers, and eventual success of co-branding alliances in the smart luxury market.

A list of important terms that are necessary to comprehend the ideas covered in this chapter is provided below. These terms shed light on the use of Neutrosophic MCDM in strategic co-branding decisions as well as the integration of smart technology with luxury branding.

1. Brand Compatibility: This factor affects the success of a co-branding partnership by determining how well two brands match in terms of values, target market, and market positioning.
2. Co-branding strategy: A marketing tactic wherein two or more brands work together to develop a new good or service, utilizing the advantages of each brand to increase its attractiveness to consumers.
3. Luxury Branding – The practice of positioning and promoting high-end products that emphasize exclusivity, premium quality, and brand prestige.
4. Market Acceptance: How well-known, embraced, and preferred a recently launched co-branded product is by consumers.
5. Multi-Criteria Decision-Making (MCDM): This framework for decision-making assesses several competing criteria to determine the best course of action.
6. Neutrosophic Sets: A mathematical method that provides a more adaptable evaluation model by taking into consideration ambiguity, indeterminacy, and contradictions in decision-making..
7. Neutrosophic Weighted Aggregation Operators – Computational tools used in decision-making to combine and analyze different weighted criteria under a neutrosophic environment.
8. Smart Luxury Products: Expensive goods that incorporate cutting-edge technology to improve functionality, exclusivity, and user experience.

This chapter explores the following key aspects:

1. Understanding how co-branding generates competitive advantage: The foundations of co-branding in smart luxury markets.
2. Smart Luxury Co-Branding Challenges: Analyzing the obstacles to fruitful brand alliances.
3. Neutrosophic MCDM Models: Presenting a strong framework for managing ambiguity in assessments of co-branding.
4. Useful Applications: Using actual cases to illustrate Neutrosophic MCDM's efficacy.
5. Strategic Suggestions and Future Research Paths: Offering perspectives on how companies can maximize co-branding tactics for sustained prosperity.

By using Neutrosophic MCDM in smart luxury co-branding, this study seeks to close the gap between technological innovation and luxury aesthetics by providing a data-driven, uncertainty-aware method for making strategic decisions in a changing digital environment.

2. INTRODUCTION TO CO-BRANDING IN SMART LUXURY GOODS

Co-branding is a strategic marketing alliance where two or more brands collaborate to create a product or service that leverages the strengths and reputations of both brands. Co-branding is essential for fostering innovation, improving market positioning, and providing distinctive customer experiences in the smart luxury market. Co-branding helps companies stand out in a crowded market by fusing the exclusivity and legacy of luxury brands with state-of-the-art technology. High-tech fashion, AI-powered watches, and intelligent cars are examples of smart luxury products that need both technical know-how and a strong luxury brand to be successful. To create a superior value proposition, this calls for strategic alliances between brands that have complementary strengths.

2.1 The Role of Co-Branding in Smart Luxury Goods

Co-branding in smart luxury products fulfills several key functions:

2.1.1 Enhancing Brand Equity and Prestige

Superior craftsmanship, heritage, and exclusivity are the hallmarks of luxury brands. These components work together with innovative technology to produce goods that appeal to tech-savvy and high-end consumers. For instance, the collaboration between Google and TAG Heuer on smartwatch development blends cutting-edge digital features with Swiss watchmaking know-how, enhancing the reputation and legitimacy of both companies.

2.1.2 Expanding Market Reach and Consumer Base

A wide range of consumers who appreciate both luxury and technology are drawn to smart luxury products. Co-branding increases market penetration by giving brands access to each other's consumer bases. A partnership between a smart wearable technology company and a luxury fashion house can appeal to younger, tech-savvy consumers while drawing in wealthy ones.

2.1.3 Driving Product Innovation and Technological Integration

Co-branding makes it easier to create cutting-edge products that raise the bar for the industry by fusing smart features with luxurious looks. Co-branding can produce innovative products that set new industry standards, as demonstrated by luxury automakers working with AI-driven companies on self-driving technology, such as the Mercedes-Benz and NVIDIA partnerships.

2.1.4 Mitigating Risk and Increasing Competitive Advantage

Technological innovations are expensive and dangerous, particularly in the luxury market where exclusivity and superior product quality are crucial. By enabling brands to split R&D expenses while maintaining a competitive edge, co-branding helps to reduce these risks. Luxury brands can access technological know-how by working with tech companies instead of having to build these capabilities internally.

Figure 1. Co-Branding Strategies in Smart Luxury Goods

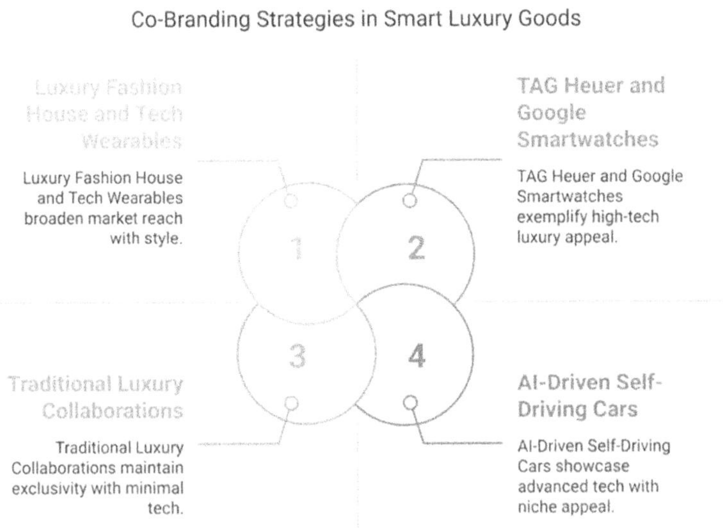

2.2 Key Elements That Affect The Success Of Co-Branding In Smart Luxury Goods

Several critical factors determine the success of co-branding initiatives in the smart luxury segment:

2.2.1 Brand Synergy and Alignment

Both brands must have comparable values, standards for quality, and market positioning for co-branding to be successful. A misalignment can drive away devoted customers and weaken brand identity. For example, unless there is a definite synergy, a luxury brand with a reputation for classic elegance might find it difficult to work with a tech company that prioritizes rapid innovation.

2.2.2 Consumer Perception and Market Acceptance

The success of co-branded smart luxury products is largely dependent on the consumers. The collaboration needs to meet customer expectations and produce perceived added value. The partnership might not succeed if it is perceived as coerced or at odds with the brand's basic values. For instance, the success of Apple and Hermès' partnership on the Apple Watch can be attributed to the smooth fusion of high-end design and innovative technology.

2.2.3 Technological Compatibility and Integration

Without sacrificing traditional luxury features, luxury brands must make sure that the technology they use in their products improves the customer experience. For a product to remain authentic, luxury materials, fine craftsmanship, and intelligent technology must all be seamlessly combined.

2.2.4 Financial and Strategic Considerations

Careful financial planning, including revenue-sharing arrangements, cost control, and investment strategies, is necessary for successful co-branding. The partnership must be profitable and strategic for both brands. The partnership might not last over time if there is no opportunity for both parties to create value.

2.2.5 Intellectual Property and Design Control

The aesthetics and design components of luxury brands are frequently strictly regulated. To avoid disputes and preserve brand integrity, co-branding partnerships must have explicit agreements on intellectual property rights and design control.

2.2.6 Marketing and Brand Communication Strategies

Promoting co-branded smart luxury products requires strong marketing and storytelling techniques. Customers must be made aware of the advantages of collaboration by brands, emphasizing the special fusion of luxury and technology. Exclusive launches, influencer partnerships, and high-end advertising all help make a successful market debut.

2.2.7 Sustainability and Ethical Considerations

Co-branding campaigns need to include sustainable practices because consumers are becoming more conscious of ethical and environmental issues. Partnerships that prioritize ethical sourcing and environmentally friendly technology have a higher chance of winning over customers' trust and loyalty.

In the smart luxury industry, co-branding has enormous potential for market expansion, consumer engagement, and innovation. However, consumer perception, technological integration, financial viability, intellectual property management, strategic marketing, and brand synergy all play a role in its success. By carefully considering these elements, brands can create pioneering smart luxury products that redefine industry standards and drive sustainable growth.

Figure 2. Success Factors in Smart Luxury Co-Branding

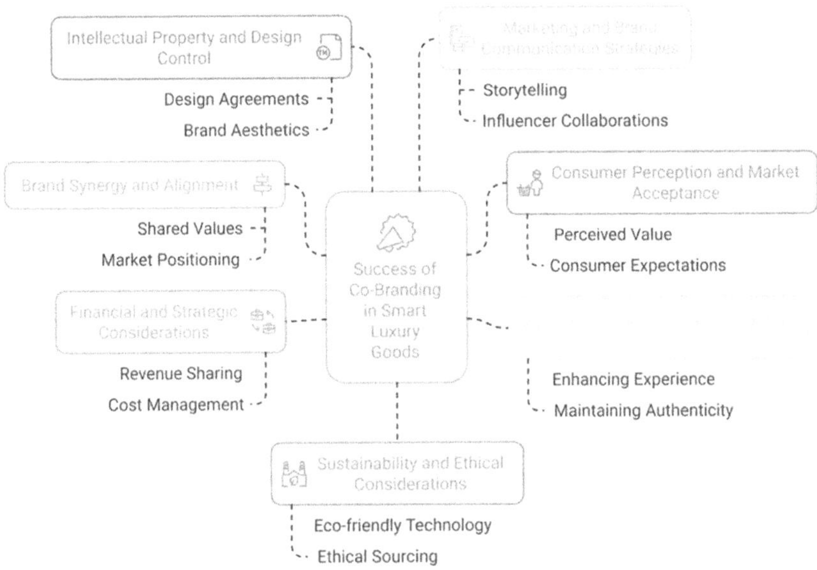

3. EXAMINING HOW CONVENTIONAL MCDM TECHNIQUES FALL SHORT WHEN IT COMES TO HANDLING UNCERTAIN DECISION-MAKING

Multiple conflicting criteria can be evaluated in decision-making processes using structured methodologies provided by MCDM techniques. The Analytical Hierarchy Process (AHP), Technique for Order of Preference by Similarity to Ideal Solution (TOPSIS), and VIKOR are examples of common traditional MCDM techniques. Despite their widespread use in decision science, these methods frequently fail to handle ambiguity, uncertainty, and incomplete data. In dynamic contexts like market analysis, strategic business decisions, and technological innovation, this limitation is especially noticeable.

3.1 Limitations Of Conventional Mcdm Techniques In Uncertain Decision-Making

Conventional MCDM approaches assume that alternatives and decision criteria can be assessed using exact numerical values, which is frequently unfeasible in practical situations. The main drawbacks of traditional MCDM techniques for managing uncertainty are as follows:

3.1.1. Inability to Capture Uncertainty in Expert Opinions

A lot of MCDM techniques ask decision-makers to give exact numerical rankings or values, which might not be feasible when experts have conflicting levels of confidence or insufficient information. For instance, experts may find it difficult to assign fixed scores when assessing co-branding options for smart luxury products because of shifting consumer preferences and market trends.

3.1.2. Rigidity in Preference Assignments

Pairwise comparisons are the foundation of techniques such as AHP, which require decision-makers to specify clear preference values between criteria and alternatives. These set preferences might not accurately represent the complexity of the real world, where people may be unsure or hesitant about their decisions. For instance, a luxury brand choosing between blockchain integration and AI-driven automation may find it challenging to establish precise preference values as a result of rapidly advancing technology.

3.1.3. Inability to Handle Indeterminate and Contradictory Data

Traditional MCDM techniques make the assumption that all available data is either completely known or can be accurately estimated. In practice, decision-makers frequently encounter contradictory information, ambiguous data, or opposing viewpoints that are difficult to handle with conventional MCDM techniques. As an illustration, reports on sustainability-driven co-branding may show that consumers are very interested in luxury eco-friendly products, but they may also express doubts about their long-term worth.

3.1.4. Limited Adaptability to Dynamic Decision Environments

A structured decision matrix at one particular moment is necessary for many static MCDM techniques. However, static models are insufficient because consumer preferences, market conditions, and technological advancements are constantly changing. As an illustration, a business assessing smart clothing technology might discover that, within months, customer preferences change, making the initial rankings of decisions no longer relevant.

3.1.5. Insufficient Representation of Linguistic and Subjective Judgments

Qualitative judgments expressed in linguistic terms (such as "highly preferred," "moderately suitable," or "low risk") are not effectively incorporated into traditional MCDM models, which rely on precise numerical inputs. As an illustration, a luxury brand working with a tech company might get individualized input from professionals in the field who characterize possible synergies using qualitative criteria as opposed to strict numerical scores.

3.1.6. Lack of Flexibility in Weight Assignment

Predefined scales and assumptions are frequently used to determine criteria weights in traditional MCDM methods. These weight distributions might not accurately represent uncertainties in the real world, where the significance of criteria can change over time. For instance, if a significant trend shift takes place, the weight given to "market perception" in co-branding decisions may rise noticeably. However, traditional models are unable to account for such dynamic shifts.

Figure 3. Limitations of Conventional MCDM in Uncertain Decision-Making

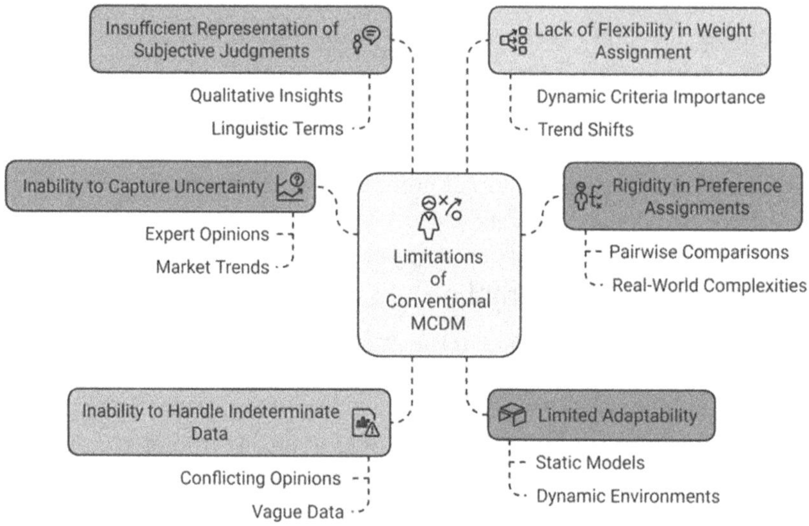

3.2. Alternative Approaches To Addressing Uncertainty In Decision-Making

Given these limitations, alternative approaches such as fuzzy MCDM and neutrosophic MCDM have been introduced to enhance decision-making under uncertainty:

Fuzzy MCDM Techniques

- Allow partial truth values instead of fixed numerical values.
- Handle subjective and imprecise judgments by using fuzzy numbers (e.g., "high," "medium," and "low" rather than exact numerical scores).

Neutrosophic MCDM Techniques

- Extend fuzzy logic by incorporating truth (T), indeterminacy (I), and falsity (F) values to capture uncertainty and contradictions.
- Provide more realistic evaluations by allowing decision-makers to express hesitation and conflicting opinions.

Traditional MCDM approaches have trouble managing uncertainty, indeterminacy, and dynamic market changes, even though they are helpful in structured decision environments. Due to these constraints, advanced decision-making frameworks that provide more flexibility, adaptability, and robustness in uncertain decision environments—like fuzzy and neutrosophic MCDM—must be adopted. Decision-makers can more effectively negotiate complex situations and make more resilient, well-informed strategic decisions by incorporating these cutting-edge techniques.

4. INTRODUCTION TO NEUTROSOPHIC SETS

To address indeterminacy and inconsistencies in decision-making problems, Smarandache (1998) developed Neutrosophic Sets (NS), an extension of classical, fuzzy, and intuitionistic fuzzy sets. NS enables decision-makers to explicitly incorporate truth, falsity, and indeterminacy components, in contrast to traditional methods that struggle with uncertainty.

A Neutrosophic Set (NS) is defined as:

$$NS = \{\langle x, T(x), I(x), F(x)\rangle \mid x \in X\}$$

where:

- T(x) represents the degree of truth (certainty)
- I(x) represents the degree of indeterminacy (uncertainty)
- F(x) represents the degree of falsity (contradiction) and all three components are independent and range within the real standard or non-standard subset of [0, 1].

4.1. Real-Life Example Of Neutrosophic Sets: Buying A New Car

Imagine Mr. Das wants to buy a new car, but Mr. Das is **unsure** about which one to choose. Mr. Das has three options:

1. **Sedan (Car A)** – Comfortable and fuel-efficient but not very spacious.
2. **SUV (Car B)** – Spacious and safe but expensive.
3. **Electric Car (Car C)** – Environmentally friendly and futuristic but has limited charging stations.

Mr. Das will evaluate these cars based on three key criteria:

- **Fuel Efficiency (C1)**
- **Comfort (C2)**
- **Affordability (C3)**

However, Mr. Das's opinions **aren't always certain** because you may have **doubts**, and different people (friends, family, and experts) may give **contradictory advice**.

Step 1: Representing Opinions in Neutrosophic Terms

For each criterion, the expert assigns three values as in Table 1, where

- **Truth (T)** → How much Mr. Das believes the car meets this criterion.
- **Indeterminacy (I)** → Mr. Das's hesitation or uncertainty.
- **Falsity (F)** → How much does Mr. Das believe the car does NOT meet the criterion?

Table 1. Neutrosophic Representation of Car Criteria

Car	Fuel Efficiency (C1)	Comfort (C2)	Affordability (C3)
Sedan (A)	(0.9, 0.1, 0.0) → Very fuel efficient	(0.7, 0.2, 0.1) → Fairly comfortable	(0.8, 0.1, 0.1) → Affordable
SUV (B)	(0.5, 0.3, 0.2) → Not very fuel efficient	(0.9, 0.1, 0.0) → Very comfortable	(0.4, 0.4, 0.2) → Expensive and uncertain
Electric Car (C)	(0.8, 0.2, 0.0) → Efficient but charging is a concern	(0.6, 0.3, 0.1) → Moderately comfortable	(0.5, 0.3, 0.2) → Uncertain due to high initial cost but low running cost

Step 2: Making a Decision Based on Neutrosophic Sets

- Sedan (A) is the best in fuel efficiency and affordability **with a** high truth value (T) **and** low falsity (F).
- SUV (B) is the most comfortable but uncertain in affordability **(high indeterminacy** I = 0.4**).**
- Electric Car (C) is fuel-efficient but has uncertainty in comfort and cost **due to** limited charging stations **and** high upfront prices.

  Final Decision (Output):
- If **fuel efficiency** and **affordability** matter most → **Choose Sedan (A)**
- If **comfort and safety** matter most → **Choose SUV (B)**

- If you are **eco-conscious and think long-term → Choose an Electric Car (C)**

4.2 Why Use Neutrosophic Sets?

Real-world decision-making is rarely simple. Conventional techniques, such as fuzzy logic (partial truth values) or binary logic (true or false), frequently fail in situations involving hesitancy, uncertainty, and divergent viewpoints. Neutrosophic Sets provide a better method in this situation. By combining three independent factors, Neutrosophic Sets offer a more flexible and realistic framework than fuzzy logic or a straightforward ranking system:

- **Truth (T)** – The degree to which something is true or favorable.
- **Indeterminacy (I)** – The degree of uncertainty, doubt, or hesitation in making a decision.
- **Falsity (F)** – The degree to which something is false or unfavorable.

4.3. How Neutrosophic Sets Improve Decision-Making

4.3.1. Consider Uncertainty in Decision-Making

Making decisions in real life frequently involves dealing with imprecise information or uncertain future circumstances. This uncertainty can be directly incorporated into the decision model with the aid of neutrophilic sets.

Example: Let's say that fuel prices are changing while you're purchasing a car. Using a conventional approach, cars would be ranked solely on their current fuel efficiency. However, you can make your choice more flexible by using Neutrosophic Sets, which allow you to assign an indeterminacy value (I) to represent the uncertainty of future fuel prices.

Another Example: Future stock performance is unpredictable if you're investing in the stock market. Instead of assuming fixed probabilities, Neutrosophic Sets let you account for market volatility, expert disagreements, and uncertain economic conditions.

4.3.2. Handle Mixed and Contradictory Opinions

When making decisions, you frequently get contradictory advice from various sources. While fuzzy logic assumes partial truth but fails to distinguish between contradictions and uncertainty, traditional ranking methods find it difficult to reconcile these divergent viewpoints.

For instance, if you have to choose between an electric car, a sedan, or an SUV, your family might favor the SUV for comfort, but experts might advise the electric car for long-term cost savings. Instead of imposing a single ranking or probability, neutrophilic sets let you give each opinion a unique T, I, and F value.

Another Example: Marketing analysts may not agree on a new branding strategy when making business decisions. Some may argue that it won't work (high F), while others may be unsure because of market trends (high I). Neutrosophic Sets enable companies to measure all viewpoints before making a strategic decision, rather than eliminating opposing viewpoints.

4.3.3. Make a Balanced Decision Considering Truth, Doubt, and False Perceptions

Decisions in the real world are not always clear-cut. While some decisions might seem favorable but have unstated risks, others might seem unfavorable but have potential benefits. Neutrosophic sets aid in creating a more realistic and nuanced assessment.

For instance, when choosing a job, you may be thrilled about a high-paying offer (high T), but you may also be concerned about the company's reputation (moderate F) and uncertain about the work-life balance (high I). Neutrosophic Sets assist you in evaluating each of the three factors independently and fairly, as opposed to relying on a binary choice or an oversimplified ranking system.

Another example is when a doctor is 80% certain that a patient has a disease ($T = 0.8$), but there is only a 10% chance that the diagnosis is incorrect ($F = 0.1$) and 30% uncertainty because of incomplete test results ($I = 0.3$). Before recommending a course of treatment, the physician can take all of these factors into account using neosophic reasoning.

Neutrosophic sets help you make more informed decisions, account for the complexity of the real world, and lower the risk of overconfidence in uncertain situations. This method works well in various fields, such as:

- Business strategy: Choosing a co-branding partner while taking consumer trend uncertainty into account.

- Finance & Investment: Assessing cryptocurrencies, stocks, or real estate with volatile market fluctuations.
- Managing contradicting data in AI-based decision-making models is the focus of artificial intelligence and machine learning.
- Healthcare: Making diagnoses for illnesses with conflicting evidence from test results and symptoms.
- Personal decision-making is the process of selecting a profession, a residence, or even a life partner while taking uncertainties and inconsistencies into account.

By enabling us to simultaneously think in terms of certainty, doubt, and contradiction, neutrophilic sets provide a more human-like decision-making process. This approach offers a dynamic, flexible, and more practical way to solve problems rather than imposing decisions into strict rankings.

4.3.4. Why Use Neutrosophic Sets in Co-Branding Decisions?

Co-branding smart luxury goods is a difficult process that calls for striking a balance between qualitative and quantitative considerations. Traditional decision-making models are insufficient because these factors are frequently ambiguous, uncertain, and the subject of differing opinions.

A better analytical framework is offered by neutrophilic sets because they can manage:

- **Truth (T)** – The extent to which a factor is favorable.
- **Indeterminacy (I)** – The level of uncertainty, doubt, or conflicting opinions.
- **Falsity (F)** – The extent to which a factor is unfavorable.

Neutrosophic MCDM allows companies to explicitly model contradiction and uncertainty when making co-branding decisions, resulting in a more thorough evaluation process.

4.4. Challenges In Smart Luxury Co-Branding

Co-branding tactics for intelligent luxury goods include a number of unknown elements, including:

- How well do the two brands complement one another in terms of prestige, identity, and values? This is known as brand reputation synergy.

- Market Perception: What impressions do customers and business leaders have of the partnership? Is the feedback contradictory?
- Compatibility of Technological Integration: Is it possible for one brand's intelligent technology to blend in with another's opulent offerings?
- Customer Emotional Engagement: Will customers develop a strong emotional bond with the co-branded product, or will they react in an ambiguous way?
- Sustainability Concerns: Does the collaboration's ethical and environmental impact raise any questions?

Indeterminacy, which includes conflicting customer perceptions, unclear market signals, and expert disagreements, is difficult for traditional MCDM models, like Fuzzy MCDM, to handle.

Figure 4. Challenges in Smart Luxury Co-Branding

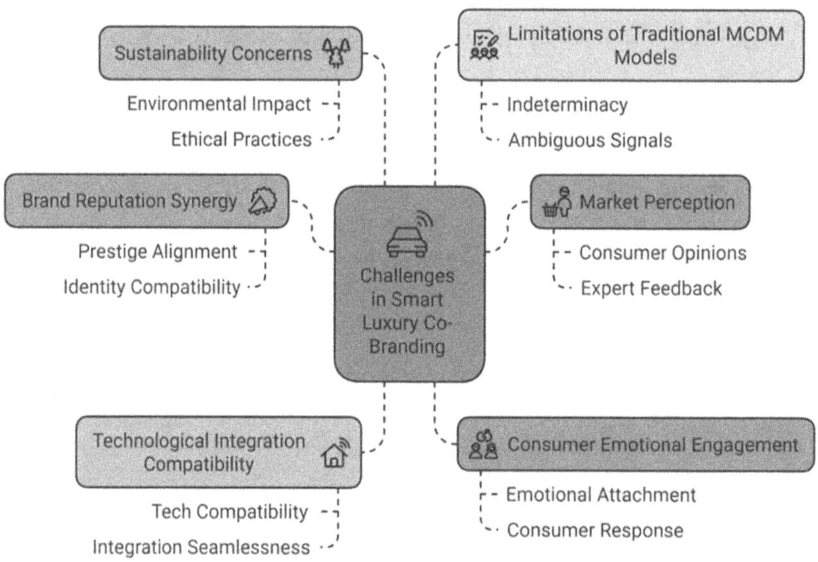

4.5. How Neutrosophic Sets Improve Co-Branding Decisions

In co-branding, brand synergy is the degree to which two partnering brands are compatible with one another in terms of identity, values, target market, and strategic objectives. This synergy is especially complicated in smart luxury co-branding since it combines the classic prestige of luxury brands with the innovative, rapidly changing nature of smart technology. How well the merged brand identities produce a unified value proposition that appeals to customers determines how successful these collaborations will be.

However, brand synergy is not always straightforward. It is influenced by various factors such as:

- Consumer trust and market perception: Do customers see the partnership as complementary or antagonistic?
- Compatibility with technology: Does the exclusivity of the luxury brand fit in with the ever-evolving landscape of smart technology?
- Cultural and emotional appeal: Is it possible for the collaboration to incorporate innovation while preserving the aspirational nature of luxury?

Traditional decision-making techniques might not adequately evaluate brand synergy in light of these uncertainties. To measure brand compatibility, Neutrosophic MCDM offers an organized method that incorporates degrees of truth (T), indeterminacy (I), and falsity (F).

4.6. Neutrosophic Approach To Brand Synergy Evaluation

Three independent values are assigned to brand synergy in Neutrosophic Decision-Making:

1. Truth (T): The degree to which the collaboration is thought to be a good strategic match.
2. Indeterminacy (I): The degree of hesitancy or uncertainty in assessing the collaboration.
3. Falsity (F): The degree to which the partnership is perceived as contradictory or incompatible.

Decision-makers can simulate real-world uncertainty, divergent expert opinions, and erratic customer reactions by taking these three factors into account.

Figure 5. Neutrosophic Approach to Evaluating Brand Synergy

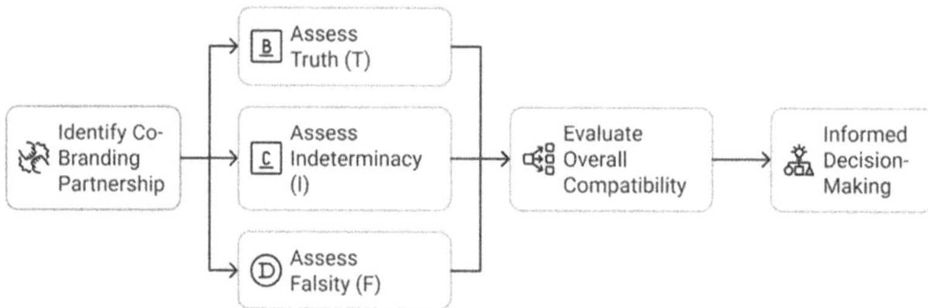

Example: Smart Luxury Watch Co-Branding (Luxury Watch + AI Technology)

Let's consider a luxury watch brand (e.g., Rolex) partnering with a smart technology company (e.g., Google) to develop an AI-powered smart watch as mentioned in Table 2.

Table 2. Luxury watch brand partnering with a smart technology company

Factor	**Neutrosophic Values**	**Explanation**
Truth (T) = 0.8	High	Consumers perceive this as an innovative collaboration that enhances luxury with smart functionalities.
Indeterminacy (I) = 0.4	Moderate	Some luxury consumers are uncertain about integrating AI with high-end watches, as they value tradition and craftsmanship.
Falsity (F) = 0.3	Low	Critics argue that AI might dilute the exclusivity of the luxury brand, making it feel less premium.

The **Neutrosophic Brand Synergy Score (NBSS)** can be calculated using a weighted aggregation of these three values to determine the overall viability of the partnership.

4.7. Factors Influencing Brand Synergy In Smart Luxury Co-Branding

4.7.1. Consumer Perception and Market Acceptance

Consumer reception plays a pivotal role in determining the success of a smart luxury collaboration. While younger, tech-savvy consumers may embrace the combination of luxury and innovation, traditional luxury buyers may resist the idea.

- **Example:** The Apple-Hermès partnership for the Apple Watch succeeded because Apple retained Hermès' exclusive craftsmanship while integrating its smartwatch technology.

To assess consumer perception using **Neutrosophic Logic**, brand managers can:

- Conduct market research to gather **T, I, and F values** based on consumer surveys.
- Evaluate whether the perceived **truth (T)** outweighs the **falsity (F)** to justify the partnership.

4.7.2. Technological Fit and Integration Challenges

Not all luxury brands can seamlessly integrate smart technology without compromising their core brand identity. If the technology feels forced rather than naturally enhancing the product, the co-branding effort may fail.

- **Example:** A high-end jewelry brand (e.g., Cartier) collaborating with a tech brand (e.g., Samsung) for a smart ring might struggle with material constraints and luxury aesthetics.

Using neutrosophic modeling:

- **Truth (T):** If the technology complements the luxury brand's existing product line, T will be high.
- **Indeterminacy (I):** If technical feasibility and execution remain uncertain, I will be high.
- **Falsity (F):** If technology disrupts the luxury experience, F will be high.

4.7.3. Brand Positioning and Strategic Alignment

Luxury brands are built on exclusivity and heritage, whereas tech companies emphasize accessibility and constant evolution. A successful co-branding initiative must find a balance between these contrasting brand identities.

- **Example:** Mercedes-Benz's collaboration with NVIDIA for AI-powered self-driving technology aligns well because both brands position themselves as industry leaders in innovation.

Using **Neutrosophic MCDM**:

- If the **perceived synergy (T)** between the brands is strong, the partnership has a high chance of success.
- If there is **market hesitation (I)** due to brand misalignment, decision-makers should analyze how to mitigate risks.
- If experts believe the collaboration damages brand reputation, the **falsity score (F)** should be carefully considered.

4.8 Neutrosophic Decision-Making For Brand Synergy Evaluation

To systematically evaluate brand synergy, decision-makers can apply a Neutrosophic Weighted Aggregation Model, where each factor is weighted based on importance:

$$NBSS = w1.T + w2.I + w3.F$$

Where:

- $w1$, $w2$, $w3$ are the assigned weights for truth, indeterminacy, and falsity, respectively.
- The final **NBSS score** determines whether the partnership should proceed.

Example Calculation (Using Weights $w_1 = 0.5$, $w_2 = 0.3$, $w_3 = 0.2$)

$$\text{NBSS} = (0.5 \times 0.8) + (0.3 \times 0.4) + (0.2 \times 0.3) = 0.4 + 0.12 + 0.06 = 0.58$$

Since the NBSS score is above 0.5, the co-branding partnership is **viable**, but decision-makers must address **indeterminacy (I = 0.4)** to ensure market acceptance.

4.9. Strategic Recommendations Based on Neutrosophic Analysis

To optimize brand synergy in smart luxury co-branding, brands should:

1. **Reduce Uncertainty (Lower Indeterminacy Score I)**
   - Conduct pilot product launches and consumer testing to gather feedback.
   - Collaborate with influencers and brand ambassadors to bridge the gap between luxury and tech audiences.
2. **Enhance Perceived Brand Synergy (Increase Truth Score T)**

- Highlight co-branding benefits through storytelling (e.g., emphasizing craftsmanship and innovation).
- Leverage limited-edition releases to maintain exclusivity.

3. **Mitigate Potential Brand Conflict (Lower Falsity Score F)**
   - Ensure seamless integration of smart features without compromising luxury aesthetics.
   - Maintain distinct luxury branding while subtly incorporating tech elements.

Evaluating brand synergy in smart luxury co-branding is complex due to the inherent uncertainties in consumer perception, technological fit, and brand alignment. By applying Neutrosophic MCDM, decision-makers can quantify truth, indeterminacy, and falsity to make informed strategic choices. This approach enhances decision accuracy, minimizes risks, and ensures that luxury-tech collaborations create meaningful and successful brand partnerships.

4.10. Assessing Consumer Perception – Addressing Mixed Or Contradictory Customer Feedback

Consumer perception is a critical determinant in the success of co-branded smart luxury products. Since luxury markets cater to highly discerning consumers, perceptions are often shaped by personal preferences, cultural influences, and evolving market trends. Unlike mass-market products, luxury brands must navigate an intricate balance between exclusivity and innovation. When luxury brands partner with technology firms, they face varied consumer reactions—ranging from enthusiasm to skepticism and outright rejection.

4.10.1. Challenges in Consumer Perception for Co-Branded Smart Luxury Products

Consumer perception is influenced by multiple factors, including brand loyalty, trust, and technological adaptability. When assessing feedback, luxury brands must consider the following challenges:

- **Divergent Consumer Segments:**
  - Traditional luxury consumers value heritage, craftsmanship, and exclusivity.
  - Tech-savvy consumers seek functionality, innovation, and digital integration.

- The intersection of these segments creates mixed reactions, making it difficult to assess a unified response.

- **Uncertainty in Market Adoption:**
  - Smart luxury products often introduce new features that may not have an immediate market precedent, causing hesitation among consumers.
  - A lack of reference points may result in uncertain responses, making it challenging to predict long-term success.
- **Perceived Brand Dilution Risks:**
  - Some consumers may feel that integrating technology undermines the brand's traditional values.
  - Others might see co-branding as a strategic move that modernizes the luxury brand without compromising its essence.
- **Influence of Media and Influencers:**
  - Consumer perception is often shaped by media reviews, influencers, and early adopters.
  - Polarizing opinions from these sources can create a fragmented market response, adding to the uncertainty.

Figure 6. Challenges in Consumer Perception for Co-Branded Smart Luxury Products

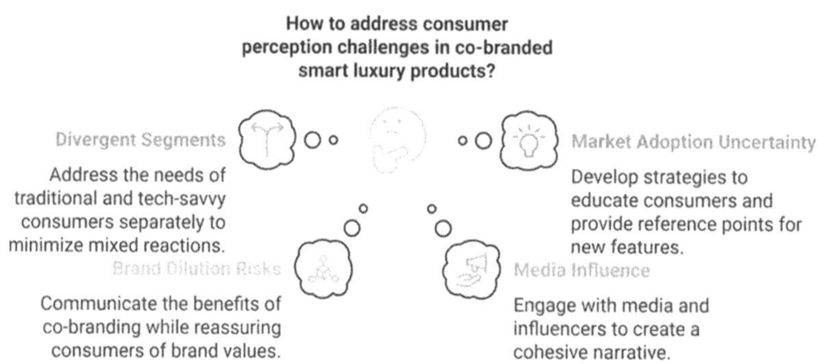

4.10.2. Neutrosophic Analysis of Consumer Feedback

Traditional decision-making models often force businesses to categorize feedback into binary (positive/negative) or scaled (good, neutral, bad) interpretations. However, real-world perceptions are rarely this straightforward. Neutrosophic sets provide a more nuanced approach by incorporating three independent components:

- **Truth (T)** → Represents the degree of positive consumer perception.
- **Indeterminacy (I)** → Captures uncertainty, hesitation, or neutral responses.
- **Falsity (F)** → Measures the degree of negative perception or rejection.

Example: High-End Fashion Brand & Tech Company Launching a Luxury Augmented Reality (AR) Shopping Experience

An Augmented Reality (AR) shopping experience is introduced by a luxury fashion house in partnership with a technology company. Customers can now virtually try on items before buying them, engaging with AI stylists, and perusing upscale fashion collections in a virtual showroom thanks to this innovation.

However, consumer perception varies:

- **Enthusiasts (T = 0.7)**
- Luxury shoppers who are tech-savvy view this as a state-of-the-art shopping experience.
- Customers who are digital natives value AR's convenience and personalization.
- Early adopters commend the combination of luxury and innovation.
- **Skeptics (I = 0.5)**
- Conventional luxury buyers are hesitant because they are worried about losing the in-store experience.
- o Some people question whether augmented reality can accurately mimic the tactile and sensory aspects of high-end shopping.
- Others are indifferent and hold off on making a firm judgment until they observe how the market responds to the technology.
- **Rejectors (F = 0.3)**
- Some consumers prefer more conventional luxury shopping experiences and think augmented reality is gimmicky and unnecessary.
- Digital interactions, according to luxury purists, lessen the exclusivity and individualized touch of luxury retail.
- Others fear that AR-driven luxury shopping could lead to over-commercialization and loss of brand mystique.

4.10.3. Leveraging Neutrosophic Analysis for Strategy Development

Instead of treating consumer feedback as binary (positive vs. negative), luxury brands can use Neutrosophic MCDM to analyze and respond to market sentiment dynamically. The strategic implications of the analysis are as follows:

1. Customizing Marketing Messages:

- Use campaigns that emphasize the innovative and futuristic appeal of augmented reality shopping to reach eager customers (T).
- Provide hybrid experiences that combine conventional in-store services with AR-enhanced personalization to allay skeptics (I).
- Assure rejectors (F) by stressing that augmented reality is a supplement to, not a substitute for, the in-store shopping experience.

2. **Adaptive Product Development:**
   - Adjust the AR platform in response to consumer feedback by including elements that replicate the in-store experience, like haptic feedback or AI-powered luxury consultation.
   - Test pilot launches in particular markets to gauge real-world adoption and make necessary adjustments to the offering based on early customer feedback.
3. **Segmented Consumer Engagement Strategies:**
   - Promote digital improvements for tech-savvy customers while providing individualized in-store experiences for conventional luxury buyers.
   - Establish influencer collaborations so that dependable luxury representatives can highlight the advantages of augmented reality without offending purists.
   - Create unique augmented reality (AR)-based VIP experiences for valuable customers, making sure the technology complements exclusivity rather than takes its place.
4. **Real-Time Sentiment Analysis for Continuous Improvement:**
   - Use sentiment tracking powered by AI to evaluate how customer perceptions are changing over time.
   - Make dynamic adjustments to strategies in response to continuous feedback, making sure that the co-branded product develops in line with consumer demands.

The way that consumers view co-branded smart luxury goods is unpredictable and differs depending on the audience. Brands can transcend conventional binary evaluations and create adaptable, data-driven strategies that serve a wide range of consumers by utilizing Neutrosophic Analysis. With this strategy, brands can:

- Reduce the risk of brand dilution and market skepticism;
- Find ways to improve and reposition co-branded products; and
- Boost engagement through tailored marketing and innovative solutions.

In order to maintain the resilience, responsiveness, and consumer-centricity of their co-branding strategies in a changing market environment, luxury brands can avoid uncertainty by adopting Neutrosophic Sets in consumer perception analysis.

Figure 7. Neutrosophic Analysis for Strategy Development

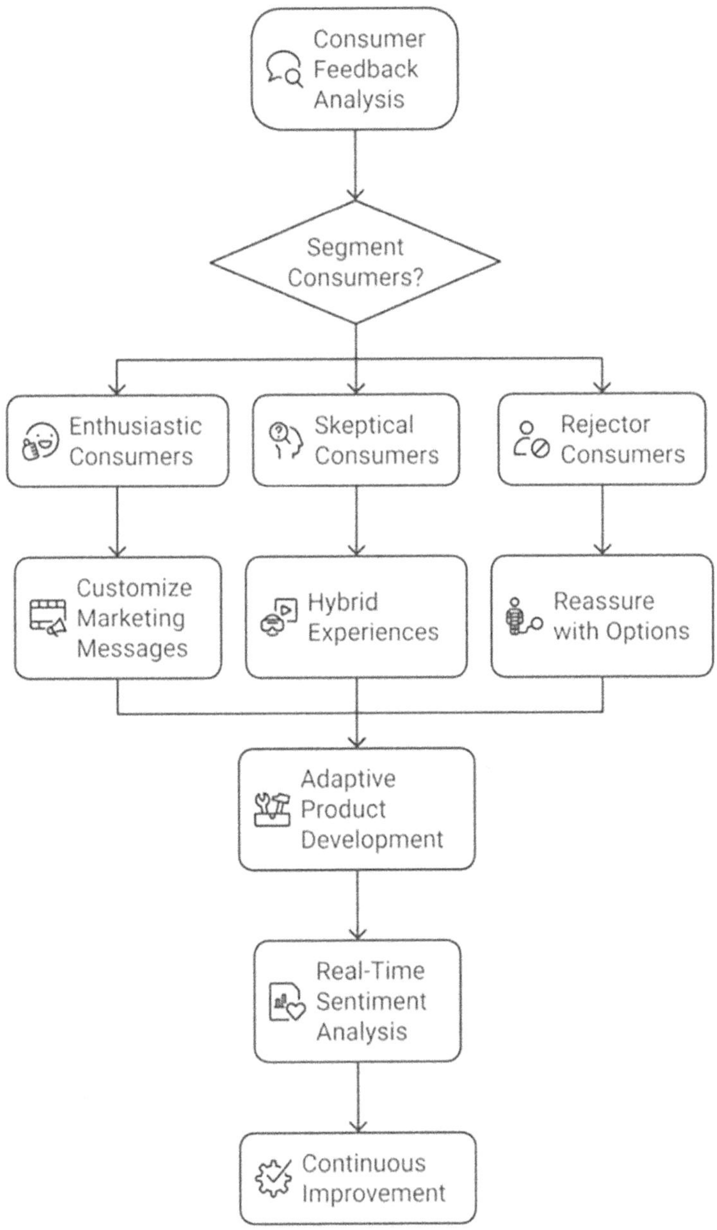

4.10.4. Quantifying Risk and Opportunity – Factoring in uncertain business risks due to technology adoption and brand collaboration

Technological co-branding is inherently **risky** due to the unpredictability of innovation adoption, competition, and economic factors. Neutrosophic Sets help businesses **model risks with uncertainty and conflicting expert opinions.**

Example: A luxury car brand co-brands with a renewable energy company to produce an **electric vehicle (EV) with AI-powered self-driving features.**

- The **business opportunity** is high due to the demand for sustainable luxury vehicles (**T = 0.85**).
- There is **uncertainty about government regulations** and public trust in self-driving technology (**I = 0.6**).
- Some investors believe the technology is too expensive and not yet viable (**F = 0.4**).

By factoring in all three aspects (T, I, F), Neutrosophic decision models can prioritize investments and mitigate risks in a more data-driven way. The comparison between Neutrosophic MCDM and Traditional MCDM Approaches is presented in Table 3.

Table 3. Neutrosophic MCDM vs. Traditional MCDM Approaches

Criteria	Traditional MCDM (e.g., Fuzzy Logic)	Neutrosophic MCDM
Handling Uncertainty	Can handle some uncertainty but assumes probability distributions.	Explicitly incorporates **indeterminate (I)** factors.
Conflicting Opinions	Difficult to model contradictory consumer and expert views.	Allows simultaneous truth, indeterminacy, and falsity values.
Decision Flexibility	Less adaptive in rapidly changing industries.	More dynamic and can adapt to evolving trends.
Real-World Applicability	Works well in well-defined, structured environments.	Works better in **complex, ambiguous decision-making scenarios** (e.g., co-branding).

Neutrosophic Sets **enhance co-branding decisions** by:

- **Providing a realistic framework** to assess both synergy and uncertainty.
- **Handling contradictory consumer and expert opinions.**
- **Modeling risks and opportunities** in a data-driven manner.

By incorporating **Neutrosophic MCDM**, luxury, and smart brands can create **stronger, more adaptive co-branding strategies** that maximize value while mitigating uncertainty.

4.11. Mathematical Representation Of Neutrosophic Mcdm In Co-Branding

In a **MCDM** framework for co-branding strategies, decision-makers need to evaluate multiple brand pairings across various criteria. However, traditional methods struggle to address **uncertainty, conflicting expert opinions, and market ambiguity**. Neutrosophic MCDM provides a more flexible and realistic decision model by incorporating **Truth (T), Indeterminacy (I), and Falsity (F)** values.

A **decision matrix** in Neutrosophic MCDM is defined as:

$$D = [(T_{ij}, I_{ij}, F_{ij})]_{m \times n}$$

where:

- Tij represents the **truth degree** of the **i-th brand alternative** under the **j-th criterion** (e.g., how well the luxury and smart brands complement each other).
- Iij represents the **indeterminacy degree**, capturing **uncertainty or hesitation** in expert evaluations (e.g., unclear brand positioning).
- Fij represents the **falsity degree**, measuring the **negative aspects** of the brand collaboration (e.g., poor brand synergy).
- m is the number of **brand alternatives**, and n is the number of **evaluation criteria** (e.g., technological compatibility, market perception, sustainability).

Each brand pairing is thus evaluated in terms of **three independent values**, rather than a single crisp score.

5. KEY NEUTROSOPHIC MCDM OPERATIONS

5.1. Single-Valued Neutrosophic Number (Svnn) In Co-Branding Decisions

A **SVNN** is a mathematical tool used to represent uncertain, imprecise, and conflicting information in decision-making. It extends traditional fuzzy logic by introducing three independent components:

- **Truth Membership (T)**: The degree to which a statement or attribute is considered true.
- **Indeterminacy Membership (I)**: The degree of hesitation, uncertainty, or vagueness regarding the statement.
- **Falsity Membership (F)**: The degree to which the statement is considered false.

These values are constrained within the range **[0, 1]** and must satisfy the following condition:

$$T + I + F \leq 3$$

This flexibility allows decision-makers to model real-world scenarios more effectively, especially in cases where uncertainty and conflicting opinions exist.

5.1.1. Why Use SVNN in Co-Branding Decisions?

In co-branding, brands must assess co-branding alliances using a variety of subjective criteria, including consumer perception, market alignment, technological compatibility, and brand reputation. The intricacy and unpredictability of expert assessments are frequently overlooked by conventional decision-making techniques. By enabling decision-makers to convey differing levels of certainty, uncertainty, and contradiction in their evaluations, SVNN gets around this.

- **For instance: Assessing a Co-Branding Partnership for Smart Luxury**

To introduce a high-end AI-powered smart watch, a luxury brand (Brand A) is thinking about forming a co-branding alliance with a smart technology company (Brand B). Using SVNN, industry experts offer their perspectives on how the market views this collaboration.

5.1.2. Expert Evaluation Using SVNN

Each expert assigns **Truth (T), Indeterminacy (I), and Falsity (F) values** based on their perception:

- **T = 0.8** → Strong positive reputation; most experts believe the partnership enhances the luxury brand's image.
- **I = 0.4** → Moderate uncertainty; some experts are unsure whether traditional luxury consumers will embrace AI-driven products.

- **F = 0.2** → Minor concerns; a few experts feel that integrating smart technology might dilute the luxury brand's exclusivity.

This evaluation can be represented as:
SVNN Market Perception = (0.8, 0.4, 0.2)
Interpretation and Decision Insights

1. **High Truth Value (0.8)**
   - Indicates a **strong overall confidence** in the co-branding strategy.
   - Experts agree that the collaboration enhances the brand image and appeals to a tech-savvy luxury audience.
2. **Moderate Indeterminacy (0.4)**
   - Suggests **uncertainty** in consumer adoption and long-term success.
   - The luxury brand should conduct further market research to reduce this ambiguity.
3. **Low Falsity Value (0.2)**
   - Implies that **few experts oppose** the partnership.
   - The main concern is the potential dilution of exclusivity, but it is not a significant deterrent.

Comparative analysis between SVNN and traditional decision-making methods is presented in Table 4.

Table 4. SVNN vs. Traditional Decision-Making Methods

Aspect	**Traditional Decision Models**	**Neutrosophic SVNN Approach**
Handling Uncertainty	Limited, assumes precise values	Explicitly models uncertainty (I)
Capturing Expert Opinions	Single numerical score	Separates truth, doubt, and falsehood
Flexibility in Decision-Making	Rigid preference rankings	Allows for hesitation and conflicting data
Real-World Applicability	Struggles with ambiguity	Ideal for strategic business decisions

5.1.3. Application of SVNN in Co-Branding Decisions

Several evaluation criteria in co-branding strategies can be added to the SVNN approach, such as:

- Technological Compatibility: Assessing if the smart technology complements the design of the luxury brand.
- Consumer Acceptance: Forecasting how the market will react given current preferences and trends.
- Financial Feasibility: Evaluating the co-branded product's profitability.

Businesses can select the ideal co-branding partner by combining SVNNs based on a variety of criteria, allowing them to make data-driven, uncertainty-aware choices.

5.2. Neutrosophic Weighted Aggregation Operators For Co-Branding Decisions

In MCDM, decision-makers evaluate multiple factors (criteria) to rank alternatives. Traditional aggregation operators, such as weighted sums or averages, work well in structured decision-making but struggle with uncertainty and conflicting expert opinions. By adding truth (T), indeterminacy (I), and falsity (F) values for every criterion, neutrophilic weighted aggregation operators go beyond conventional techniques. This enables decision-makers to account for hesitancy and uncertainty, resulting in more thorough and trustworthy choices.

5.2.1. Mathematical Formulation of the Neutrosophic Weighted Averaging Operator

The Neutrosophic Weighted Averaging (NWA) operator is formulated as:

$$NWA\left(Ai\right) = \sum_{j=1}^{n} wj \cdot (Tij, Iij, Fij)$$

where:

- Ai is the i-th alternative (e.g., a co-branding partnership).
- j represents the j-th criterion in the decision-making process.
- wj is the weight assigned to the j-th criterion, where $\sum$wj=1.
- (Tij,Iij,Fij) represents the neutrosophic values assigned to the i-th alternative under the j-th criterion.

The NWA operator calculates an aggregated score that accounts for:

- **Truth (T)**: The degree to which the alternative meets the criterion.
- **Indeterminacy (I)**: The level of hesitation or uncertainty in expert judgments.
- **Falsity (F)**: The extent to which the alternative does not meet the criterion.

This approach ensures that decision-making incorporates **uncertainty and contradictions**—a limitation of traditional MCDM methods.

5.2.2. Application of NWA in Smart Luxury Co-Branding

To illustrate the utility of NWA in co-branding, consider two brand partnerships evaluated across three key criteria:

- **Technological Innovation (C1)**: How well the smart luxury product integrates new technologies.
- **Market Compatibility (C2)**: The level of synergy between brand identities and target consumers.
- **Brand Prestige (C3)**: The strength of each brand's reputation in the luxury sector.

5.2.3. Neutrosophic Decision Matrix for Two Brand Pairings

The expert evaluations for two brand partnerships (Brand A & B vs. Brand C & D) are represented as Neutrosophic Numbers as in Table 5.

Table 5. The expert evaluations for two brand partnerships (Brand A & B vs. Brand C & D)

Brand Pairing	Technological Innovation (C1)	Market Compatibility (C2)	Brand Prestige (C3)	Aggregated Score
Brand A & B	(0.7, 0.3, 0.2)	(0.6, 0.4, 0.3)	(0.8, 0.2, 0.1)	(0.7, 0.3, 0.2)
Brand C & D	(0.6, 0.4, 0.3)	(0.5, 0.5, 0.4)	(0.7, 0.3, 0.2)	(0.6, 0.4, 0.3)

Weight Assignment:

- Technological Innovation (w1) = 0.4
- Market Compatibility (w2) = 0.3
- Brand Prestige (w3) = 0.3

Now, applying the NWA operator:

$$NWA(Ai) = \sum wj(Tij, Iij, Fij)$$

For **Brand A & B:**

NWA(A1) = (0.4×0.7, 0.4×0.3, 0.4×0.2) + (0.3×0.6, 0.3×0.4, 0.3×0.3) + (0.3×0.8, 0.3×0.2, 0.3×0.1)

= (0.28 + 0.18 + 0.24, 0.12 + 0.12 + 0.06, 0.08 + 0.09 + 0.03)

= (0.70, 0.30, 0.20)

For **Brand C & D**:

NWA(A2) = (0.4 × 0.6, 0.4 × 0.4, 0.4 × 0.3) + (0.3 × 0.5, 0.3 × 0.5, 0.3 × 0.4) + (0.3 × 0.7, 0.3 × 0.3, 0.3 × 0.2)

= (0.24 + 0.15 + 0.21, 0.16 + 0.15 + 0.09, 0.12 + 0.12 + 0.06)

= (0.60, 0.40, 0.30)

Brand A & B (0.70, 0.30, 0.20) is the better choice for co-branding since it has a higher truth value (T) and a lower falsity value (F). The indeterminacy value (I) for brands C & D (0.60, 0.40, 0.30) is higher, suggesting greater ambiguity in the expert assessments. Both options might be ranked similarly in traditional MCDM, but by taking uncertainty (I-values) into account, neutrophilic MCDM offers a more thorough analysis. This is especially helpful when making strategic branding decisions in an environment where consumer preferences and market conditions are always changing.

5.2.4. Advantages of Neutrosophic Weighted Aggregation Operators

1. **Handles Uncertainty and Contradictions**
   - o Neutrosophic MCDM incorporates indeterminacy (I) to capture hesitancy in expert opinions, in contrast to conventional approaches that depend on precise values.
2. **Better Risk Management**
   - o Decision-makers can spot possible hazards in co-branding alliances by specifically taking falsity (F) into account.
3. **Improves Decision Robustness**
   - o By combining truth, uncertainty, and falsehood, neurosophic MCDM produces more robust rankings that are flexible enough to adjust to changing markets.
4. **Flexible and Scalable Approach**

- The NWA operator can be used in various decision-making areas, such as product innovation, sustainability, and finance.

By specifically taking uncertainty and contradictions into account, the Neutrosophic Weighted Aggregation Operator (NWA) improves on conventional decision-making models. NWA helps brands make well-informed and risk-aware strategic alliances by offering a more nuanced, data-driven approach to partnership evaluation when it comes to co-branding smart luxury products.

5.3. Neutrosophic Similarity Measures

The degree to which two brands align concerning their target market, technological compatibility, and market positioning is assessed using neutrophilic similarity measures (NSM). To effectively handle uncertainty and contradictory evaluations, neutrosophic similarity takes into account three independent factors: truth (T), indeterminacy (I), and falsity (F). This is in contrast to traditional similarity measures that rely on binary or fuzzy logic.

5.3.1. Mathematical Formulation

The similarity between two brand alternatives A and B across n criteria is calculated using the following formula:

$$NSM(A,B) = \sum_{j=1}^{n}(1 - | TAj - TBj | -| IAj - IBj | -| FAj - FBj |)$$

where:

- TAj, TBj = Truth values of brands A and B under criterion j (how well they meet the criterion).
- IAj, IBj = Indeterminacy values of brands A and B under criterion j (uncertainty in meeting the criterion).
- FAj, FBj = Falsity values of brands A and B under criterion j (how much they fail to meet the criterion).
- n = Total number of criteria considered in the co-branding evaluation.

Interpreting NSM Values

- **NSM(A, B) close to 1** → Strong compatibility between the two brands.
- **NSM(A, B) close to 0** → Weak compatibility, indicating misalignment in brand identity, target market, or technological integration.

5.3.2. Example: Evaluating Co-Branding Suitability

Consider a luxury fashion brand exploring partnerships with two potential smart technology brands:

1. **Smart Jewelry Brand (Brand X)**
   - o T=0.8 (High compatibility with luxury aesthetics)
   - o I=0.3 (Some uncertainty about consumer response)
   - o F=0.1 (Minimal mismatch)
2. **Smart Home Tech Brand (Brand Y)**
   - o T=0.5 (Moderate compatibility with luxury aesthetics)
   - o I=0.5 (Higher uncertainty about how well the luxury market will accept the tech)
   - o F=0.3 (Higher likelihood of brand misalignment)

To compute the **NSM** between **Brand X** and **Brand Y**, we apply the formula:

$$NSM(X, Y)=1-(|0.8-0.5|+|0.3-0.5|+|0.1-0.3|) =1-(0.3+0.2+0.2)=1-0.7=0.3$$

NSM(X, Y) = 0.3 indicates that, due to its higher falsity and indeterminacy values, Brand X (Smart Jewelry) is a better candidate for co-branding than Brand Y (Smart Home Tech). Neutrosophic similarity measures are more realistic in complex decision-making because traditional similarity models would not account for indeterminacy, whereas Brand X's lower falsity (F) indicates a lower chance of market rejection than Brand Y.

5.3.3. Why Use Neutrosophic Similarity Measures in Co-Branding?

1. **Handles Uncertainty and Contradictory Data**
   - o Unlike conventional similarity measures, NSM can account for **uncertain, ambiguous, or conflicting brand evaluations.**
2. **Accounts for Market and Technological Alignment**
   - o Evaluates brand fit not just based on direct compatibility but also based on **uncertainty and potential risks.**
3. **Provides More Granular Insights for Decision-Making**
   - o Instead of giving a binary or fuzzy similarity score, NSM provides a **structured measure of alignment** across multiple factors.

5.3.4. Potential Applications in Co-Branding

- Luxury Automotive & Smart Tech: Assessing the parallels between smart tech companies (like NVIDIA) and luxury automakers (like Bentley) in terms of AI-assisted driving.
- Luxury Fashion & AI: Assessing how well luxury fashion brands (like Gucci) work with wearables that use artificial intelligence (like Apple).
- Sustainable Co-Branding: Evaluating how well luxury brands and green tech companies align to develop sustainable products.

By combining truth, indeterminacy, and falsity, NSM offers a potent tool for evaluating brand compatibility in co-branding decisions. This strategy aids decision-makers in navigating volatile, changing markets and choosing the most advantageous alliances for astute luxury partnerships.

5.4. Why Neutrosophic Mcdm Is Essential For Co-Branding

- Captures Market Uncertainty: In contrast to conventional approaches, it takes into account conflicting expert opinions and changing consumer trends.
- Handles Conflicting Data: Capable of combining brand partnerships' uncertain, negative, and positive elements into a single model.
- Enhances Decision Accuracy: Businesses make better data-driven co-branding decisions by incorporating aggregation measures and neutrophic similarity.

5.4.1. Example of Neutrosophic Sets in Co-Branding for Smart Luxury Products

A co-branding initiative to introduce a Tesla x Rolex Smart Luxury Watch is being considered by Tesla, a pioneer in electric cars and cutting-edge technology, and Rolex, a representation of classic luxury and accuracy. The objective is to create an AI-powered, eco-friendly, and high-end smart watch by fusing Rolex's craftsmanship with Tesla's cutting-edge technology.

However, because of variables like consumer perception, brand fit, and market reaction, this choice is fraught with uncertainty, conflicting expert opinions, and insufficient data. We employ Neutrosophic Sets in an MCDM technique to address these complexities.

Step 1: Define Decision Criteria

The **co-branding decision** is evaluated based on **four key criteria**:

1. **Brand Synergy (C1)** – Compatibility of Tesla and Rolex's brand identities.
2. **Market Demand (C2)** – Expected consumer acceptance and demand.
3. **Technological Feasibility (C3)** – Compatibility of Tesla's tech with Rolex's traditional watchmaking.
4. **Sustainability & Innovation (C4)** – Use of sustainable materials and innovative features.

The **decision alternatives** are:

- A1: Co-branding that is fully integrated (Tesla-Rolex smart luxury watch).
- A2: Technology licensing (Tesla supplies the battery and AI technology for the Rolex watch).
- A3: Independent development: Every company creates its own smart watch.

Step 2: Expert Opinions in Neutrosophic Form

Industry experts provide evaluations using **Neutrosophic Numbers** for each criterion **(T, I, F)**:

- T (Truth) → Confidence in the success of the alternative.
- I (Indeterminacy) → Uncertainty or hesitation about the alternative.
- F (Falsity) → Rejection or contradiction of the alternative.
- Neutrosophic Decision Matrix

Each alternative is evaluated under each criterion using **SVNNs** as presented in Table 6.

Table 6. Neutrosophic representation of co-branding strategies

Alternatives	**Brand Synergy (C1)**	**Market Demand (C2)**	**Tech Feasibility (C3)**	**Sustainability (C4)**
A1 (Full Co-branding)	(0.85, 0.10, 0.05)	(0.80, 0.15, 0.05)	(0.70, 0.20, 0.10)	(0.75, 0.15, 0.10)
A2 (Licensing)	(0.70, 0.20, 0.10)	(0.60, 0.25, 0.15)	(0.85, 0.10, 0.05)	(0.65, 0.20, 0.15)
A3 (Independent Dev.)	(0.50, 0.30, 0.20)	(0.55, 0.35, 0.10)	(0.60, 0.25, 0.15)	(0.50, 0.30, 0.20)

Step 3: Neutrosophic Decision-Making Process

We apply a **Neutrosophic Weighted Aggregation** approach to rank the alternatives:

Step 3. Compute the Aggregated Neutrosophic Scores

Using a **weighted aggregation method**, we assign relative weights to each criterion:

W = (w1, w2, w3, w4) = (0.3, 0.3, 0.2, 0.2)

We then calculate the **Neutrosophic Score** for each alternative using a weighted sum:

$$NS(Ai) = \left(\sum_{j=1}^{n} wj \cdot Tij,\ \sum_{j=1}^{n} wj \cdot Iij,\ \sum_{j=1}^{n} wj \cdot Fij\right)$$

For **A1 (Full Co-branding)**:

$$TA1 = (0.3 \times 0.85) + (0.3 \times 0.80) + (0.2 \times 0.70) + (0.2 \times 0.75) = 0.79$$

$$IA1 = (0.3 \times 0.10) + (0.3 \times 0.15) + (0.2 \times 0.20) + (0.2 \times 0.15) = 0.14$$

$$FA1 = (0.3 \times 0.05) + (0.3 \times 0.05) + (0.2 \times 0.10) + (0.2 \times 0.10) = 0.07$$

Similarly, for **A2 (Licensing)** and **A3 (Independent Dev.)**, we get:

- **A2**: T = 0.72, I = 0.19, F = 0.10
- **A3**: T = 0.55, I = 0.30, F = 0.15

Step 4: Decision Ranking Based on Neutrosophic Scores

Neutrosophic Similarity Measures are used to choose the best option:

- The option with the lowest F (falsity) and the highest T value (truth) is favored.
- The choice is deemed risky and requires more research if I (indeterminacy) is high.

Final Ranking:

1. **A1 (Full Co-Branding)** → (0.79, 0.14, 0.07) → **Best choice**
2. **A2 (Licensing)** → (0.72, 0.19, 0.10)→ **Moderate choice**
3. **A3 (Independent Development)** → (0.55, 0.30, 0.15) → **Least preferred**

Step 5: Business Interpretation

- The best strategy is A1 (Tesla x Rolex Smart Luxury Watch), which shows strong brand synergy and market demand with the highest truth value (0.79) and lowest falsity value (0.07).
- Because it has a moderate degree of truth and falsity, A2 (Technology Licensing) is a substitute if Rolex is apprehensive about full integration.
- A3 (Independent Development) is a riskier and less practical approach because it has the highest uncertainty (I = 0.30).

Businesses can systematically quantify uncertainties, contradictions, and expert hesitations when making co-branding decisions by using Neutrosophic Sets. Compared to conventional techniques, this approach offers a more flexible and realistic framework for decision-making, resulting in data-driven co-branding strategies for astute luxury goods.

5.5. Application Of Neutrosophic Sets In Smart Luxury Co-Branding

Luxury fashion brand Louis Vuitton (LV) and technology giant Apple are considering a co-branding venture to develop a high-end smart luxury smart watch. The objective is to develop a product that combines LV's history of exclusivity and craftsmanship with Apple's technological know-how.

Four primary criteria are used by decision-makers to assess the co-branding partnership's viability:

1. Brand Image Alignment: The degree to which the values, aesthetics, and market positioning of the two brands mesh well.
2. Technological Innovation Fit: How well LV's opulent craftsmanship blends with Apple's state-of-the-art technology.
3. Market Demand: The expected reaction and demand from consumers for a luxury smart watch bearing a co-brand.

4. Sustainability Considerations: The cooperation's effects on the environment and moral issues.

Industry experts provide their assessments for each criterion using Neutrosophic Numbers (T, I, F) as presented in Table 7, where:

- **T(x)** (Truth) represents the degree to which the criterion is satisfied.
- **I(x)** (Indeterminacy) accounts for uncertainty or conflicting opinions.
- **F(x)** (Falsity) measures the degree to which the criterion is not satisfied.

Table 7. Industry experts provide their assessments for each criterion using Neutrosophic Numbers (T, I, F),

Criteria	T (Truth)	I (Indeterminacy)	F (Falsity)
Brand Image Alignment	0.85	0.2	0.05
Technological Innovation Fit	0.9	0.15	0.05
Market Demand	0.75	0.3	0.1
Sustainability Considerations	0.7	0.35	0.15

Analysis of Neutrosophic Values

- **Brand Image Alignment (T = 0.85, I = 0.2, F = 0.05):** The high Truth value (0.85) indicates a strong brand matching between Apple and LV. Both brands are known for their inventiveness and premium image. There is little brand identity conflict, as indicated by the low Falsity value (0.05). Nonetheless, the Indeterminacy (0.2) indicates a lack of certainty regarding the potential response of conventional LV clients to a technologically advanced luxury timepiece.
- **Technological Innovation Fit (T = 0.9, I = 0.15, F = 0.05):** LV's craftsmanship and Apple's long history of innovation complement each other nicely. This synergy is confirmed by the extremely high Truth value (0.9). The Falsity value (0.05) implies that there are no serious worries regarding technological compatibility, while the Indeterminacy (0.15) is comparatively low, suggesting little uncertainty.
- **Market Demand (T = 0.75, I = 0.3, F = 0.1):** Despite the high anticipated demand, the Indeterminacy (0.3) is higher than the other criteria, suggesting a range of consumer responses. While some luxury buyers might be excited about a smartwatch bearing the LV brand, others might favor more conven-

tional timepieces. The product may still be rejected by a portion of the market, according to the Falsity value of 0.1.

- **Sustainability Factors (F = 0.15, I = 0.35, T = 0.7):** Particularly in the luxury market, sustainability is a crucial factor. The Indeterminacy value (0.35) is the highest of all the criteria, even though the Truth value (0.7) is comparatively strong. This raises questions regarding supply chain ethics, battery life, and the smartwatch's materials' effects on the environment. Concerns that certain elements of the partnership might not be in line with sustainable practices are reflected in the Falsity value (0.15).

5.6. Decision-Making Using Neutrosophic Mcdm Techniques

MCDM techniques like VIKOR or TOPSIS can be used to analyze the best course of action for this co-branding initiative.

Step 1: Construct a Neutrosophic Decision Matrix

Each criterion is assigned a weight based on its importance. Suppose the weights are:

- Brand Image Alignment (W1) = 0.3
- Technological Innovation Fit (W2) = 0.3
- Market Demand (W3) = 0.2
- Sustainability Considerations (W4) = 0.2

Step 2: Compute the Weighted Neutrosophic Scores

Using a weighted sum method, we compute the aggregated scores for each criterion.
Truth Score (T):

$$T_{total} = (0.3 \times 0.85) + (0.3 \times 0.9) + (0.2 \times 0.75) + (0.2 \times 0.7) = 0.815$$

Indeterminacy Score (I):

$$I_{total} = (0.3 \times 0.2) + (0.3 \times 0.15) + (0.2 \times 0.3) + (0.2 \times 0.35) = 0.23$$

Falsity Score (F):

$$F_{total} = (0.3 \times 0.05) + (0.3 \times 0.05) + (0.2 \times 0.1) + (0.2 \times 0.15) = 0.07$$

Step 3: Interpret the Results

- The co-branding initiative is promising, as evidenced by the high truth value (0.815), which shows a strong alignment between brand identity and technological compatibility.
- The moderate indeterminacy (0.23) points to certain uncertainties that require more investigation, mainly in the areas of market demand and sustainability.
- The partnership has few risks or contradictions, as indicated by the low falsity score (0.07).

Step 4: Compare with Other Co-Branding Strategies

Other co-branding options, such as Apple and Gucci or Louis Vuitton and Samsung, can be assessed using the same Neutrosophic MCDM technique to make sure the LV-Apple partnership is the best choice. The most practical approach would be the one with the highest truth value and the lowest falsity value.

Using Neutrosophic MCDM to assess co-branding choices:

- Indeterminacy values are used to quantify consumer response uncertainty.
- Falsity values are used to capture potential risks.
- High Truth values indicate strong brand synergies.

With its strong technological compatibility, market appeal, and brand alignment, the Louis Vuitton x Apple Smartwatch shows great promise as a co-branding strategy. However, targeted marketing tactics and sustainable product development should be used to address market acceptance and sustainability uncertainties.

The efficacy of Neutrosophic MCDM in strategic decision-making for astute luxury co-branding is illustrated in this case study. Brands can maximize their partnerships and guarantee long-term success by combining fuzzy logic, professional assessments, and real-world uncertainties.

6. CONCLUSION

Co-branding is now a crucial strategy for maintaining a competitive advantage since the luxury industry's adoption of smart technology has transformed conventional branding tactics. To assess co-branding tactics for intelligent luxury goods, this chapter presented a Neutrosophic MCDM framework that takes into account hesitancy, uncertainty, and divergent expert opinions. When it comes to handling the inherent complexity of consumer perception, brand synergy, and technological integration, traditional decision-making models frequently fail. A more thorough, data-driven method of evaluating the viability of co-branding is made possible by the use of neutrophilic sets, neutrophilic weighted aggregation operators, and neutrophilic similarity measures.

This study showed that, in comparison to conventional approaches, Neutrosophic MCDM offers a more flexible and accurate framework for decision-making through computational models and real-world case studies. To maximize both brand equity and innovation potential, the results emphasize the significance of assessing truth (T), indeterminacy (I), and falsity (F) values in co-branding evaluations. By making strategic decision-making easier, the suggested model enables luxury brands to collaborate with tech companies in ways that improve long-term sustainability, consumer engagement, and product differentiation.

Future Work

While the current study provides a strong foundation for decision-making in smart luxury co-branding, several areas warrant further exploration:

1. To dynamically evaluate customer perception in real time, future research can investigate sentiment analysis and AI-driven predictive modeling. By continuously updating weight assignments based on changing market trends, machine learning algorithms can improve the accuracy of Neutrosophic decision-making models.
2. The Neutrosophic framework can incorporate the use of blockchain technology for intellectual property protection, authenticity verification, and supply chain transparency. Co-branding agreements might be automated by smart contracts, guaranteeing smooth cooperation between luxury and tech companies.
3. Before releasing joint products, creating a Digital Twin framework for intelligent luxury co-branding could offer virtual simulations to gauge consumer reaction and brand compatibility. Further research should examine how Neutrosophic MCDM can incorporate ethical, circular economy, and environmental considerations into smart luxury co-branding, as sustainability becomes a crucial determinant in consumer decision-making.
4. Extending the current model to examine how regional market dynamics and cultural preferences affect the effectiveness of astute luxury co-branding alliances.

Future studies can further improve decision-making frameworks by taking these developments into account. This will guarantee that luxury-tech partnerships continue to be creative, customer-focused, and strategically viable in a market that is becoming more digital and dynamic.

REFERENCES

Anuar, N. S. K., Abidin, I. H. Z., Samsudin, A., Adzmy, A., & Zulkornain, L. H. (2021). Revisiting the Issues of Co-Branding. In Jurnal Intelek (Vol. 16, Issue 2, p. 11). https://doi.org/DOI: 10.24191/ji.v16i2.391

Büyüközkan, G., & Görener, A. (2015). Evaluation of product development partners using an integrated AHP-VIKOR model. In Kybernetes (Vol. 44, Issue 2, p. 220). Emerald Publishing Limited. https://doi.org/DOI: 10.1108/K-01-2014-0019

Das, A. K. (2016). On partially included intuitionistic fuzzy rough relations. *Afrika Matematika*, *27*(5-6), 993–1001. DOI: 10.1007/s13370-016-0395-2

Das, A. K. (2018). Weighted fuzzy soft multiset and decision-making. *International Journal of Machine Learning and Cybernetics*, *9*(5), 787–794. DOI: 10.1007/s13042-016-0607-y

Das, A. K., & Granados, C. (2021). An advanced approach to fuzzy soft group decision-making using weighted average ratings. *SN Computer Science*, *2*(6), 471. DOI: 10.1007/s42979-021-00873-5

Das, A. K., & Granados, C. (2022a). A new fuzzy parameterized intuitionistic fuzzy soft multiset theory and group decision-making. *Journal of Current Science and Technology*, *12*(3), 547–567. DOI: 10.14456/jcst.2022.42

Das, A. K., & Granados, C. (2022b). FP-intuitionistic multi fuzzy N-soft set and its induced FP-Hesitant N soft set in decision-making. *Decision Making: Applications in Management and Engineering*, *5*(1), 67–89. DOI: 10.31181/dmame181221045d

Das, A. K., & Granados, C. (2022c). IFP-intuitionistic multi fuzzy N-soft set and its induced IFP-hesitant N-soft set in decision-making. *Journal of Ambient Intelligence and Humanized Computing*, *14*(8), 10143–10152. DOI: 10.1007/s12652-021-03677-w

Das, A. K., & Granados, C. (2023). Preference intuitionistic fuzzy rough relation and its theoretical approach. *Journal of the Indian Mathematical Society*, *90*(3-4), 199–212.

Das, A. K., Granados, C., & Bhattacharya, J. (2022). Some new operations on fuzzy soft sets and their applications in decision-making. *Songklanakarin Journal of Science and Technology*, *44*(2), 440–449. DOI: 10.14456/sjst-psu.2022.61

Das, A. K., Gupta, N., & Granados, C. (2023). Weighted hesitant bipolar-valued fuzzy soft set in decision-making. *Songklanakarin Journal of Science and Technology, 45*(6), 681–690. https://sjst.psu.ac.th/journal/45-6/10.pdf

Das, A. K., Gupta, N., Mahmood, T., Tripathy, B. C., Das, R., & Das, S. (2024). An innovative fuzzy multi-criteria decision making model for analyzing anthropogenic influences on urban river water quality. *Iran Journal of Computer Science*. Advance online publication. DOI: 10.1007/s42044-024-00211-x

Das, A. K., Gupta, N., Mahmood, T., Tripathy, B. C., Das, R., & Das, S. (2025a). An efficient water quality evaluation model using weighted hesitant fuzzy soft sets for water pollution rating. In Kumar, A., Kumar, P., Rathee, S., & Kumar, B. (Eds.), *Mechatronics: Concepts, Tools, Applications, and New Trends* (1st ed., p. 17). CRC Press., DOI: 10.1201/9781003494478-10

Das, A. K., Gupta, N., Patra, S., Tripathy, B. C., Das, R., & Datta, R. (2025b). An effective fuzzy soft decision-making model for rating water pollution: A case study of Manu, the longest river in Tripura, India, Vietnam. *Journal of Science and Technology*.

Dey, P. P., Pramanik, S., & Giri, B. C. (2016). TOPSIS for solving multi-attribute decision-making problems under bi-polar neutrosophic environment. In Smarandache, F., & Pramanik, S. (Eds.), *New trends in neutrosophic theory and applications* (pp. 65–77). Pons Editions.

Drakou, A., Papachristou, E., Tarnanidis, T., & Bilalis, N. (2024). Transforming fashion education: The impact of ai-driven 3d design on learning, creativity, and sustainability Proceedings of the 13th Hellenic Conference on Artificial Intelligence, https://doi.org/DOI: 10.1145/3688671.3688794

Fakhravar, H., & Tahami, H. (2022). International Co-Branding and Firms Finance Performance. In arXiv (Cornell University). Cornell University. https://doi.org/DOI: 10.48550/arXiv.2202

Gkiouzepas, L., & Tarnanidis, T. (2024). Revisiting Visual Communication: Conceptualizing How Extremity of Claims Work in the Context of Advertising Visuals. In Tarnanidis, T., & Sklavounos, N. (Eds.), *New Trends in Marketing and Consumer Science* (pp. 337–359). IGI Global., DOI: 10.4018/979-8-3693-2754-8.ch018

. Granados, C., Das, A.K., Das, B. (2022). Some continuous neutrosophic distributions with neutrosophic parameters based on neutrosophic random variables. Advances in the Theory of Nonlinear Analysis and its Applications, 33(2): 293-308. https://doi.org/DOI: 10.31197/atnaa.1056480

. Granados, C., Das, A.K., Osu, B. (2023). Weighted neutrosophic soft multiset and its application to decision making. Yugoslav journal of operations research, 33(2): 293-308. https://doi.org/DOI: 10.2298/YJOR220915034G

Halwani, L. (2020). The Online Experience of Luxury Consumers: Insight into Motives and Reservations. In International Journal of Business and Management (Vol. 15, Issue 11, p. 157). Canadian Center of Science and Education. https://doi.org/DOI: 10.5539/ijbm.v15n11p157

Huang, R. (2023). Marketing Strategy and Operation Mode of Luxury Goods Lead to Brand Premium. *BCP Business & Management*, *47*, 49–54. DOI: 10.54691/bcpbm.v47i.5168

Ismyrlis, V., Tarnanidis, T., & Moschidis, E. (Eds.). (2020). *Knowledge Management Practices in the Public Sector*. IGI Global., DOI: 10.4018/978-1-7998-1940-0

Javornik, A., Duffy, K., Rokka, J., Scholz, J., Nobbs, K., Motala, A., & Goldenberg, A. (2021). Strategic approaches to augmented reality deployment by luxury brands. *Journal of Business Research*, *136*, 284–292. DOI: 10.1016/j.jbusres.2021.07.040

Jitta, A. S., Manda, V. K., & Tarnanidis, T. (2025). Geospatial Technologies for Smart Cities. In Darwish, D., & Chemingui, H. (Eds.), *Recent Trends in Geospatial AI* (pp. 103–136). IGI Global Scientific Publishing., DOI: 10.4018/979-8-3693-8054-3.ch004

Kamalakannan, R., Ramesh, C., Shunmugasundaram, M., Sivakumar, P., & Mohamed, A. (2020). Evaluvation and selection of suppliers using TOPSIS. In *Materials Today Proceedings* (Vol. 33, p. 2771). Elsevier BV., DOI: 10.1016/j.matpr.2020.02.105

Kampakaki, E., Papachristou, E., Tarnanidis, T., & Bilalis, N. (2022). What are the most important digital qualifications of the employees in the clothing and fashion companies? In *Proceedings of the 12th Hellenic Conference on Artificial Intelligence* (pp. 1-7). https://doi.org/DOI: 10.1145/3549737.3549804

Karypidis, M., Tarnanidis, T., & Papachristou, E. (2024). Treating Textile Effluents for Sustainable Fashion and Green Marketing. In Tarnanidis, T., Papachristou, E., Karypidis, M., & Ismyrlis, V. (Eds.), *Driving Green Marketing in Fashion and Retail* (pp. 166–179). IGI Global., DOI: 10.4018/979-8-3693-3049-4.ch010

Khaliq, L. N., & Manda, V. K. (2024). Customer Experience in the Web 3.0 Era: The Meeting of Blockchain and the Metaverse. In Majeed, M., Ofori, K., Amoako, G., Alolo, A., & Awini, G. (Eds.), *The Rise of Blockchain Applications in Customer Experience* (pp. 258–274). IGI Global., DOI: 10.4018/978-1-6684-7649-9.ch015

Liu, Y. (2023). Digital Transformation in the Luxury Industry: Strategies, Challenges, And Opportunities. In Highlights in Business Economics and Management (Vol. 23, p. 650). DOI: 10.54097/xgftec52

Manda, V. K., Bikkina, A., & Tarnanidis, T. (2025b). Future trends in Virtual Reality (VR). In *Virtual Reality Technologies and Real Life Applications* (pp. 83–96). Deep Science Publishing., DOI: 10.70593/978-81-982935-1-0_5

Manda, V. K., Manda, V. K., & Katneni, V. (2022). Blockchain for the asset management industry. *World Review of Science, Technology and Sustainable Development, 19*(1-2), 170–185. DOI: 10.1504/WRSTSD.2023.127266

Manda, V. K., Sagi, S., & Yadav, A. (2024). Blockchain in Advertising and Marketing: Revolutionizing the Industry Through Transparency and Trust. In Tarnanidis, T., & Sklavounos, N. (Eds.), *New Trends in Marketing and Consumer Science* (pp. 89–112). IGI Global., DOI: 10.4018/979-8-3693-2754-8.ch005

Manda, V. K., Tarnanidis, T., & Majeed, M. (2025a). Advanced topics in robotics. In D. Darwish (Ed.), Advances in Robots Technologies and Implementations (pp. 96-112). Deep Science Publishing. https://doi.org/DOI: 10.70593/978-81-983916-1-2_5

Milanesi, M., Guercini, S., & Runfola, A. (2022). Let's play! Gamification as a marketing tool to deliver a digital luxury experience. In Electronic Commerce Research (Vol. 23, Issue 4, p. 2135). Springer Science+Business Media. https://doi.org/DOI: 10.1007/s10660-021-09529-1

Mukherjee, A., & Das, A. K. (2015a). Application of interval valued intuitionistic fuzzy soft set in investment decision making, *2015 Fifth International Conference on Advances in Computing and Communications (ICACC)*, Kochi, India, 2015, pp. 61-64, https://doi.org/DOI: 10.1109/ICACC.2015.37

Mukherjee, A., & Das, A. K. (2015b). Relations on intuitionistic fuzzy soft multi sets. In Kim, K. (Ed.), *Information Science and Applications. Lecture Notes in Electrical Engineering* (Vol. 339). Springer., DOI: 10.1007/978-3-662-46578-3_71

Mukherjee, A., & Das, A. K. (2015c). Interval valued intuitionistic fuzzy soft multi set theoretic approach to decision making problems, 2015 International Conference on Computer, Communication and Control (IC4), Indore, India, 2015, pp. 1-5, https://doi.org/DOI: 10.1109/IC4.2015.7375640

Mukherjee, A., & Das, A. K. (2016a). Application of fuzzy soft multi sets in decision-making problems. In: Nagar, A., Mohapatra, D., Chaki, N. (eds) Proceedings of 3rd International Conference on Advanced Computing, Networking and Informatics. Smart Innovation, Systems and Technologies, vol. 43. Springer, New Delhi. https://doi.org/DOI: 10.1007/978-81-322-2538-6_3

Mukherjee, A., & Das, A. K. (2016b). Topological structure formed by intuitionistic fuzzy rough relations. *Journal of the Indian Mathematical Society*, *83*(1-2), 135–144.

Mukherjee, A., & Das, A. K. (2022). Einstein operations on fuzzy soft multisets and decision making. *Boletim da Sociedade Paranaense de Matematica*, *40*, 1–10. DOI: 10.5269/bspm.32546

Mukherjee, A., & Das, A. K. (2023). Essentials of fuzzy soft multisets: Theory and applications, Springer Nature Singapore Pte Ltd., 1-151. ISBN: 978-981-19-2759-1, https://doi.org/DOI: 10.1007/978-981-19-2760-7

Mukherjee, A., Das, A. K., Gupta, N., Patra, S., Mahmood, T., Das, S., & Das, R. (2025). *A novel generalized interval-valued neutrosophic rough soft set framework for enhanced decision-making: Application in water quality assessment.* Transactions on Fuzzy Sets and System.

Pramanik, S., Dalapati, S., Alam, S., & Roy, T. K. (2018). VIKOR-based MAGDM strategy under bipolar neutrosophic set environment. *Neutrosophic Sets and Systems*, *19*, 57–69.

Pramanik, S., Das, S., Das, R., & Tripathy, B. C. (2023). Neutrosophic BWM-TOPSIS strategy under SVNS environment. *Neutrosophic Sets and Systems*, *56*, 178–189.

Rathi, R., Garg, R., Kataria, A., & Chhikara, R. (2022). Evolution of luxury marketing landscape: A bibliometric analysis and future directions. *Journal of Brand Management*, *29*(3), 241–257. DOI: 10.1057/s41262-022-00273-x

Remy, N., Catena, M., & Durand-Servoingt, B. (2023). Digital inside: Get wired for the ultimate luxury experience. https://www.mckinsey.com/capabilities/growth-marketing-and-sales/our-insights/digital-inside-get-wired-for-the-ultimate-luxury-experience

Smarandache, F. (2005). Neutrosophic set: A generalization of the intuitionistic fuzzy sets. *International Journal of Pure and Applied Mathematics*, *24*, 287–297.

Smarandache, F. A. (1998). unifying field in logics: Neutrosophic logic. Neutrosophy, neutrosophic set, neutrosophic probability and statistics. Rehoboth: American Research Press.

Tang, Y. (2023). Analysis of the Current Chinese Consumer for Luxury Goods - Generation Z. *Lecture Notes in Education Psychology and Public Media*, *4*(1), 531–537. DOI: 10.54254/2753-7048/4/2022171

Tarnanidis, T. (2022). Measuring Sustainable Entrepreneurship in the Food Sector. [IJSECSR]. *International Journal of Sustainable Entrepreneurship and Corporate Social Responsibility*, *7*(1), 1–8. https://doi.org/DOI:10.4018/IJSECSR.314157. DOI: 10.4018/IJSECSR.314157

Tarnanidis, T. (2023). Understanding reference points to make purchase decisions: Overview, phases, and time-features. *Marketing Science & Inspirations*, *18*(4), 2–13. DOI: 10.46286/msi.2023.18.4.1

Tarnanidis, T. (2024a). Exploring the Impact of Mobile Marketing Strategies on Consumer Behavior: A Comprehensive Analysis. *International Journal of Information, Business and Management*, *16*(2), 1–17.

Tarnanidis, T. (Ed.). (2024b). *Consumer Experience and Decision-Making in the Metaverse*. IGI Global., DOI: 10.4018/979-8-3693-4167-4

Tarnanidis, T. (2024e). What Lies Ahead for Marketing in Wholesale and Retailing. In Tarnanidis, T. (Ed.), *Reshaping Marketing Science in Wholesaling and Retailing* (pp. 354–363). IGI Global., DOI: 10.4018/979-8-3693-6145-0.ch015

Tarnanidis, T., & Gkiouzepas, L. (2024). Leveraging Marketing Strategies to Help Consumers Make Decisions Using Metaverse Technologies. In Tarnanidis, T. (Ed.), *Consumer Experience and Decision-Making in the Metaverse* (pp. 158–171). IGI Global., DOI: 10.4018/979-8-3693-4167-4.ch009

Tarnanidis, T., Karypidis, M., Manda, V. K., Bilali, S., & Papachristou, E. (2025b). Discovering the Patterns of Slow Fashion Consumption in the Rich Market. In Tarnanidis, T., Papachristou, E., Karypidis, M., & Manda, V. (Eds.), *Sustainable Practices in the Fashion and Retail Industry* (pp. 159–178). IGI Global Scientific Publishing., DOI: 10.4018/979-8-3693-9959-0.ch007

Tarnanidis, T., Papachristou, E., Karypidis, M., & Ismyrlis, V. (Eds.). (2023a). *Social Media and Online Consumer Decision Making in the Fashion Industry*. IGI Global., DOI: 10.4018/978-1-6684-8753-2

Tarnanidis, T., Papachristou, E., Karypidis, M., & Manda, V. K. (2025e). Luxury Fashion Consumption and Branding in the Affluent Market. In Tarnanidis, T. (Ed.), *Cross-Cultural Perspectives on Luxury Marketing* (pp. 161–172). IGI Global., DOI: 10.4018/979-8-3693-7535-8.ch008

Tarnanidis, T., Papathanasiou, J., Ismyrlis, V., & Kittu Manda, V. (Eds.). (2025c). *Multiple-Criteria Decision-Making (MCDM) Techniques and Statistics in Marketing*. IGI Global., DOI: 10.4018/979-8-3693-9122-8

Tarnanidis, T., Papathanasiou, J., Mareschal, B., & Vlachopoulou, M. (2025f). MCDM Review in marketing and managerial decisions: Practical implications and Future research. *Management Science Letters*, *15*(1), 45–54. Advance online publication. DOI: 10.5267/j.msl.2024.3.004

Tarnanidis, T., Papathanasiou, J., Osei-Frimpong, K., & Owusu-Frimpong, N. (2024b). How young consumers are influenced by the valence of positive and negative frames: A cross-cultural perspective. *International Journal of Information and Decision Sciences*, *16*(2), 109–133. DOI: 10.1504/IJIDS.2024.139825

Tarnanidis, T., Papathanasiou, J., & Subeniotis, D. (2023e). How can we teach young consumers to become entrepreneurs? Application of specific instruments. *World Review of Entrepreneurship, Management and Sustainable Development*, *19*(6), 431–453. DOI: 10.1504/WREMSD.2023.133735

Tarnanidis, T., Papathanasiou, J., Vlachopoulou, M., & Mareschal, B. (2023d). Review and the Use of PROMETHEE Methods in Marketing (Problems). In Tarnanidis, T., Vlachopoulou, M., & Papathanasiou, J. (Eds.), *Influences of Social Media on Consumer Decision-Making Processes in the Food and Grocery Industry* (pp. 196–212). IGI Global., DOI: 10.4018/978-1-6684-8868-3.ch009

Tarnanidis, T., & Tarnanidis, J. (2024). Analyzing the SERVQUAL method for senior high school education's gaps and factors. *European Journal of Educational Management*, *7*(4), 227–243. DOI: 10.12973/eujem.7.4.227

Tarnanidis, T., Vlachopoulou, M., Mareschal, B., & Papathanasiou, J. (2025d). Marketing in the Metaverse Through MCDM Methods and Statistics. In Tarnanidis, T., Papathanasiou, J., Ismyrlis, V., & Kittu Manda, V. (Eds.), *Multiple-Criteria Decision-Making (MCDM) Techniques and Statistics in Marketing* (pp. 249–266). IGI Global Scientific Publishing., DOI: 10.4018/979-8-3693-9122-8.ch011

Tarnanidis, T., Vlachopoulou, M., & Papathanasiou, J. (Eds.). (2023c). *Influences of social media on consumer decision-making processes in the food and grocery industry*. IGI Global., DOI: 10.4018/978-1-6684-8868-3

Tarnanidis, T. K. (Ed.). (2024d). *Reshaping Marketing Science in Wholesaling and Retailing*. IGI Global., DOI: 10.4018/979-8-3693-6145-0

Tarnanidis, T. K. (Ed.). (2025). *Cross-Cultural Perspectives on Luxury Marketing*. IGI Global., DOI: 10.4018/979-8-3693-7535-8

Tarnanidis, T. K., Papachristou, E., Karypidis, M., & Ismyrlis, V. (2023b). How social media affects consumer behavior in the fashion industry. In *Social Media and Online Consumer Decision Making in the Fashion Industry* (pp. 324–337). IGI Global., DOI: 10.4018/978-1-6684-8753-2.ch017

Tarnanidis, T. K., Papachristou, E., Karypidis, M., & Ismyrlis, V. (2024a). *Driving Green Marketing in Fashion and Retail. Driving Green Marketing in Fashion and Retail.* IGI Global., DOI: 10.4018/979-8-3693-3049-4

Tarnanidis, T. K., Papachristou, E., Karypidis, M., & Manda, V. (Eds.). (2025a). *Sustainable Practices in the Fashion and Retail Industry*. IGI Global., DOI: 10.4018/979-8-3693-9959-0

Tarnanidis, T. K., & Sklavounos, N. (Eds.). (2024). *New Trends in Marketing and Consumer Science*. IGI Global., DOI: 10.4018/979-8-3693-2754-8

Tarnanidis, T. K. (2024c). Fashion consumers' response to the COVID-19 pandemic and today's future. *International Journal of Science and Research Archive, 12*(02), 648–655. DOI: 10.30574/ijsra.2024.12.2.1286

Tsaples, G., & Tarnanidis, T. (2020). A System Dynamics Model and Interface for the Simulation and Analysis of Milk Supply Chains. In I. Management Association (Ed.), Supply Chain and Logistics Management: Concepts, Methodologies, Tools, and Applications (pp. 108-135). IGI Global Scientific Publishing. https://doi.org/ DOI: 10.4018/978-1-7998-0945-6.ch006

Wittstruck, D., & Teuteberg, F. (2012). Integrating the concept of sustainability into the partner selection process: a fuzzy-AHP-TOPSIS approach. In International Journal of Logistics Systems and Management (Vol. 12, Issue 2, p. 195). Inderscience Publishers. https://doi.org/DOI: 10.1504/IJLSM.2012.047221

Xu, Q. (2020). A Study of the Marketing Strategies of High-End Luxury Brands: The Case of Hermes. In *Advances in economics, business and management research/ Advances in Economics, Business and Management Research*. Atlantis Press., DOI: 10.2991/aebmr.k.201128.088

Zadeh, L. A. (1965). Fuzzy sets. *Information and Control, 8*(3), 338–353. DOI: 10.1016/S0019-9958(65)90241-X

Chapter 7
Connected Couture: Co-Branding in the Era of Smart Luxury Fashion

Sushila Sahani
https://orcid.org/0000-0002-5739-144X
Lovely Professional University, India

ABSTRACT

The luxury fashion industry is undergoing a major transformation driven by digital innovation and tech partnerships. This chapter explores how collaborations between fashion brands and tech companies are reshaping the meaning of luxury for modern consumers. It highlights the growing demand for personalization, immersive digital experiences, and sustainability as key influences on brand strategies. It discusses the rise of smart luxury products from sensor-embedded clothing to AI-powered personalization. It also delves into how technologies such as artificial intelligence, IoT, and data analytics are driving new fashion experiences. Digital fashion's role is explored through examples like Balenciaga's Fortnite collaboration and Gucci's presence on Roblox, showcasing new forms of digital ownership and exclusivity. The chapter also examines how technology is helping brands meet ethical and sustainability goals, while noting concerns like privacy and tech's environmental footprint.

INTRODUCTION

The luxury fashion industry, long defined by exclusivity, craftsmanship, and heritage, is undergoing a transformative shift driven by technological advancements and changing consumer preferences. Co-branding, a strategy where two or more brands collaborate to create a product bearing both names, has become a cornerstone for luxury brands seeking to innovate and expand their market reach. In recent years,

DOI: 10.4018/979-8-3373-3010-5.ch007

the emergence of "smart luxury fashion", luxury products enhanced with technologies like blockchain, augmented reality (AR), virtual reality (VR), and the Internet of Things (IoT) has redefined co-branding, enabling partnerships with technology companies to create unique, digitally enhanced offerings.

In fashion, luxury hinges on uniqueness and a bold brand identity that sets a label apart. Successful luxury branding relies on exclusivity, sharp communication, and precise positioning. A brand's identity, rooted in history, culture, and reputation drives its appeal. Yet, traditional strategies often clash with luxury's focus on prestige over mass appeal. Enhancing brand perception often involves establishing flagship stores in prestigious locations and participating in major fashion shows, which reinforce the brand's aspirational image. To preserve exclusivity, luxury brands rely on premium pricing and carefully controlled distribution, while also navigating the need for differentiation across various market segments to drive growth without compromising their elite status. While luxury branding emphasizes exclusivity and a distinct brand identity, modern market dynamics often encourage collaborations as a strategy for growth and innovation. One such approach is co-branding, which allows luxury and non-luxury brands alike to expand their reach. It allows brands create unique product experiences without compromising individual brand value. Such partnership model offers both opportunities and challenges, shaping how consumers perceive and interact with the combined brand offering.

Co-branding unites two brands to create a single product, blending their strengths. It's a powerful tool for growth, pooling resources and expanding market presence. The goals? Wider exposure, shared costs, and a stronger competitive edge. For success, brands must align, sharing similar value and complementary strengths. Defining the target audience and roles clearly is critical. While co-branding boosts visibility and reach, it risks brand confusion or internal rivalry. Additionally, consumer perception plays a vital role, as attitudes toward co-branded products are shaped by how favorably the partner brands are viewed. Collaborations between high-equity brands generally generate more positive consumer responses, whereas significant disparities in brand equity can lead to uncertainty and skepticism about the co-branded offering.

Luxury brands can't rely solely on status and exclusivity anymore. The 2008 financial crisis proved even affluent consumers feel economic shocks, with luxury demand dropping 8% (Danziger, 2015). Co-branding emerged as a vital tool for growth and resilience. Partnering with high-street retailers lets luxury brands reach broader audiences while preserving prestige. However, while co-branding can be lucrative, it must be approached cautiously, as excessive or poorly aligned collaborations risk diluting a brand's luxury image. This chapter explores the dynamics of co-branding in the era of smart luxury fashion, focusing on how luxury fashion brands collaborate with technology firms to produce innovative products. Through theoretical insights, case studies, and an analysis of benefits and challenges, this chapter aims to provide a comprehensive understanding of this evolving landscape.

Connected Couture: A New Paradigm

Connected couture represents a revolutionary evolution in luxury fashion, merging traditional artisanal craftsmanship with cutting-edge digital technologies to redefine exclusivity, personalization, and sustainability for today's tech-savvy, socially conscious consumers. This innovative approach integrates Internet of Things (IoT)-enabled garments, artificial intelligence (AI)-driven personalization, and blockchain-based authentication to create a seamless blend of heritage and innovation. IoT-infused couture incorporates sensors, microchips, and conductive fabrics into high-end clothing, enabling garments to interact in real time with the wearer and their environment enhancing comfort, functionality, and uniqueness without compromising on aesthetics. AI is also transforming couture by enabling hyper-personalized fashion experiences. Machine learning algorithms analyze consumer data ranging from style preferences and sizing to social media behavior to generate bespoke designs tailored to individual lifestyles. Blockchain technology plays a crucial role in ensuring authenticity, ethical sourcing, and transparency. By recording a garment's lifecycle from raw materials to final purchase, blockchain acts as a digital passport, offering traceability and reassurance.

As connected couture continues to evolve, it caters to a generation that values both innovation and heritage. Looking forward, advancements in flexible electronics, generative AI, and decentralized blockchain networks are expected to make connected couture more accessible and sustainable. With the global *smart fabrics market* projected to reach around USD 48.34 billion by 2032, growing at a CAGR of 30.9% (Beyond Market Insights. (2024), brands that successfully balance craftsmanship with technology are poised to lead the next wave of luxury fashion. In this digital era, connected couture is not just redefining luxury; it's shaping its future.

THE EVOLUTION OF LUXURY IN THE DIGITAL AGE

Historically, luxury was defined by rare, handcrafted items such as fine jewelry, bespoke clothing, and high-end cars available only to the elite. These possessions served as status symbols, often driven by the "Veblen effect," where visible consumption signaled wealth and social standing (Bagwell et al., 1996). In the early 20th century, luxury was scarce and rooted in exclusivity and craftsmanship. By the late 20th century, globalization and brand expansion began to change this dynamic. Luxury brands like LVMH and Kering broadened their reach through marketing, mass production of entry-level goods (like perfumes and accessories), and global distribution. This led to the rise of "mass luxury," making designer products more accessible to the middle class. As a result, luxury shifted from being purely about rarity to also emphasizing brand image and perception.

In the 21st century, the digital age brought further transformation. Social media amplified aspirational lifestyles and fueled trends like "quiet luxury," where minimalist, logo-free designs subtly signal wealth to those in the know. At the same time, experiential luxury began to gain traction offering exclusive travel, wellness retreats, and curated cultural events. This marks a shift from owning physical items to valuing unique, personalized experiences that reflect changing consumer priorities like authenticity and emotional connection.

Smart luxury also emerged in the late 1990s and early 2000s with early experiments in wearable technology (King, 2011). Though basic at first, these innovations began blending fashion with functionality, laying the groundwork for what would become "connected couture" a new form of luxury that integrates technology, design, and lifestyle.

One of the earliest milestones was Philips' Lumalive Textile Displays. This innovative technology embedded flexible LED displays into fabric, allowing garments to showcase dynamic visuals, patterns, and text. Demonstrated at the IFA trade show in Berlin, Lumalive jackets and skirts transformed clothing into wearable screens, hinting at a future where garments could entertain or communicate. Although originally designed for promotional and advertising purposes, Lumalive's potential inspired designers to imagine couture that could respond to its environment. However, limitations like bulky electronics and high production costs hindered its commercial viability challenges that continue to influence the development of modern IoT-enabled fashion (El-Sayed & El-Fanagely, 2022). At the same time, brands like CuteCircuit emerged as trailblazers in fashion tech and introduced emotionally responsive garments such as the Hug Shirt (Cranny-Francis, 2013), which used embedded sensors and actuators to simulate the sensation of a hug via Bluetooth. By 2008, the brand unveiled the Galaxy Dress an elegant piece adorned with 24,000 full-color LEDs layered beneath silk, powered by discreet microelectronics. Displayed at the Museum of Science and Industry in Chicago, the dress exemplified how technology could elevate, rather than disrupt, luxury fashion (Li, 2018).

Academic and artistic innovators also contributed to the early landscape of smart fashion. MIT's Media Lab, through its Wearable Computing initiative (1997), explored responsive garments using conductive fabrics and microcontrollers technologies that would later influence high-end applications (Berzowska, 2005). Meanwhile, designer Hussein Chalayan's 2007 collection incorporated motorized dresses that changed shape on the runway, merging mechanical engineering with haute couture to craft futuristic fashion narratives. These avant-garde experiments demonstrated that technology could enrich storytelling, exclusivity, and experiential depth in luxury fashion (Yeo, 2021). Together, these early efforts marked a critical turning point, showing that technology could amplify the emotional, functional, and aesthetic qualities of luxury clothing. Yet despite their vision, widespread adoption was slowed by technical limitations such as bulky components, limited battery life, and high production costs as well as resistance from a luxury market still rooted in tradition. Nonetheless, these foundational explorations paved the way for today's more refined, integrated, and consumer-ready connected couture.

UNDERSTANDING CO-BRANDING IN FASHION

Co-branding, also known as brand alliance, involves two or more brands partnering to create a product or service that leverages the strengths of each (Keller, 2003). In the luxury fashion industry, co-branding has historically been used to enhance brand image, access new consumer segments, and create exclusive products that generate significant media attention. A notable example is the 2008 collaboration between Bugatti and Hermès, which produced the Bugatti Veyron FBG Hermès supercar, blending automotive engineering with luxury fashion aesthetics (Hypebeast, 2008). Such partnerships capitalize on shared values and complementary expertise, resulting in products that are greater than the sum of their parts.

Academic research highlights the effectiveness of co-branding in luxury fashion. For instance, co-branding with high street retailers, such as H&M's collaborations with Versace and Balmain, has been shown to be a more effective "masstige" (mass prestige) strategy than downward brand extensions, eliciting favorable consumer responses without diluting brand prestige (Shan et al., 2022) The rise of technology has expanded co-branding opportunities, with luxury brands now partnering with tech firms to create smart luxury products that align with the digital preferences of younger consumers.

Co-branding in fashion serves several strategic purposes. Brands can borrow positive associations from one another. A fashion brand partnering with a technology brand, for instance, may benefit from associations with innovation and modernity. Collaborations help brands reach new demographics or geographic markets. For example, a luxury brand collaborating with a streetwear label may tap into younger, urban audiences. The scarcity and exclusivity of co-branded products often create a "drop culture" effect, generating hype and encouraging immediate purchase behavior. In a saturated market, co-branding offers a way to differentiate offerings, creating buzzworthy products that stand out through their novelty. Co-branding in fashion is no longer limited to simple logo placement or joint promotions; it is a complex, strategic practice that shapes brand identity, consumer perception, and market reach.

THE RISE OF SMART LUXURY FASHION

Smart luxury fashion refers to luxury products that incorporate advanced technologies to enhance functionality, interactivity, and exclusivity. These technologies include blockchain for product authenticity and traceability, IoT for connected wearable, AR/VR for immersive retail experiences, and artificial intelligence (AI) for personalized design. The adoption of smart technologies is driven by the need to remain relevant in a digital-first world and to cater to tech-savvy consumers, particularly millennial and Generation Z, who value innovation, sustainability, and personalized experiences.

The luxury fashion industry has seen significant growth in recent years, with leading firms experiencing double-digit growth (Cabigiosu, 2020). This growth is partly fueled by digital innovations that allow brands to maintain their exclusivity while embracing new business models. For example, blockchain technology enables brands to create digital passports for products, ensuring authenticity and enhancing resale value (Kuusela, 2024). Similarly, NFTs have emerged as a new form of luxury collectible, offering exclusivity in the digital realm Joy, (Abeyatunge, 2024) notes that luxury brands are leveraging digital technologies to make operations more agile and responsive without compromising their core values of craftsmanship and personalization

KEY TECHNOLOGIES POWERING SMART LUXURY FASHION

The integration of advanced technologies into luxury fashion is redefining the industry, enabling brands to create innovative, interactive, and exclusive products that resonate with tech-savvy consumers. These technologies not only enhance functionality and personalization but also reinforce the exclusivity and craftsmanship that define luxury. Below, we explore the key technologies driving smart luxury fashion, supported by examples of collaborations between fashion brands and technology firms.

Blockchain for Authenticity and Sustainability

Blockchain is a decentralized, tamper-proof digital ledger that records transactions across a network of computers, ensuring data integrity and transparency. In luxury fashion, it tracks a product's lifecycle from raw material sourcing to manufacturing, retail, and resale using immutable records. Each product is assigned a unique digital identifier, often linked to a QR code or NFC chip, creating a "digital passport" that consumers can access to verify authenticity and ethical sourcing (Shee Weng, 2024).

Applications

- **Authenticity verification:** Blockchain ensures products are genuine, critical for high-value items like handbags or watches. The Aura Blockchain Consortium (LVMH, Prada, Cartier) uses blockchain to create digital certificates that verify product origin and ownership, enhancing trust in both primary and secondary markets (Ballini et al. 2025).
- **Supply chain transparency:** The success of the luxury fashion industry largely relies on safeguarding brand value, ensuring quality and authenticity, and maintaining efficient supply chains . Achieving this requires a shift toward genuine luxury brands that prioritize sustainability at their core.Stella McCartney's is one of the Brands which trace raw materials to ensure ethical sourcing, such as organic cotton or recycled polyester, providing consumers with verifiable data on environmental impact. Blockchain allows real-time tracking of goods throughout the supply chain, reducing delays, fraud, and inefficiencies. Brands can demonstrate responsible sourcing and sustainability claims with verifiable data, strengthening their luxury narrative (Tsicalos, 2023).
- **Resale and circular economy:** Blockchain supports the secondhand luxury market by authenticating pre-owned goods, increasing their resale value. Platforms like Arianee, partnered with Richemont, issue blockchain-based certificates for watches and jewelry, facilitating secure resale transactions (Herinckx & Ghislain, 2022).

Sustainability impact

- **Reducing counterfeiting:** By ensuring authenticity, blockchain minimizes the environmental waste caused by counterfeit goods, which often use low-quality, non-sustainable materials. Blockchain platforms like Ethereum or Hyperledger are commonly used, with smart contracts automating verification processes. For example, a smart contract can trigger a digital certificate upon product sale, ensuring traceability. Vacca et al., 2021).
- **Ethical sourcing:** Transparent supply chains enable brands to verify eco-friendly practices, such as using low-carbon materials or fair-trade labor, aligning with consumer demand. Integration with IoT devices, such as RFID tags, enhances real-time tracking, ensuring data accuracy across global supply chains. Al Mahmud et al. (2025)
- **Circular economy:** Blockchain extends product lifecycles by supporting resale, reducing the need for new production and associated emissions.

Challenges

- **High costs:** Developing and maintaining blockchain infrastructure requires significant investment in software, hardware, and partnerships, which may be prohibitive for smaller brands.
- **Consumer education:** Luxury consumers may lack familiarity with blockchain, necessitating user-friendly interfaces (e.g., mobile apps) to present traceability data clearly.
- **Scalability:** Blockchain systems must handle large transaction volumes across global supply chains, requiring robust network design to avoid latency or cost inefficiencies.
- **Energy concerns:** While mitigated by newer protocols, blockchain's environmental footprint remains a concern, requiring brands to adopt low-energy solutions to align with sustainability goals. Energy consumption is a concern, as some blockchains (e.g., Ethereum pre-2022) rely on energy-intensive proof-of-work mechanisms. However, newer protocols like proof-of-stake or private blockchains (used by Aura) reduce environmental impact (Platt et al, 2021).

While blockchain improves authenticity and transparency, it comes with trade-offs. The high costs of setting up and maintaining it can be a burden, especially for smaller luxury brands, limiting its use to expensive items or big companies. Even newer blockchain systems still use energy, which may clash with sustainability goals. Brands need to choose eco-friendly options, which might be slower or harder to scale. Plus, since the technology is complex, customers need to be educated to use it otherwise, they may not engage with features like digital passports. Overall, brands must balance these challenges with the long-term benefits of building trust and supporting circular practices.

3D Printing for Waste Reduction and Customization

3D printing, or additive manufacturing, builds objects layer by layer from digital designs, using materials like polymers, metals, or bio-based composites. In luxury fashion, it enables rapid prototyping and bespoke production, producing intricate designs with minimal material waste compared to traditional subtractive methods (e.g., cutting fabric) (Faludi et al, 2017).

Applications

- **Bespoke accessories:** Stratasys collaborations with renowned designers like Karim Rashid, KAIMIN and Ganit Goldstein produced Digitally embroidered garment, Luxury dresses and handbags, combining intricate designs with sustainable production. These products highlight the blend of **technology, sustainability, customization**, and **cultural storytelling** in modern fashion design through 3D printing (Saunders, 2022).
- **Couture garments:** Iris van Herpen's couture collections feature 3D-printed fabrics that mimic organic forms, such as skeletal or floral structures, offering unique designs that traditional methods cannot replicate (Smelik, 2017).
- **On-demand production:** 3D printing supports made-to-order models, minimizing overstock and unsold inventory, a major contributor to fashion waste (Clero, 2023).

Sustainability impact

- **Waste reduction:** Unlike traditional manufacturing, which generates up to 20% fabric waste, 3D printing uses precise material amounts, reducing environmental impact (Nyika et al., 2022).
- **Material innovation:** Brands can use sustainable materials like recycled plastics or bio-based resins, lowering reliance on non-renewable resources. Sustainable options include PLA (plant-based plastic), recycled PET, and bio-composites, which reduce carbon footprints compared to traditional leather or synthetics (Agrawal & Bhat, 2025). Fused Deposition Modeling (FDM) and Stereolithography (SLA) 3D printing technology are common for fashion, with SLA offering high precision for intricate designs like jewelry (Chakraborty & Biswas, 2019). CAD software like Autodesk Fusion 360 enables designers to create complex patterns, optimized for minimal material use and structural integrity (Land, 2023).
- **Circular economy:** 3D-printed products can be designed for durability or recyclability, extending lifecycles and supporting resale markets (Ruiz et al, 2022).

Challenges

- **High costs:** 3D printers and sustainable materials are expensive, limiting adoption to high-value, bespoke items. Scaling for broader collections requires cost reductions.

- **Scalability:** Current 3D printing speeds and material limitations hinder mass production, making it suitable primarily for limited-edition or couture pieces.
- **Quality assurance:** Ensuring 3D-printed products meet luxury standards for durability, texture, and finish requires rigorous testing and material advancements.
- **Consumer perception:** Some consumers may view 3D-printed items as less artisanal, necessitating branding efforts to emphasize their exclusivity and sustainability.

3D printing offers strong sustainability and customization benefits, but it also comes with challenges. The high cost of machines and materials limits its use to expensive, niche products, making it less accessible for smaller brands or large collections. Its slow speed and limited materials also make it hard to scale for mass production. Some consumers may see 3D-printed items as less handcrafted, which can clash with the luxury image. Brands will need to market these products carefully to emphasize their exclusivity and eco-friendly value. Overall, they must weigh these downsides against the technology's potential for reducing waste and enabling unique designs.

Artificial Intelligence (AI) for Supply Chain Optimization and Sustainable Design

AI encompasses machine learning, predictive analytics, and generative algorithms that analyze data to optimize processes and create innovative designs. In luxury fashion, AI streamlines supply chains, predicts consumer demand, and enables sustainable design by simulating eco-friendly materials.

Applications

- **Supply chain optimization:** LVMH uses machine learning to optimize inventory and logistics, reducing energy consumption and overproduction. AI analyzes sales data, weather patterns, and trends to align production with demand This supports LVMH's sustainability goals while maintaining operational efficiency (Ji et al., 2023).
- **Predictive analytics:** Ralph Lauren's partnership with Amazon Web Services employs AI to forecast trends and manage inventory, minimizing unsold stock and associated waste (Ramirez, 2024).
- **Generative design:** Balmain's experiments with generative AI create unique textile patterns based on consumer inputs, reducing physical prototyping and authenticated via blockchain for exclusivity (Balmain, 2023)

Sustainability impact

- **Reduced overproduction:** AI-driven demand forecasting minimizes excess inventory, which accounts for significant waste in fashion (up to 30% of garments go unsold globally) (Younus, 2025).
- **Energy efficiency:** Optimized supply chains reduce transportation and storage emissions, lowering the industry's carbon footprint (Ojadi et al., 2024).
- **Sustainable design:** Generative AI can help create eco-friendly designs with minimal resource use, while material simulations promote low-impact fabrics (Shabi et al. 2025).

Challenges

- **Privacy concerns:** AI relies on consumer data, raising concerns about compliance with regulations like GDPR. Brands must ensure transparent data practices (Mbah, 2024).
- **Integration with craftsmanship:** AI-driven design must complement human artistry to preserve luxury's artisanal essence, requiring careful oversight.
- **Cost and expertise:** Implementing AI requires investment in software, cloud infrastructure, and skilled data scientists, which may be challenging for smaller brands.
- **Consumer trust:** Transparency about AI's role in design is crucial to avoid perceptions of in-authenticity in luxury products.

AI can improve supply chains and support sustainable design, but it also has drawbacks. Using consumer data raises privacy concerns, which require strict rules and can add cost and complexity. The high price of AI tools and expertise can be a barrier for smaller brands. Relying too much on AI for design may also reduce the handcrafted feel of luxury items, which could turn off customers who value human creativity. Brands need to use AI in a way that supports, not replaces, traditional methods striking a balance between innovation, efficiency, and authenticity.

Smart Textiles and Sustainable Materials

Smart textiles integrate technologies like sensors, conductive fibers, or responsive materials into fabrics, offering functionality like health monitoring or environmental adaptability. Sustainable materials, such as bio-based leather or recycled fibers, reduce environmental impactThey play a key role in combining innovation with eco-conscious design, appealing to luxury consumers who prioritize sustainability (Dutta & Bansal 2024).

Applications

- **Health-monitoring fabrics:** Ralph Lauren's exploration with biotech firms aims to integrate biometric sensors into clothing, monitoring heart rate or stress, enhancing wellness-focused luxury (Tschorn, 2015).
- **Adaptive fabrics:** Google's Project Jacquard, initially with Levi's, is evolving to include temperature-regulating or self-cleaning fabrics, reducing environmental impact (Bohn, 2019).
- **Bio-based materials:** Hermès' partnership with MycoWorks develops mushroom-based leather (Sylvania), a sustainable alternative to animal leather, used in products like the Victoria bag (Mallon, 2024).

Sustainability impact

- **Reduced environmental footprint:** Bio-based materials like mycelium leather require less water and energy than traditional leather. Mycelium leather (from mushroom roots) and lab-grown collagen are grown in controlled environments, reducing land use. Recycled PET and organic cotton are processed to meet luxury standards (Hildebrandt & Bezama 2021).
- **Extended lifecycles:** Smart textiles with self-cleaning or durable properties reduce washing and replacement needs, lowering resource use (Cheung & Li 2018).
- **Cruelty-free alternatives:** Bio-based and recycled materials eliminate reliance on animal products, aligning with ethical consumer values (Nejadrezaei et al., 2024).

Challenges

- **Scalability:** Producing smart textiles and bio-based materials at scale is costly and technically complex, limiting their use to high-value items.
- **Consumer education:** Luxury consumers may need education to appreciate the texture or performance of sustainable materials compared to traditional leather or silk.
- **Integration:** Embedding sensors without compromising aesthetics or comfort requires advanced design, as bulky tech can detract from luxury appeal.
- **Cost:** R&D and production of smart textiles are expensive, requiring brands to balance sustainability with profitability.

Smart textiles and sustainable materials support eco-friendly luxury, but they come with challenges. High production costs and technical difficulties make it hard to scale, often limiting their use to exclusive, high-end items. Adding sensors or smart features can affect how the product looks or feels, so careful design is needed to keep the luxury feel. Some consumers may also be unsure about new bio-based materials due to their texture or performance, which means brands need to invest in education and marketing. Overall, brands must balance these issues with the benefits of innovation and sustainability while maintaining their image and customer expectations.

CASE STUDIES IN TECH-FASHION CO-BRANDING

The synergy of connectivity and couture defines the new era of smart luxury fashion, where co-branding between fashion houses and technology firms creates products that transcend traditional boundaries. Connectivity—enabled by technologies like IoT, blockchain, and AI—fosters seamless interactions between consumers, brands, and digital ecosystems, while couture preserves the artistry, exclusivity, and heritage of luxury craftsmanship. These collaborations produce connected couture, where innovative functionality meets timeless elegance, redefining luxury for a digitally native audience. Below, we explore key case studies that exemplify this fusion, highlighting how strategic partnerships amplify brand value, enhance consumer engagement, and bridge physical and digital realms.

Levi's x Google: Jacquard Smart Denim

Launched in 2016 and expanded in subsequent years, the Levi's x Google Jacquard project integrates Google's Jacquard technology into Levi's Commuter Trucker Jacket. The jacket features touch-sensitive fabric on the sleeve, allowing users to perform actions like answering calls, controlling music, or navigating maps through gestures. Priced at approximately $200–$350, the jacket targets urban, tech-savvy consumers, blending Levi's iconic denim aesthetic with Google's innovative IoT platform (Arthur, 2016).

Impact

The collaboration broadened Levi's appeal to younger, digitally native consumers, positioning the brand as a pioneer in accessible smart fashion. While not a luxury brand, Levi's leveraged the premium perception of smart technology to enhance its market positioning, aligning with the smart luxury trend. The project also showcased Google's ability to partner with fashion brands to create functional, stylish wearables.

Consumer Reception

The Jacquard jacket received positive feedback for its innovative functionality and seamless integration of technology, though some consumers noted the limited gesture controls and high price point relative to standard denim jackets. The collaboration generated significant media buzz, reinforcing both brands' innovative images.

The Jacquard platform exemplifies connectivity by integrating IoT into everyday wear, allowing the jacket to sync seamlessly with smartphones via a Bluetooth-enabled cufflink. This creates a dynamic interface where physical gestures translate into digital actions, fostering a fluid connection between the wearer and their digital ecosystem. The technology's open API encourages developers to expand functionality, ensuring the jacket evolves with consumer needs. While Levi's is not a traditional luxury brand, the jacket's bespoke integration of technology elevates its craftsmanship. The touch-sensitive fabric is woven with precision to maintain denim's tactile allure, reflecting a modern reinterpretation of couture through functional artistry. Limited-edition releases enhance exclusivity, aligning with luxury's emphasis on scarcity.

Apple x Hermès: Apple Watch Hermès

Introduced in 2015 and updated annually, the Apple Watch Hermès is a collaboration between Apple and Hermès, combining Apple's smartwatch technology with Hermès' luxury craftsmanship. Priced from $1,200 to over $1,500, the watch features exclusive Hermès watch faces, handcrafted leather bands, and signature designs like the Double Tour and Single Tour. The collection is sold through Apple and Hermès channels, emphasizing exclusivity and prestige (Spangler, 2023).

Impact

The partnership elevated the Apple Watch from a tech gadget to a luxury fashion accessory, attracting affluent consumers who value both technology and heritage craftsmanship. It strengthened Apple's position in the luxury wearable market and reinforced Hermès' relevance among tech-savvy buyers. The collaboration also set a benchmark for tech-fashion alliances, demonstrating how functionality and aesthetics can coexist.

Consumer reception

The Apple Watch Hermès was widely praised for its elegant design and seamless integration of technology, though some criticized the high price point compared to standard Apple Watches. The exclusive branding and limited-edition bands drove strong demand among luxury consumers, with waitlists reported for certain models.

Louis Vuitton's Via Treasure Trunks

Launched in June 2023, the Via Treasure Trunks collection consists of NFTs priced at €39,000 each, each accompanied by a made-to-order physical trunk. Owners can purchase digital "keys" to unlock additional bespoke products, some customized to the owner. The NFTs are soul-bound, non-transferable, and limited to several hundred pieces, available in the US, Canada, France, UK, Germany, Japan, and Australia (Hirschmiller, 2023). The NFTs, while digital, embody couture through their scarcity and bespoke design, offering a modern interpretation of luxury ownership.

Impact

This initiative has positioned Louis Vuitton as a leader in the luxury NFT space, attracting attention from both traditional luxury consumers and crypto enthusiasts. The high price point and limited availability align with the brand's exclusivity, while the 'phygital' approach bridges physical and digital luxury.

Consumer reception

The collection generated significant buzz due to its exclusivity and innovative approach, though the soul-bound nature sparked debate among NFT collectors who value tradability. The waitlist process and selective invitation system further enhanced the perception of exclusivity.

Gucci's SuperGucci NFTs

In February 2022, Gucci partnered with Superplastic to launch the SuperGucci NFT collection, featuring 200 publicly available NFTs and additional pieces for Superplastic's community. Each of the 10 unique NFTs included a handmade Italian ceramic sculpture, combining digital art with physical luxury. The NFTs were sold on Superplastic's website and Gucci's Vault platform, with a raffle system implemented to counter bot issues (Ledger Insights. 2022).The NFTs, while digital, carry the same exclusivity, with each piece reflecting Gucci's heritage of innovation and artistry.

Impact

The collection sold out quickly, demonstrating strong demand for luxury-digital crossovers. It helped Gucci engage with a younger, digitally native audience, reinforcing its position as an innovative luxury brand.

Consumer reception

The blend of digital and physical products appealed to collectors and fashion enthusiasts, though some criticized the high prices and bot-related issues. The inclusion of physical ceramics added tangible value, enhancing consumer satisfaction.

Balenciaga's Cristóbal Balenciaga: To the Moon

Launched on July 21, 2022, this collection featured 10,000 unique NFTs inspired by Cristóbal Balenciaga's original drawings, minted on Crypto.com NFT for $100 each. It included 50 exclusive one-of-one "haute couture" NFTs, with rewards such as Balenciaga gear, gift cards, and 70-year-old drawings. The project involved multiple collaborators, including Art Consulting, Art Vein, Animal Concerts, Cointelegraph, and Crypto.com NFT(Cointelegraph, 2022). The collection pays homage to Cristóbal Balenciaga's architectural silhouettes and meticulous craftsmanship, reimagining his sketches as digital art. The one-of-one NFTs embody haute couture's exclusivity, with their rarity and historical significance elevating them to the status of digital heirlooms.

Impact

The collection successfully merged fashion history with modern technology, creating a unique offering that honored Balenciaga's heritage while embracing digital innovation. It attracted both fashion and crypto communities, broadening the brand's appeal.

Consumer reception

The collection was well-received, with consumers appreciating the historical significance and the innovative use of technology. The rewards system added value, though the mystery box format sparked mixed reactions among buyers seeking specific NFTs.

These case studies illustrate how co-branding in smart luxury fashion harnesses connectivity to create dynamic, interactive experiences and couture to preserve the artistry and exclusivity of luxury. From IoT-enabled denim to blockchain-backed NFTs, these partnerships redefine what it means to be luxurious in a digital world, blending innovation with heritage to captivate modern consumers.

FUTURE TRENDS IN CO-BRANDING FOR SMART LUXURY FASHION

The future of co-branding in smart luxury fashion is poised for further innovation, driven by emerging technologies and evolving consumer expectations. Key trends include:

AI Personalization at Scale

Artificial intelligence is revolutionizing how luxury brands deliver personalized experiences, from tailored product recommendations to bespoke designs. Collaborations with AI-driven platforms are enabling brands to create hyper-personalized fashion experiences that resonate with individual consumer preferences. For example, Stitch Fix's partnership with AI algorithms to curate personalized clothing boxes has inspired luxury brands to explore similar models .In the future, luxury brands like Dior or Chanel could partner with generative AI companies, such as xAI, to co-create unique patterns or accessories based on customer inputs, enhancing exclusivity through one-of-a-kind digital or physical products. AI-powered styling apps are also reducing return rates by up to 35% through precise fit predictions, fostering consumer trust and loyalty (Twarogal & Puczyk, 2025). These collaborations will require brands to balance data-driven personalization with privacy concerns, ensuring seamless integration into the luxury experience.

Smart Textiles and Adaptive Fabrics

The development of smart textiles, fabrics embedded with technologies like sensors, conductive fibers, or responsive materials is redefining functionality in luxury fashion. Collaborations with tech firms specializing in material science are driving this trend. For instance, Google's Project Jacquard, initially launched with Levi's, is evolving to include more advanced applications, such as temperature-regulating fabrics or health-monitoring fibers. Luxury brands like Prada could partner with biotech firms like Bolt Threads to create adaptive fabrics that adjust to environmental conditions or user needs, combining sustainability with cutting-edge design. Such collaborations could produce garments that monitor vital signs or change color based on mood, appealing to consumers seeking both innovation and exclusivity. However, brands must address challenges like production scalability and consumer education to ensure adoption.

Sustainability and Circular Economy Integration

As consumer demand for eco-friendly luxury grows, co-branding with tech companies focused on sustainable materials, production methods, or circular economy models is becoming a strategic priority. Hermès' partnership with MycoWorks to develop mushroom-based leather alternatives exemplifies this trend, offering a sustainable yet luxurious material. Blockchain technology is also enhancing transparency in sustainable supply chains, with platforms like the Aura Blockchain Consortium (LVMH, Prada, Cartier) enabling digital product passports for traceability. Looking ahead, luxury brands could collaborate with tech startups like Circularise to create blockchain-enabled resale platforms, allowing consumers to verify and trade pre-owned luxury goods with guaranteed authenticity. Challenges include high development costs and ensuring sustainability claims are credible to avoid greenwashing accusations.

Metaverse and Spatial Computing Integration

The metaverse continues to redefine digital luxury, with luxury brands co-branding with virtual platforms to create immersive fashion experiences. Gucci's Vault in The Sandbox and Nike's Nikeland on Roblox demonstrate how brands are leveraging virtual worlds to engage digitally native audiences (Beunardeau, 2024). In the future, luxury brands like Balenciaga could partner with spatial computing firms, such as Magic Leap or Meta, to develop augmented reality (AR) showrooms or virtual reality (VR) fashion shows that blend physical and digital craftsmanship. These collaborations could enable consumers to "wear" digital couture in virtual spaces or customize avatars with exclusive NFT-based accessories, reinforcing brand exclusivity. However, brands must navigate consumer fatigue with digital products and ensure virtual experiences maintain the tactile allure of luxury.

Generative AI for Co-Created Designs

Beyond personalization, generative AI offers opportunities for luxury brands to co-create designs with consumers or tech partners, blurring the lines between creator and customer. Luxury brands like Versace could collaborate with AI platforms to allow customers to input design preferences, generating unique garments or accessories authenticated via blockchain. These partnerships would enhance consumer engagement and create new revenue streams through limited-edition, AI-crafted products. The challenge lies in maintaining brand control over aesthetics and ensuring AI outputs align with the heritage and quality expectations of luxury consumers.

Advanced Biometrics in Wearables

The integration of advanced biometrics into luxury wearables is an emerging frontier, driven by consumer interest in health and wellness. Apple's collaboration with Hermès for the Apple Watch has set a precedent, but future partnerships could involve medical tech firms like Fitbit or Oura to create luxury wearables that monitor advanced health metrics, such as blood oxygen levels or stress indicators. For instance, collaboration between Bulgari and a biotech firm could produce a smart jewelry line that combines gemstone craftsmanship with health-tracking sensors, appealing to affluent consumers prioritizing holistic well-being. These partnerships require careful design to ensure functionality does not overshadow aesthetics, as well as robust data privacy measures to protect sensitive health information.

Web3 and Decentralized Fashion Ecosystems

The rise of Web3 technologies, including decentralized platforms and blockchain-based ownership, is creating new opportunities for co-branding. Luxury brands could partner with Web3 platforms like Decentraland or The Sandbox to create decentralized fashion ecosystems where consumers own, trade, and showcase digital luxury assets. For example, Dolce & Gabbana's NFT initiatives could evolve into a broader collaboration with a Web3 marketplace, enabling community-driven design or tokenized loyalty programs. These partnerships offer brands access to crypto-native audiences but face challenges like regulatory uncertainty and the need to educate consumers about decentralized technologies (Ledger Insights, 2021).

These trends suggest that co-branding will continue to evolve, with luxury brands balancing innovation with their core values of exclusivity and craftsmanship. However, brands must navigate challenges such as consumer fatigue with digital products and the need to maintain authenticity in virtual environments.

MANAGERIAL STRATEGIES FOR CO-BRANDING SUCCESS

In the era of smart luxury fashion, co-branding has emerged as a transformative strategy that seamlessly blends the timeless allure of luxury with the dynamic capabilities of cutting-edge technology, redefining exclusivity, craftsmanship, and consumer engagement. By forging strategic partnerships with technology firms, luxury fashion brands are not only enhancing product functionality through innovations like blockchain for authenticity, AI-driven personalization, 3D printing for sustainable customization, and smart textiles for adaptive wearability but also addressing modern consumer demands for sustainability, transparency, and immersive digital experiences. To maximize the potential of co-branding in smart luxury fashion, brands must adopt strategic approaches that align technological innovation with their core values of exclusivity and craftsmanship. The following managerial strategies provide a roadmap for successful collaborations with technology firms:

Strategic Partner Selection

Choosing the right technology partner is critical to ensuring alignment in values, brand equity, and consumer perception. Luxury brands should prioritize tech firms that share similar ethos, such as a commitment to sustainability for eco-conscious brands like Stella McCartney, which has partnered with Bolt Threads to develop bio-based materials like Mylo (a mycelium-based leather alternative). For instance, a brand emphasizing ethical sourcing might collaborate with a blockchain platform like the Aura Blockchain Consortium to ensure supply chain transparency, as seen with LVMH and Prada. Compatibility in innovation goals is equally important Hermès' partnership with Apple for the Apple Watch Hermès succeeded because both brands valued precision craftsmanship and cutting-edge design. Misaligned partnerships, such as a luxury brand collaborating with a low-cost tech firm, risk diluting brand prestige. To mitigate this, brands should conduct thorough due diligence, assessing the partner's reputation, technological capabilities, and alignment with target audience expectations.

Consumer Engagement through Storytelling

Effective communication of technological benefits is essential to engage consumers and justify premium pricing. Luxury brands should leverage storytelling to highlight how technologies enhance exclusivity, authenticity, or sustainability. For example, blockchain's role in ensuring product authenticity can be communicated through immersive campaigns, such as QR code-enabled digital passports that allow consumers to trace a product's journey from raw materials to retail, as implemented by Stella McCartney. Augmented reality (AR) experiences can further enhance engagement Burberry's AR partnership with Google Lens allows consumers to explore product stories through interactive digital overlays. Social media platforms like Instagram and TikTok are powerful channels for these campaigns, enabling brands to showcase tech-driven innovations through visually compelling narratives. For instance, Gucci's Roblox presence uses virtual storytelling to immerse users in brand heritage, reinforcing the value of digital luxury assets like NFTs.

Risk Management and Gradual Integration

Integrating advanced technologies carries risks, including consumer skepticism, technical failures, or brand dilution. To mitigate these, brands should pilot tech integrations in limited-edition lines to test consumer response before scaling. For example, Ralph Lauren's exploration of biometric-sensing fabrics was initially tested in niche products like the PoloTech shirt, allowing the brand to gauge demand and refine technology without over-committing resources. Similarly, luxury brands experimenting with smart textiles, such as temperature-regulating fabrics, can launch capsule collections to assess functionality and aesthetics. This approach minimizes the risk of alienating core customers who value traditional craftsmanship while allowing brands to build credibility in the smart luxury space. Additionally, brands must ensure that tech integrations align with their aesthetic standards bulky sensors or obtrusive designs can detract from the luxury appeal, as seen in early wearable tech iterations.

Cost Optimization through Shared Resources

Developing and implementing technologies like blockchain, AI, or 3D printing requires significant investment in research and development (R&D), infrastructure, and talent. To offset these costs, luxury brands should share R&D expenses with tech partners, leveraging their expertise and resources. For instance, the Levi's x Google Jacquard project benefited from Google's investment in IoT technology, reducing Levi's financial burden while enabling innovative product development. Similarly,

LVMH's collaboration with Amazon Web Services for AI-driven supply chain optimization allows the luxury conglomerate to access cutting-edge analytics without building proprietary systems from scratch. Strategic cost-sharing agreements can also include joint marketing efforts, as seen in the Apple x Hermès collaboration, where both brands promoted the Apple Watch Hermès through their respective channels, amplifying reach and reducing individual costs. Smaller brands, with limited budgets, can explore consortiums like the Aura Blockchain Consortium to pool resources with other luxury players, making advanced technologies more accessible.

Navigating Consumer Challenges

Consumer challenges, such as fatigue with digital products and privacy concerns, also require careful management. For instance, the proliferation of NFTs has led to skepticism among some consumers who question their long-term value. Brands can address this by pairing digital assets with physical products, as seen in Gucci's SuperGucci collection, which included handmade ceramics. Privacy concerns, particularly with AI and biometric wearables, necessitate transparent data practices and compliance with regulations like GDPR. Brands should communicate how consumer data is protected, using storytelling to build trust. To succeed, brands must balance technological innovation with the artisanal essence of luxury. Overreliance on technology risks alienating consumers who value heritage, while underinvestment in innovation could cede market share to competitors. Strategic co-branding, guided by careful partner selection, engaging storytelling, risk management, and cost optimization, ensures that luxury brands remain resilient and relevant.

IMPACT OF NEW TARIFF IMPOSITIONS ON CURRENT LUXURY MARKET

The Trump administration's tariff policies, implemented in April 2025, have introduced significant disruptions to the luxury market, exacerbating existing challenges. On April 5, 2025, a 10% baseline tariff was imposed on all U.S. imports, with additional tariffs of 20% on European Union goods and up to 145% on Chinese imports effective April 9. These tariffs, aimed at addressing trade imbalances, have disrupted the luxury sector, which relies heavily on global supply chains (Huld, 2025).

Increased Costs and Price Hikes

The EU, the world's largest supplier of luxury goods (accounting for 70% of the global market), faces 20% tariffs, impacting brands such as Prada (which derives 17% of its revenue from the Americas), Gucci (24% of Kering's revenue comes from North America), and Hermès (20% of its sales are from the Americas). France alone exported $5 billion in luxury goods to the U.S. in 2024. These tariffs are likely to increase costs, forcing brands to raise prices or absorb losses, which could dampen demand. For example, a Forbes report from April 2025 noted that luxury brands like LVMH and Hermès are bracing for price hikes, potentially alienating aspirational consumers (Danziger, 2025)

Consumer Confidence and Market Slowdown

Tariffs have shaken consumer confidence, with affluent U.S. shoppers pausing purchases amid economic uncertainty. Bain & Company's Claudia D'Arpizio warned that tariffs could exacerbate an already weakened luxury market, forecasting a potential 2% sales decline in 2025 (Reuters, 2025). Posts on X reflect this sentiment, with users noting a "freeze" in the multi-billion-dollar U.S. luxury market and a valuation crash for brands like Louis Vuitton and Dior due to trade war fallout.

Supply Chain Disruptions

The EU's dominance in luxury production means tariffs will disrupt supply chains. Brands may face higher costs for raw materials and finished goods, particularly leather goods, jewelry, and apparel. Meanwhile, China's 84% retaliatory tariffs on U.S. goods and exposure of luxury brand manufacturing sources have further eroded brand exclusivity, as consumers' access high-quality replicas at lower prices (Pound, 2025).

Regional Impacts

While LVMH is partially shielded by U.S.-based production, smaller brands without local facilities are more vulnerable. The closure of the de minimis loophole, which allowed tariff-free imports of low-value goods from China, has also raised costs for e-commerce platforms, impacting accessible luxury brands. In response, some brands are exploring alternative sourcing from countries like Vietnam, though this shift is costly and time-intensive

Brand Strategies and Co-Branding Implications

Luxury brands are adopting varied strategies to mitigate tariff impacts. LVMH, partially shielded by U.S. production facilities (e.g., Louis Vuitton factories in Texas), may absorb some costs to maintain market share. Brands like Gucci and Hermès, with stronger pricing power, are passing costs to consumers, risking pushback from price-sensitive buyers. Others, like BMW, are temporarily covering auto tariffs to delay price hikes. Sourcing shifts are also emerging, with brands exploring production in Vietnam or Indonesia to bypass Chinese tariffs, though this increases complexity for tech-integrated products like smart textiles (Sutar, 2025)

For co-branding in smart luxury fashion, tariffs complicate partnerships with tech firms. Collaborations like Apple-Hermès or Google-Levi's rely on global supply chains, and tariff-induced cost increases could delay innovation or raise prices for IoT wearables and AR apps. Blockchain-based authentication, as used by the Aura Consortium, may gain traction to reassure consumers of value amid price hikes, but development costs could strain smaller brands. Luxury brands need to focus on core values, prioritize key clients, and invest in tech-enabled experiences to rebuild consumer connections, aligning with co-branding trends like AI personalization and metaverse integration.

CONCLUSION

The luxury fashion industry stands at a pivotal crossroads, where heritage and innovation converge to redefine the essence of exclusivity, craftsmanship, and consumer engagement. The rise of smart luxury fashion, fueled by strategic co-branding between fashion houses and technology firms, has ushered in a new era of digitally enhanced products and experiences that resonate with the values of modern, tech-savvy consumers. This chapter has explored how technologies such as blockchain, artificial intelligence, 3D printing, and smart textiles are not only enhancing functionality and personalization but also addressing pressing demands for sustainability and ethical production. Through case studies like Levi's x Google, Apple x Hermès, and Louis Vuitton's Via Treasure Trunks, we have seen how these collaborations create synergies that amplify brand value, broaden market reach, and redefine luxury as a blend of tangible craftsmanship and intangible digital experiences.

The integration of advanced technologies into luxury fashion has expanded the boundaries of what constitutes a luxury product, moving beyond physical goods to include digital assets like NFTs and immersive virtual experiences in the metaverse. These innovations cater to a new generation of consumers who prioritize personalization, authenticity, and sustainability alongside traditional markers of prestige.

Blockchain, for instance, ensures traceability and authenticity, reinforcing consumer trust while supporting circular economy models. AI-driven personalization and predictive analytics optimize supply chains and reduce waste, aligning with environmental goals. Meanwhile, smart textiles and 3D printing offers sustainable alternatives that merge cutting-edge design with eco-conscious production, addressing the industry's environmental footprint.

However, the path forward is not without challenges. The recent imposition of tariffs, as seen in the Trump administration's policies of April 2025, has introduced significant disruptions, increasing costs and threatening consumer confidence in an already volatile luxury market. These economic pressures underscore the importance of strategic co-branding as a tool for resilience, enabling brands to innovate and differentiate in a competitive landscape. Yet, brands must navigate risks such as consumer fatigue with digital products, privacy concerns, and the potential dilution of exclusivity through poorly aligned partnerships. Balancing technological integration with the artisanal essence of luxury remains critical to maintaining brand integrity.

Looking ahead, the future of co-branding in smart luxury fashion is bright, with emerging trends like AI-driven personalization, advanced biometrics, and Web3 ecosystems poised to further transform the industry. Collaborations with tech firms will continue to drive innovation, enabling brands to create hyper-personalized, sustainable, and immersive experiences that resonate with evolving consumer expectations. For instance, partnerships with generative AI platforms could empower consumers to co-create bespoke designs, while blockchain-enabled resale platforms could strengthen circular economy initiatives. The metaverse and spatial computing offer opportunities to redefine fashion shows and retail experiences, blending physical and digital realms to enhance exclusivity.

Ultimately, the success of co-branding in smart luxury fashion hinges on a delicate balance: leveraging technology to enhance, rather than overshadow, the core values of craftsmanship, exclusivity, and heritage that define luxury. As brands navigate economic uncertainties and technological advancements, those that prioritize authenticity, sustainability, and meaningful consumer engagement will thrive. The era of connected couture is not merely a trend but a transformative shift, promising a future where luxury is as dynamic, innovative, and inclusive as the consumers it serves. By embracing strategic partnerships and cutting-edge technologies, luxury fashion is poised to redefine itself, blending timeless elegance with the limitless possibilities of the digital age.

REFERENCES

Abeyatunge, W. G. N. N. (2024). Crafting Tomorrow's Elegance Through the Evolution of Luxury Marketing in a Digital World. *Cross-Cultural Perspectives on Luxury Marketing*, 173.

Agrawal, K., & Bhat, A. R. R. (2025). Advances in 3D Printing with Eco-Friendly Materials: A Sustainable Approach to Manufacturing. *RSC Sustainability*.

Al Mahmud, M. A., Hossan, M. Z., Tiwari, A., Khatoon, R., Sharmin, S., Hosain, M. S., & Ferdousmou, J. (2025). Reviewing the Integration of RFID and IoT in Supply Chain Management: Enhancing Efficiency and Visibility. *Journal of Posthumanism*, *5*(3), 409–437. DOI: 10.63332/joph.v5i3.746

Bagwell, L. S., & Bernheim, B. D. (1996). Veblen effects in a theory of conspicuous consumption. *The American Economic Review*, •••, 349–373.

Ballini, E., Battaglia, L., & Cedrola, E. (2025). Blockchain in luxury fashion: a collaborative approach to innovation and trust. In *24° International Marketing Trends Conference-What Managers Want and Real-World Needs* (Vol. 1, pp. 1-9). International Conference Marketing Trends.

Balmain. (2023, December). Balmain x Ant Kai: Balmain continues its trailblazing journey with the Unicorn sneakers. https://int.balmain.com/en/experience/balmain-x-antkai

Berzowska, J. (2005). Electronic textiles: Wearable computers, reactive fashion, and soft computation. *Textile (Oxford)*, *3*(1), 58–75. DOI: 10.2752/147597505778052639

Beunardeau, P. (2024, August 11). The future of luxury: How Web3 is elevating high-end brands. LinkedIn. https://www.linkedin.com/pulse/future-luxury-how-web3-elevating-high-end-brands-pierre-beunardeau-pmjoe

Beyond Market Insights. (2024, October). Smart fabrics market size, share, and trends analysis report by product (passive, active, very smart), by functionality (sensing, energy harvesting, luminescence & aesthetics), by end-use, by region, and segment forecasts, 2024–2032 (Report No. 6300). https://beyondmarketinsights.com/report/smart-fabrics-market/

Bohn, D. (2019, September 30). Google's Project Jacquard is available on new Levi's jackets. The Verge. https://www.theverge.com/2019/9/30/20888909/google-levi-project-jacquard-available-trucker-sherpa-jackets-price-design

Cabigiosu, A. (2020). *Digitalization in the luxury fashion industry*. Springer International Publishing. DOI: 10.1007/978-3-030-48810-9

Chakraborty, S., & Biswas, M. C. (2019). *Fused deposition modeling 3D printing technology in textile and fashion industry: Materials and innovation. Modern Concepts in Material Science*. Iris Publishers.

Cheung, T. W., & Li, L. (2018). Sustainable development of smart textiles: A review of 'self-functioning'abilities which makes textiles alive. *Journal of textile engineering & fashion technology, 4*(2), 151-156.

Clero, L. M. (2023). *Deadstock in the Fashion Industry: Is the On-Demand Business Model a Viable and Sustainable Solution?* (Master's thesis, Universidade Catolica Portuguesa (Portugal)).

Cointelegraph. (2022, July 19). Fashion meets NFTs: Original Cristóbal Balenciaga designs released in NFT drop slated for July 21. https://cointelegraph.com/press-releases/fashion-meets-nfts-original-cristobal-balenciaga-designs-released-in-nft-drop-slated-for-july-21

Cranny-Francis, A. (2013). Smart Textiles: Giving a New Sense to the Notion That Clothing Makes the (Hu) man. In *Technology and Touch: The Biopolitics of Emerging Technologies* (pp. 151–171). Palgrave Macmillan UK. DOI: 10.1057/9781137268310_8

Danziger, P. (2015). Five luxe trends for 2015. Unity Marketing. https://unitymarketingonline.com/wp-content/uploads/Five-Key-Luxury-Market-Trends-FINAL-NEW.pdf

Danziger, P. N. (2025, April 11). *As luxury brands brace for tariffs, affluent consumers hit pause*. Forbes. https://www.forbes.com/sites/pamdanziger/2025/04/11/as-luxury-brands-brace-for-tariffs-affluent-consumers-hit-pause/

Dutta, S., & Bansal, P. (2024). *Evolution of sustainable wearables: integrating cutting-edge techniques for future textile innovation*. Research Journal of Textile and Apparel.

El-Sayed, A., & El-Fanagely, B. (2022). The importance of digital lightening fashion technique in.

Faludi, J., Cline-Thomas, N., & Agrawala, S. (2017). *3D printing and its environmental implications. The Next Production Revolution Implications for Governments and Business, Organization for Economic Cooperation and Development*. OECD.

Herinckx, J., & Ghislain, R. (2022). The use of blockchain to fight counterfeiting in the second-hand luxury fashion market. *cit. a*, 48.

Hildebrandt, J., Thrän, D., & Bezama, A. (2021). The circularity of potential bio-textile production routes: Comparing life cycle impacts of bio-based materials used within the manufacturing of selected leather substitutes. *Journal of Cleaner Production*, *287*, 125470. DOI: 10.1016/j.jclepro.2020.125470

Hirschmiller, S. (2023, June 7). Unpacking the Web3 strategy behind Louis Vuitton's €39k digital Treasure Trunk NFTs. Forbes. https://www.forbes.com/sites/stephaniehirschmiller/2023/06/07/unpacking-the-strategy-behind-louis-vuittons-39k-digital-treasure-trunk-nfts/

Huld, A. (2025, April 11). Trump raises tariffs on China to 145% – Overview and trade implications. China Briefing. https://www.china-briefing.com/news/trump-raises-tariffs-on-china-to-125-overview-and-trade-implications/

Hypebeast. (2008, March 4). Hermès x Bugatti Veyron FBG. https://hypebeast.com/2008/3/hermes-x-bugatti-veyron-fbg

Ji, Y., Sun, Z., & Xia, Y. (2023). Leveraging Digital Transformation: Analysis of LVMH's Impact on Economic, Social, and Environmental Performance. *Advances in Economics. Management and Political Sciences*, *33*, 189–193.

Joy, A., Zhu, Y., Peña, C., & Brouard, M. (2022). Digital future of luxury brands: Metaverse, digital fashion, and non-fungible tokens. *Strategic Change*, *31*(3), 337–343. DOI: 10.1002/jsc.2502

Keller, K. L. (2003). Strategic branding management: Building measuring, and managing brand equity.

King, M. F. (2011). *Fashion, the body and technology: Tracing early 20th century techno-utopian ideas, aesthetics and impulses in 21st century wearable technology* (Doctoral dissertation, Queensland University of Technology).

Kuusela, M. (2024). Reimagining luxury: The potential of digital product passports in accelerating circular business models In high-end fashion.

Land, K. M. (2023). *Improving CAD Designs with Autodesk Fusion 360: A project-based guide to modelling effective parametric designs*. Packt Publishing Ltd.

Ledger Insights. (2021, July 15). *Luxury brand Dolce & Gabbana to launch NFT collection*. https://www.ledgerinsights.com/luxury-brand-dolce-gabbana-to-launch-nft-collection-blockchain/

Ledger Insights. (2022, January 18). Gucci to launch NFT collection with Superplastic brand. https://www.ledgerinsights.com/gucci-to-launch-nft-collection-with-superplastic-brand/

Li, J. (2018). *The Lighting Technology of Fashion Designs in the Contemporary Social Context* (Master's thesis, State University of New York at Stony Brook).

Mallon, S. (2024). Materialities of 'Mushroom-Leather': A Critical Appraisal of the Material Dimensions and Sustainability of Fungus-Based Alternatives to Leather for use in Fashion and Textiles. *Ethnologia Fennica*, *51*(1), 81–110. DOI: 10.23991/ef.v51i1.141682

Mbah, G. O. (2024). Data privacy in the era of AI: Navigating regulatory landscapes for global businesses.

Nejadrezaei, N., Henchion, M., & O'Neill, E. (2024). Consumers Perspectives on Biobased Products: A Scoping Review. *Cleaner Production Letters*, 100073.

Nyika, J., Mwema, F. M., Mahamood, R. M., Akinlabi, E. T., & Jen, T. C. (2022). Advances in 3D printing materials processing-environmental impacts and alleviation measures. *Advances in Materials and Processing Technologies, 8*(sup3), 1275-1285.

Ojadi, J. O., Odionu, C., Onukwulu, E., & Owulade, O. (2024). Big Data Analytics and AI for Optimizing Supply Chain Sustainability and Reducing Greenhouse Gas Emissions in Logistics and Transportation. *International Journal of Multidisciplinary Research and Growth Evaluation*, *5*(1), 1536–1548. DOI: 10.54660/.IJMRGE.2024.5.1.1536-1548

Platt, M., Sedlmeir, J., Platt, D., Xu, J., Tasca, P., Vadgama, N., & Ibañez, J. I. (2021, December). The energy footprint of blockchain consensus mechanisms beyond proof-of-work. In *2021 IEEE 21st International Conference on Software Quality, Reliability and Security Companion (QRS-C)* (pp. 1135-1144). IEEE. DOI: 10.1109/QRS-C55045.2021.00168

Pound, J. (2025, April 9). China slaps 84% retaliatory tariffs on U.S. goods in response to Trump. CNBC. https://www.cnbc.com/2025/04/09/china-slaps-retaliatory-tariffs-of-84percent-on-us-goods-in-response-to-trump.html

Ramirez, M. (2024, May 16). Ralph Lauren Corporation: AI use cases 2024. PitchGrade. https://pitchgrade.com/companies/ralph-lauren-corporation-ai-use-cases

Reuters. (2025, May 15). Luxury sector faces more gloom as Bain cuts sales forecast. ET BrandEquity. https://brandequity.economictimes.indiatimes.com/news/marketing/luxury-sector-faces-more-gloom-as-bain-cuts-sales-forecast/121184224

Ruiz, L. E., Pinho, A. C., & Resende, D. N. (2022). 3D printing as a disruptive technology for the circular economy of plastic components of end-of-life vehicles: A systematic review. *Sustainability (Basel)*, *14*(20), 13256. DOI: 10.3390/su142013256

Saunders, S. (2022, June 9). Stratasys 3D printed fashion collection debuts at Milan Design Week. 3DPrint.com. https://3dprint.com/291745/collaborative-stratasys-3d-printed-fashion-collection-debuts-at-milan-design-week/

Shabi, S. B. M., Bhuvaneshwari, M., Kanimozhi, S., & Pavithra, S. (2025, February). Artificial Intelligence, Intelligent Manufacturing, and Sustainable Futures in Next-Gen Textiles: A Comprehensive Approach. In *2025 3rd International Conference on Intelligent Data Communication Technologies and Internet of Things (IDCIoT)* (pp. 2144-2149). IEE

Shan, J., Lu, H., & Cui, A. P. (2022). 1+ 1> 2? Is co-branding an effective way to improve brand masstige? *Journal of Business Research*, *144*, 556–571. DOI: 10.1016/j.jbusres.2022.01.058

Shee Weng, L. (2024). Digital Product Passports: Transforming Industries Through Transparency, Circularity, and Compliance. *Digital Product Passports: Transforming Industries Through Transparency, Circularity, and Compliance (November 10, 2024).*

Smelik, A. M. (2017). *Cybercouture: the fashionable technology of Pauline van Dongen.* Iris van Herpen and Bart Hess.

Spangler, A. R. (2023). *Integrating Fashion and Technology in Wearable Devices: A Study on Co-Branding Effects* (Doctoral dissertation, University of Pennsylvania).

Sutar, S. (2025, April 16). Case study: The impact of tariff wars on luxury fashion brands and their pricing models. LinkedIn. https://www.linkedin.com/pulse/case-study-impact-tariff-wars-luxury-fashion-brands-pricing-sutar-mvaxc

Tschorn, A. (2015, August 25). Ralph Lauren's biometric-sensing smart shirt hits retail Thursday. Los Angeles Times. https://www.latimes.com/fashion/alltherage/la-ar-ralph-lauren-s-biometric-sensing-smart-shirt-hits-retail-aug-27-20150825-story.html

Tsicalos, Z. (2023). The Fashion Industry, the Environment & Sustainability: A Call for an International Treaty and What This Means for the Future of the Fashion Industry Notes and Student Works. *Journal of International Business and Law*, *23*(1), 7.

Twarogal, P., & Puczyk, A. (2025, July 4). AI in fashion retail industry: Why retailers must break free from old habits. Neontri. https://neontri.com/blog/ai-fashion-retail/

Vacca, A., Di Sorbo, A., Visaggio, C. A., & Canfora, G. (2021). A systematic literature review of blockchain and smart contract development: Techniques, tools, and open challenges. *Journal of Systems and Software*, *174*, 110891. DOI: 10.1016/j.jss.2020.110891

Yeo, P. (2021). Exhibiting Transformative Fashion: Digital Interventions to Enhance Display and Interpretation. *Costume*, *55*(1), 97–120. DOI: 10.3366/cost.2021.0184

Younus, M. (2025). The Economics of A Zero-Waste Fashion Industry: Strategies To Reduce Wastage, Minimize Clothing Costs, And Maximize & Sustainability. *Strategic Data Management and Innovation*, *2*(01), 116–137. DOI: 10.71292/sdmi.v2i01.15

Chapter 8 The Fusion of Co-Branding and Digital Fashion: Transforming Smart Luxury Products in the Digital Era

Sunanda Vincent Jaiwant
https://orcid.org/0000-0002-1161-5970
Christ University, India

ABSTRACT

In the evolving landscape of luxury products, co-branding strategies and digital fashion integration have become essential tools for enhancing brand value and consumer engagement. Smart luxury products, which incorporate advanced technology, require a unique blend of traditional craftsmanship and digital innovation to maintain exclusivity while catering to tech-savvy consumers. This chapter explores the strategic collaborations between luxury brands and technology firms, analysing their impact on brand equity, consumer perception, and market performance. It examines case studies of successful co-branding initiatives and evaluates the role of digital fashion in enhancing personalization and sustainability in luxury consumption.

INTRODUCTION

Luxury companies are navigating an ever-changing terrain in the digital age, where co-branding meets with cutting-edge technology. This convergence has resulted in "smart luxury" products, which effortlessly combine aesthetic sophistication with technical innovation. These products meet consumers' needs for exclusivity, identity

DOI: 10.4018/979-8-3373-3010-5.ch008

expression, and individualized digital experiences. The traditional concept of luxury, which was formerly limited to tangible objects, is moving into the digital sphere. Brands are increasingly selling virtual fashion goods, NFTs, and immersive digital experiences that appeal to a tech-savvy audience. This trend is seen in collaborations such as Dolce & Gabbana's Collezione Genesi NFT collection, which merged physical and digital works, indicating a big step toward incorporating blockchain technology into fashion. Furthermore, the combination of artificial intelligence (AI) and augmented reality (AR) improves personalization in the luxury industry. Companies like FFFACE.ME have pioneered semi-digital apparel and AR mirrors, allowing customers to virtually try on things and interact with brands in novel ways. These technology improvements not only improve the customer experience, but they also create new opportunities for brand involvement and storytelling.

The rise of virtual influencers and digital fashion platforms exemplify this tendency. Tribute Brand, for example, sells digital clothing that can be worn in augmented reality, demonstrating a growing need for virtual fashion that is consistent with personal identity and environmental concerns. In this dynamic climate, co-branding allows luxury brands to collaborate with tech businesses, artists, and digital platforms. Such collaborations allow brands to build on each other's strengths, broaden their reach, and co-create value in both physical and digital locations. As the physical and digital worlds become increasingly intertwined, the merging of co-branding and digital fashion represents a watershed moment in the luxury business.

THEORETICAL FRAMEWORKS

- **Brand Equity Theory**

Brand equity theory is a key marketing idea that explains the manner in which the worth of a company results from customer impressions, attitudes, and behaviors. Two seminal models, David Aaker's Brand Equity Model and Kevin Lane Keller's Customer-Based Brand Equity (CBBE) Model, provide comprehensive frameworks for evaluating and maintaining brand equity.

- **Aaker's Brand Equity Model**

According to David Aaker (1991), brand equity is a collection of assets and liabilities associated with a brand's name and symbol that add to or detract from the value supplied by a product or service. His model identified five critical dimensions:

Brand loyalty is a customer's attachment to a brand, which influences repeat purchases and reduces marketing expenses. Brand awareness is the ability of customers to recognize or recall a brand, which can impact attitudes and decisions. Perceived Quality: A consumer's assessment of the overall quality or superiority of a product or service. Brand associations are mental connections and attributes that shape a brand's identity and value. Other proprietary brand assets include patents, trademarks, and channel partnerships that offer a competitive advantage. Aaker's model highlights that these characteristics contribute to brand equity by increasing customer value and offering benefits to the company, such as higher margins and brand extensions.

- **Keller's Customer-Based Brand Equity (CBBE) Model.**

Kevin Lane Keller (1993) takes a consumer-centric perspective, defining brand equity as the differential effect that brand knowledge has on customer responses to brand marketing. His CBBE model is designed as a pyramid with four increasing levels: Brand identity (Who are you?) which entails establishing brand salience in order to assure identification and recall. Brand Meaning (Who Are You?) which is establishing brand connotations through performance and visuals. Brand Response (What about you?) that is gathering consumer opinions and feelings about the brand. Brand Resonance (What about you and me?) that is creating deep psychological relationships that lead to loyalty and advocacy. Keller's concept emphasizes the necessity of creating a great brand by molding consumer perceptions and experiences.

- **Impact of Co-Branding on Brand Equity**

Co-branding, a strategic relationship between two or more brands, can have a major impact on brand equity. Effective co-branding can enhance brand associations by combining strengths and creating new positive connotations. Partnerships with established brands can improve quality perceptions. Broaden brand awareness is established by connecting with partner audiences. However, co-branding carries hazards, such as the possibility for brand identity dilution if the relationship lacks strategic alignment. Therefore, careful evaluation of brand compatibility and shared values is critical. Recognizing and managing brand equity using models such as Aaker's and Keller's is critical for capitalizing on co-branding opportunities while preserving brand integrity.

- **Rogers' Diffusion of Innovation Theory**

Rogers (2003) describes the process by which new ideas and technologies spread across cultures. Adoption of digital fashion and smart luxury products is extremely important for innovators and early adopters—typically younger, affluent, and tech-savvy consumers. This theory explains market segmentation, technology acceptability, and the rate of adoption among luxury audiences.

- **Resource-Based View (RBV) of the Firm**

According to Barney's RBV, organizations obtain a competitive edge by leveraging distinctive resources, both tangible (technology) and intangible (brand legacy). Co-branding between luxury and technology firms generates a hybrid resource that can provide exclusivity, innovation, and cultural capital, making it more difficult for competitors to replicate.

- **Consumer Culture Theory (CCT)**

CCT investigates how people use things to express themselves and navigate social meanings. Digital fashion, particularly in luxury, enables symbolic consumption, status signalling, and customisation in both physical and virtual environments. Co-branded smart devices reflect changing customer desires to portray identity via aesthetics and technology.

LITERATURE REVIEW

- Co-Branding in Luxury Markets

Co-branding has evolved as a strategic tactic for premium firms that want to remain culturally relevant. Successful collaborations such as Apple x Hermès and Balenciaga x Crocs demonstrate how co-branding can blend innovation and tradition. Hillyer & Tikoo (1995) and Park et al. (1996) found that brand image congruence is critical for consumer acceptability. When brands with similar values align, consumer trust and brand attraction increase.

- **Digital Fashion: Origins and Evolution**

Digital fashion, which includes virtual clothes, augmented reality try-ons, and blockchain-authenticated gear, began in experimental design spaces but has since become a viable commercial and branding tool. Scholars such as Rissanen and McQuillan (2016) highlight the environmental and artistic potential of digital fashion. Virtual collections (e.g., from The Fabricant or DressX) allow firms to engage younger, technologically savvy customers while decreasing material waste.

- **Smart Luxury Products and Experience Innovation**

Smart luxury products are designed to be both functional and aesthetically premium. These include smartwatches, connected accessories, and IoT-embedded apparel. Research by Kim & Ko (2012) and Kapferer (2015) confirms that smart features, when combined with exclusivity and design, appeal to Millennial and Gen Z consumers. These products also create new forms of experiential engagement, allowing brands to build emotional and practical value.

- **Personalization, artificial intelligence, and predictive analytics**

Luxury brands are increasingly using artificial intelligence (AI) to provide individualized consumer experiences. Predictive analytics provides for personalized product recommendations, whilst AR allows customers to visually try things. Pine & Gilmore (1999) and Deloitte (2020) found that customisation improves consumer satisfaction, retention, and emotional brand attachment in the luxury industry.

- **Blockchain and authentication assurance**

Authenticity is essential in luxury branding, and blockchain technology offers a way to authenticate product origin, ownership, and supply chain integrity. Tapscott and Tapscott (2018) describe how blockchain can prevent counterfeiting and increase consumer trust. Co-branded digital fashion combined with blockchain enables NFTs and digiAuthenticity is essential in luxury branding, and blockchain technology offers a way to authenticate product origin, ownership, and supply chain integrity. Tapscott and Tapscott (2018) describe how blockchain can prevent counterfeiting and increase consumer trust. Co-branded digital fashion combined with blockchain enables NFTs and digital ownership, increasing product exclusivity and secondary market value.

- **Digital Fashion and Sustainability**

Environmental concerns have prompted luxury firms to adopt more sustainable strategies. Digital fashion dramatically minimizes environmental impact by minimizing fabric waste, shipping, and overproduction. Fletcher and Tham (2019) found that digital clothes provide circular solutions and lessen environmental impact while keeping a strong brand story and visual presence.

- **Challenges to Brand Authenticity and Over-Digitalization**

While innovation is welcome, it must be consistent with the main brand story. Beverland (2006) contends that retaining authenticity is critical for luxury brands. Over-digitalization, or cosmetic tech add-ons, can diminish brand equity if not done properly. Consumer distrust may develop if the digital change appears unconnected from brand legacy.

- **Social Media, Virtual Influencers, and Digital Engagement**

Luxury businesses are transitioning to immersive digital engagement methods that include Instagram, TikTok, and the metaverse. Virtual influencers (e.g., Lil Miquela) and gamified experiences (e.g., Gucci Garden in Roblox) enable marketers to connect with digital natives. According to studies, engagement-driven campaigns increase brand virality, emotional connection, and community-based loyalty, all of which are crucial for long-term brand health.

MARKETING STRATEGY AND CONSUMER BEHAVIOR

Strategic Alignment in Co-Branding

Successful co-branding is dependent on the strategic alignment of the partnered brands. This alignment includes shared values, complementary target audiences, and a consistent market positioning. For example, Apple and Hermès' partnership displays synergy. Apple's focus on innovation and design perfection complements Hermès' tradition of craftsmanship and elegance. They collaborated to create the Apple Watch Hermès, which combines cutting-edge technology with timeless elegance, appealing to customers who respect both functionality and distinction.

Similarly, Balenciaga's partnership with Fortnite exemplifies strategic alignment by combining high fashion and digital game culture. This collaboration aims to reach a younger, tech-savvy clientele that is interested in both virtual and actual fashion. By incorporating Balenciaga's designs into Fortnite's virtual environment, the company efficiently targets people who may not normally associate with high fashion, broadening its audience base. These examples highlight the necessity of co-branding campaigns that link brand identities and values in order to generate authentic and appealing offers for consumers.

Leveraging Digital Marketing Tools

In the digital age, brands use a variety of marketing techniques to enhance their co-branding efforts. Social media platforms, influencer collaborations, and virtual events all play important roles in reaching and engaging audiences.

Gucci's relationship with Roblox exemplifies the effective use of digital channels to reach out to younger audiences. Gucci used Roblox's metaverse to create virtual experiences that not only showed its products but also engaged people in interactive and engaging ways. This strategy allows Gucci to engage with the gaming community, increasing brand exposure and loyalty among Gen Z consumers.

Furthermore, influencer relationships increase brand visibility and credibility. Collaborating with influencers that resonate with the target audience can help to magnify marketing messaging and increase customer engagement. These collaborations allow firms to reach a larger audience and establish trust through genuine endorsements.

Virtual events, like as online product debuts and digital fashion shows, help to enhance co-branding campaigns. These events generate buzz and excitement by giving consumers exclusive access to new items and experiences. Brands may use digital marketing technologies to successfully deliver their messages and engage with customers in relevant ways.

Omnichannel Integration

Integrating online and offline channels guarantees a consistent customer experience, which is critical for current marketing tactics. Balenciaga's cooperation with Fortnite illustrates successful omnichannel integration. The cooperation expanded beyond the virtual realm to include physical products, allowing customers to buy real-world counterparts of the in-game clothes. This seamless transition between digital and physical items improved the consumer experience by offering additional touchpoints for involvement.

Omnichannel integration also requires uniform branding across all platforms. Whether a consumer engages with a business online via social media or in-store, the messaging, images, and overall experience should be consistent in order to establish brand identification. This consistency fosters consumer trust and loyalty since customers can rely on a consistent brand experience regardless of channel. Moreover, information acquired from numerous sources can be used to personalize marketing activities. By studying consumer behavior across touchpoints, organizations can customize their goods and messaging to individual tastes, increasing customer happiness and driving sales.

Data Analytics and AI Personalization

The use of data analytics and artificial intelligence (AI) allows firms to tailor their services and forecast consumer behavior. Large datasets can be analyzed to gain insights about consumer preferences, purchase patterns, and levels of engagement.

Gucci's use of big data exemplifies this methodology. The brand uses data analytics to improve its digital marketing methods, improving campaigns to reach its target demographic. This data-driven approach enables Gucci to create individualized information and experiences, forging deeper connections with customers. AI enhances personalization efforts by enabling predictive analytics. Machine learning algorithms can examine past consumer behavior to predict future behaviors, allowing firms to adjust their marketing efforts more effectively. For example, AI can offer items based on individual tastes, raising conversion rates and improving customer happiness.

Furthermore, AI-powered chatbots and virtual assistants offer consumers real-time assistance by answering questions and guiding them through the shopping process. These solutions improve the customer experience by providing immediate support and personalized recommendations, which drives engagement and loyalty. Effective digital co-branding marketing techniques include strategic alignment between collaborating businesses, the use of digital marketing technologies, the integration of online and offline channels, and the use of data analytics and artificial intelligence for personalization. By using these strategies, brands can increase consumer engagement, establish loyalty, and drive business success in the ever-changing digital market.

Challenges and Risks in Digital Co-Branding

While digital co-branding provides countless potential for innovation and market expansion, it also introduces a number of hurdles and risks that companies must overcome in order to maintain their integrity and consumer trust.

Authenticity Dilution: One of the key problems with co-branding is the potential loss of brand legitimacy. When brands with different beliefs, styles, or target audiences collaborate, customers may perceive the collaboration as inauthentic or opportunistic. This image can destroy brand equity and drive away loyal customers. For example, a luxury fashion business that collaborates with a fast-food chain may suffer backlash from its affluent consumer base, who may regard the partnership as a deviation from the brand's premium image. Such collaborations might lead to consumer distrust and a loss of brand confidence.

Mitigation strategies can be conducting thorough brand compatibility assessments to ensure potential partners align with similar values and brand identities. Transparent communication can be ensured with consumers, stressing shared values and mutual benefits. Selective collaborations can be established with brands that align with their heritage and target demographic.

Technological Overreach: In the quest of innovation, brands may exaggerate their technological skills or introduce innovations that alienate their core client base. Overemphasis on cutting-edge technology without regard for the preferences of all consumer categories might result in disengagement and discontent. For **example, A** premium watchmaker offering a smartwatch with too many functions may confuse traditional clients who value simplicity and quality. If the smartwatch's appearance and functionality depart too far from the brand's primary offers, it may fail to appeal to its existing customers.

Mitigation strategies can be designed with the consumer in mind, allowing technological developments to enhance rather than undermine the primary brand experience. Incremental integration leads to the progressive introduction of new technology, allowing consumers to adjust and providing options to meet a variety of tastes. input Loops can be implemented on a regular basis to collect and analyze user input in order to gauge reactions to technology improvements and make required adjustments.

Digital Divide: The digital divide is the disparity between people who have access to contemporary information and communication technology and those who don't. This barrier can hinder the reach and effectiveness of digital co-branding projects, especially when targeting underserved or rural groups. For example, a fashion firm offering an exclusive online collection may unintentionally exclude clients who lack dependable internet access or digital literacy, limiting the campaign's success in specific groups. plans such as Inclusive Campaign Design will create marketing plans that take into account different levels of access, providing alternative methods for consumers to interact with the business. Community outreach would entail collaborating with local organizations to bridge the digital divide and providing resources to underserved communities. Multiplatform Presence To reach a larger audience, make campaigns accessible across several platforms, including traditional media. Bridge the digital divide and provide services to underserved populations.

Data Privacy Concerns: In the digital age, co-branding initiatives that use customer data must prioritize data protection in order to maintain confidence and comply with legislation. Collecting personal information through co-branded platforms without clear consent channels can result in customer backlash and legal action, particularly if it is mishandled or improperly protected. To reduce these risks, organizations should use transparent data policies that clearly tell consumers about data collecting and acquire explicit consent. Strong security measures, such as encryption and two-factor authentication, are required to protect user data from breaches. Furthermore, staying current with data protection legislation such as the General Data Protection Regulation (GDPR) and ensuring that all procedures comply with legal standards are critical measures in preserving consumer trust and avoiding legal ramifications.

Integration Complexity: Integrating Challenges in Co-Branding Campaigns are very much prevalent. Co-branding efforts face substantial obstacles when attempting to merge disparate brand systems, organizational cultures, and operational processes. Such integration issues might result in operational inefficiencies, misaligned goals, and a fragmented customer experience. For example, a collaboration between a luxury fashion brand and a technological company may face challenges in aligning supply chains, customer care norms, and brand message, potentially leading to discrepancies and consumer confusion. To mitigate these issues, detailed preparation is required. Creating extensive integration plans that include all parts of the cooperation, from logistics to branding, guarantees a consistent approach. Cross-functional teams made up of individuals from both brands can help to ensure smooth integration and quick resolution of issues. These teams can use their different abilities and viewpoints to effectively synchronize tactics. Continuous monitoring of the integration process enables timely assessments and adjustments, maintaining alignment and operational efficiency throughout the co-branding

While digital co-branding offers exciting opportunities for innovation and market expansion, it is essential for brands to be aware of and address the associated challenges and risks. By implementing strategic measures and maintaining a consumer-centric approach, brands can navigate these challenges effectively and build successful, sustainable partnerships. While digital co-branding presents exciting prospects for innovation and market expansion, businesses must be aware of and handle the related challenges and risks. Brands can effectively traverse these hurdles and form successful, long-term partnerships by applying strategic strategies and remaining consumer-centric.

STRATEGIC IMPLICATIONS FOR BRAND MANAGERS

In the evolving landscape of digital co-branding, brand managers must navigate complex dynamics to ensure successful partnerships and sustained brand equity. This section delves into key strategic considerations that are pivotal for brand managers in the realm of co-branding.

Frameworks for Brand Alignment: Developing strong standards for assessing brand affinity is critical to the success of co-branding activities. These structures ensure that collaborations are strategically aligned, authentically resonate with customers, and meet mutual goals. These frameworks' key components include examining brand values, target markets, and long-term objectives. Prada and Adidas' cooperation is a prime example of strategic brand alignment. Both firms are committed to sustainability and innovation, as evidenced by their combined Re-Nylon line. This alliance harmonizes their company values and appeals to environmentally conscious customers.

A comprehensive brand audit examines each brand's underlying values, mission, vision, and market positioning. This assessment helps identify whether the brands have complementary ideologies and can collaborate to build a unified brand narrative. For example, both Prada and Adidas value innovation and sustainability, so their cooperation on the Re-Nylon line was a natural fit. This line combines Prada's luxury workmanship with Adidas' sporting innovation, incorporating recycled materials to appeal to environmentally concerned customers. Collecting and evaluating customer reviews is critical to ensuring that the collaborative branding effort is consistent with the target audience's expectations and preferences. Considering consumer perceptions can help influence the development of products and marketing tactics that appeal to both firms' client bases. In the example of Prada and Adidas, their collaboration capitalized on consumers' common interest in environmentally friendly clothing, increasing the perceived value of their collaborative offers. Developing combined strategies that capitalize on the strengths of both brands is critical for accomplishing common goals. This includes coordinating product development, marketing campaigns, and distribution methods. The Prada-Adidas collaboration shows this strategy, combining Prada's design knowledge with Adidas' technological capabilities to develop goods that reflect both companies' identities while also addressing current sustainability problems. Implementing these principles allows brand managers to guarantee that co-branding agreements are cohesive, strategically aligned, and authentically resonate with customers, boosting brand equity and achieving long-term success.

Sustainable Co-branding Practices: Incorporating sustainability into co-branding projects appeals to environmentally conscientious customers and boosts company reputation.

For example, the Prada x Adidas Re-Nylon line demonstrates a dedication to sustainability by employing Prada's Re-Nylon fabric, which is manufactured from recycled materials. This effort not only encourages environmental responsibility, but it also addresses the growing consumer demand for sustainable products.Prada's Implementation Strategies:

- Prioritize sustainable and recyclable materials for product development.
- Ensure transparency by clearly communicating sustainable efforts to consumers through marketing and product labeling.

- Collaborate with groups that promote environmental sustainability.

Managing Digital Innovation: Luxury brands must embrace technological innovations proactively in order to remain competitive and relevant in today's quickly expanding digital landscape. Integrating technologies like artificial intelligence (AI) into design and marketing processes allows organizations to tailor customer experiences, streamline operations, and stimulate creativity. Adidas, for example, uses 3D printing and artificial intelligence to create unique shoe designs that are suited to each customer's foot form and size, increasing both innovation and effectiveness.

Risk Mitigation Strategies: Although technological advancement has many advantages, it also poses potential hazards to brand image and confidence among customers. The cancellation of Nike's RTFKT virtual shoe project acts as a cautionary tale. Despite the undertaking's conclusion, Nike faced a group of lawsuits from customers who claimed that they had been misled, saying the NFTs were unregistered securities and that they would not have bought them if they had been informed about the planned closure. This event emphasizes the value of good communication and risk evaluation in digital ventures. To minimize these hazards, companies ought to carry out periodic assessments to recognize possible obstacles connected to collaborative branding campaigns, create exhaustive emergency strategies for dealing with recognized hazards, and preserve honest dialogue with all stakeholders throughout the co-branding process to regulate requirements while establishing credibility.

FUTURE DIRECTIONS IN SMART LUXURY CO-BRANDING

The future of co-branding in the luxury fashion industry is being shaped by technological advancements and evolving consumer expectations. This section explores emerging trends that are poised to redefine co-branding strategies.

- **AI-Powered Customization:** Artificial intelligence enables personalized consumer experiences, from product recommendations to virtual try-ons, enhancing customer satisfaction. Generative AI is transforming the way luxury fashion is conceived, produced, and personalized. Through algorithms capable of learning consumer preferences and generating unique design variations, brands can offer hyper-personalized smart products. For example, brands like Gucci and Nike have explored AI-assisted design for creating custom sneakers or digital outfits tailored to individual style profiles. In the context of co-branding, AI enables each partner brand to contribute its strengths: the tech brand provides the AI infrastructure while the fashion house offers heritage-rich design sensibilities. This synergy results in unique product offerings that can adapt to the consumer's digital identity, mood, or environment in real-time. For instance, a smart luxury watch co-designed by a tech brand and a fashion label could change its interface aesthetics based on mood-sensing AI or fashion choices. AI also enables predictive customization, where future consumer trends are anticipated and integrated into the design pipeline—minimizing inventory risk while maximizing brand relevance. Example: AI-driven tools allow brands to analyze consumer data and preferences, enabling the creation of customized products and marketing strategies.

- **AI-driven personalization strategies:** Incorporating artificial intelligence (AI) into fashion retail has transformed how firms understand and interact with customers. By analyzing massive volumes of data, AI allows firms to acquire insights into consumer behavior and preferences, resulting in highly tailored purchasing experiences. For example, AI-powered recommendation engines can assess browsing history and previous purchases to promote products that match individual preferences, increasing consumer happiness and loyalty. This level of personalization not only enhances the shopping experience, but it also increases sales by providing customers with things they are more likely to buy. Furthermore, AI can help businesses optimize inventory management by predicting trends and consumer demand, ensuring that popular items are appropriately stocked. Feedback methods are also essential for improving personalization strategies. Implementing tools for collecting and analyzing consumer feedback enables organizations to constantly enhance their offers and customize experiences to changing customer needs. For example, services like Stitch Fix use AI algorithms that incorporate client feedback to improve stylist recommendations, balancing creative styling and data-driven insights. By using AI in this way, fashion firms can create a dynamic, responsive shopping environment that resonates with customers and drives long-term engagement.
- **Web3 and Decentralization:** Adopting Web3 innovations enables brands to establish centralized systems, giving customers more control and encouraging community interaction. Web3 proposes decentralized methods of ownership and community governance via NFTs, DAOs, and blockchain-based economies. In the luxury fashion industry, this results in new forms of offering value. Luxury NFTs, for example, can serve as digital authentication certificates or provide access to special events, drops, or metaverse places. Co-branding in this ecosystem can combine fashionable designs and blockchain functionality. A partnership between a premium brand and a blockchain start-up could end up in limited-edition virtual handbags that exist only in the metaverse, with a value for sale maintained on the blockchain. Gucci's Vault

project and Prada's Timecapsule NFT releases were initial instances of such innovation.

Decentralized Autonomous Organizations (DAOs) are blockchain-based corporations governed by token-holding members rather than a centralized authority, and they are rapidly changing the way luxury companies interact with their audiences. In these systems, consumers have voting rights proportional to their token holdings, allowing them to influence design decisions, governance regulations, and even environmental activities (for example, voting on eco-friendly materials). Websites like MetaFactory and Zora demonstrate this trend by allowing "consumer-creators" to co-own community-driven brands and share income earned from collaboratively created fashion goods.

In a similar manner Red DAO, an ensemble of fashion fans, made news by paying more than $1.3 million for Dolce & Gabbana's "Doge Crown" NFT, illustrating DAO communities' purchasing power and influence in the luxury industry. Luxury companies such as LVMH have also experimented with DAO-style groups via their Aura blockchain, which provides Digital Product Passports and encourages consumer participation in authenticity certification and lineage tracing. Other than possession, DAOs promote transparency and trust. Participants are able to monitor a garment's entire lifecycle—from raw materials to retail—via blockchain-recorded provenance, as envisioned in DAO-governed sustainable fashion models, during which identifying a QR code provides source, labor circumstances, and carbon footprint. Successful integration of DAOs into luxury co-branding necessitates meticulous architecture and community design. Brands must first integrate blockchain, selecting scalable and secure platforms—such as Ethereum or specialized enterprise chains—to host DAO smart contracts that transparently record governance rules and token-based voting procedures. Next, creating user-friendly community platforms (e.g., Discord channels or bespoke portals) is necessary for onboarding token holders, facilitating debates, and guiding newbies through participation processes. Embedding smart contracts automates membership management, proposal submissions, and vote counting, guaranteeing that decisions—such as picking a limited-edition co-branded digital wardrobe—are implemented without manual interference or tampering. To maintain momentum and trust in an active DAO community, enterprises should hold regular digital events, such as AMAs with designers, virtual studio tours, or co-design workshops, and transparently report on decision results and token economics. Finally, ongoing legal and regulatory compliance is critical: businesses must collaborate with legal specialists to navigate the securities legislation surrounding token sales and establish clear guidelines for intellectual property rights when community-generated designs enter the commercial arena. Luxury companies may employ these integrated strategies to not only improve consumer interaction, but also contribute to developing new collaboration company structures in the digital fashion frontier.

• **Sustainable Digital Twins and Eco-Conscious Smart Fashion:** Digital twins provide virtual replicas of physical products, facilitating virtual interactions and enhancing the shopping experience. Sustainability remains a central concern in the luxury industry, and digital twins—virtual replicas of physical products—are becoming key tools for environmental tracking and control. By linking each physical item to its digital twin, brands can monitor lifecycle data, carbon emissions, and even ownership history. For example, a co-branded luxury jacket embedded with sensors could send environmental performance data to its digital counterpart on the blockchain. Such applications enable transparent communication of sustainability metrics, fostering consumer trust and ethical brand positioning. Brands like Stella McCartney and PANGAIA are already exploring bio-fabrication and digital material passports that align with this model. From a co-branding perspective, IT companies can provide the infrastructure for digital twin technology, while luxury brands provide the product and narrative. This strengthens brand equity while meeting the aware consumer's desire for transparency and environmental integrity. For example, brands like H&M are producing digital twins of human models to showcase apparel goods, allowing shoppers to see how products fit and look in a virtual setting.

Digital twins are virtual replicas of physical products that enable virtual interactions and enhance the shopping experience. Sustainability remains a big problem in the luxury industry, and digital twins (virtual replicas of actual products) are becoming important tools for environmental monitoring and control. Companies may track lifecycle statistics, carbon emissions, along with property history by connecting every physical thing to its digital equivalent. For example, a co-branded designer clothing outfitted with sensors may provide ecological sustainability data to its blockchain counterpart. Such solutions allow for the clear sharing of sustainability characteristics, fostering customer trust and ethical brand positioning. Stella McCartney and PANGAIA are already looking on bio-fabrication and digitally generated credentials that are sustainable with this theoretical framework.

- **Virtual Influencers:** As luxury businesses navigate the increasingly digital-first market, the intersection of co-branding and digital fashion is primed for transformational changes. Emerging technologies such as generative AI, Web3, digital twins, and synthetic influencers are shaping future trajectories. These advances not only change the way luxury is made and consumed, but they also necessitate rethinking brand strategy, identity, and customer involvement. The development of computer-generated influencers such as Lil Miquela, Shudu, and Noonoouri has marked a significant shift in digital marketing. These synthetic personas provide brands complete control over their branding, beauty, and communicating, without the unpredictable nature of human influencers.

Virtual influencers can function as branded avatars in luxury co-branding, embodying both partners' ethos. For example, a collaboration between a French couture house and a Silicon Valley AI firm may result in the creation of a virtual brand ambassador who appears in metaverse fashion shows or TikTok commercials wearing AI-generated smart clothing. These virtual beings serve as immersive storytellers and product demonstrations, flawlessly blending brand storytelling throughout physical and virtual platforms.

Virtual influencers have additional ethical and legal benefits—they can be trained to demonstrate diverse standards, prevent controversies, and ensure constant brand consistency. Example: Brands like Prada and Nars have collaborated with computer-generated influencers to promote products, creating a futuristic and engaging brand presence. Content production is essential for drawing virtual influencers to reality. Brands use modern CGI, motion capture, and AI technologies to create entertaining video for a variety of digital platforms, including Instagram, TikTok and YouTube. These virtual influencers connect with real individuals in digital settings, resulting in a captivating blend of fantasy and realism for consumers. Cultivating relationships between virtual influencers and their followers increases audience engagement while also creating a sense of community and trust. Virtual influencers can create meaningful relationships with consumers by continuously producing honest and relevant material, resulting in increased brand loyalty and digital presence.

- **NFTs and Digital Ownership:** Non-fungible tokens (NFTs) have transformed digital ownership by allowing customers to own unique digital assets and creating new opportunities for brand engagement and revenue streams. Luxury fashion businesses are at the forefront of this innovation, using NFTs to create special digital collectibles that reflect their brand identity and consumer preferences. For example, Gucci's cooperation with Superplastic produced the "SuperGucci" NFT line, which combined digital art with the brand's distinctive aesthetics. Similarly, Prada's Timecapsule NFT line includes limited-edition digital content linked to physical products, creating a tangible link between the digital and physical domains.

Using NFTs in luxury fashion necessitates numerous strategic measures. To begin, brands focus on NFT creation by creating one-of-a-kind digital assets that reflect their brand and appeal to consumer interests. These assets frequently include aspects of the brand's legacy and design philosophy, ensuring authenticity and exclusivity. Second, marketplace integration is critical; placing NFTs on trustworthy platforms like as OpenSea or Rarible allows for easier buying and selling, broadening the brand's digital reach. Finally, providing ownership benefits boosts value and engagement; NFT holders may receive benefits like unique content, early access to new collections, or invitations to virtual events. This technique not only promotes a sense of community, but it also strengthens brand loyalty in the digital age.

- **The Blending of Phygital Experiences:** Future luxury co-branding will most likely take place in the "phygital" environment, which merges the physical and digital worlds. Smart mirrors, holographic fashion presentations, AR-enabled product interfaces, and haptic feedback in wearable technology will all become part of luxury's immersive retail story. A co-branded experience may include entering a flagship store and using a smart mirror powered by a tech partner to preview how a digital-only coat will appear on the shopper's avatar in the metaverse. The actual goods may include an integrated chip that connects to its NFT certificate, which the consumer can then show in a digital fashion museum or game environment. The convergence of physical and digital will need new design thinking, legal frameworks, and user experience methods, as well as opportunities for deeper brand involvement and post-purchase connection.

METHODOLOGICAL FRAMEWORK

This study employs a qualitative case study approach to investigate the dynamics of co-branding collaborations within the digital fashion landscape. The qualitative methodology is chosen for its strength in exploring complex phenomena within their real-life contexts, allowing for an in-depth understanding of the strategies, consumer perceptions, and outcomes associated with digital co-branding initiatives.

The collected data were subjected to thematic analysis to identify recurring patterns and themes related to co-branding strategies and consumer responses. Content analysis was employed to systematically code and interpret textual and visual data from marketing materials and consumer feedback. Sentiment analysis tools were utilized to quantify consumer attitudes expressed in social media content. This study uses a qualitative, interpretive research approach to investigate the role of co-branding and digital fashion in converting smart luxury products. The analysis attempts to provide a thorough understanding of how these collaborations influence consumer engagement and brand evolution in the digital age by combining multiple data sources such as academic literature, brand case studies, industry reports, and digital artifacts. A selective selection strategy was used to find co-branding relationships of strategic value in the luxury sector. Notable cases include Apple x Hermès, Balmain x Pokémon, and Gucci x Roblox, which were chosen based on factors like as prominence, brand alignment, and technology innovation. Each collaboration was evaluated using the criteria of value co-creation, brand synergy, and customer engagement.

The Apple x Hermès advertisement stressed the seamless blending of technology and traditional craftsmanship. Visuals showcased the elegance of Hermès' leather straps with Apple's sleek wristwatch design, underlining the image of a perfect collaboration between the two businesses. Social media content emphasized the product's versatility and appeal to a discerning audience. Balmain's campaign with Pokémon included vivid visuals and whimsical elements to convey the essence of both brands. The marketing materials included models dressed in Balmain fashion alongside iconic Pokémon characters, resulting in a visually striking juxtaposition that appealed to a wide range of consumers. The campaign effectively conveyed a theme of inclusion and creativity.

Gucci's digital advertising on Roblox featured deep storytelling and interactive components. Users could walk through themed rooms that represented various aspects of the brand's history and aesthetic. These virtual settings not only kept users engaged, but also allowed them to express themselves through customisable avatars and digital fashion items. Data from reliable sources such McKinsey, Business of Fashion, WGSN, and Bain & Co. were reviewed to situate these case studies within broader industry trends. These publications shed light on the acceptance and efficacy of digital co-branding tactics. In addition, social media analytics, app-based virtual try-on measurements, and NFT sales data were used to evaluate customer involvement and the impact of digital initiatives.

For example, surveys revealed an increasing consumer interest in personalized and immersive digital experiences, prompting luxury brands to investigate virtual platforms and technologies. Gucci's success in the metaverse, as indicated by strong user engagement on Roblox, demonstrated the power of digital spaces in expanding brand reach and establishing community. Additionally, the increasing popularity of NFTs and digital collectibles indicated a shift in consumer behavior, with people wanting unique digital assets that provide exclusivity and status. Balmain capitalized on this trend by incorporating digital elements into their collaborations, increasing consumer interaction and bringing up new revenue streams.

CONTENT ANALYSIS OF DIGITAL CO-BRANDING CAMPAIGNS

The content research focuses on four major digital co-branding collaborations: Apple x Hermès, Balenciaga x Fortnite, Gucci x Roblox, and Prada x Adidas. These examples were chosen for their unique use of fashion and technology, as well as their major impact on customer engagement and brand impression.

- **Apple x Hermès: Fusion of Technology and Luxury**

The Apple x Hermès cooperation brought together Apple's technological power and Hermès' luxurious workmanship to offer a premium wristwatch experience. Marketing materials emphasized the product's rarity and refinement, using high-quality graphics and simple design features. Consumer reaction praised the flawless blending of usefulness and design, however some critics cited the expensive price as a barrier to accessibility.

- **Balenciaga x Fortnite: Bridging Fashion and Gaming**

Balenciaga's collaboration with Fortnite brought high-fashion elements to the gaming world, with players able to wear virtual costumes and accessories. The campaign used immersive narrative and interactive experiences to engage a younger, digitally native audience. Social media study found a good response, with people expressing excitement about the new merging of fashion and gaming. The cooperation has spurred debate over the future of digital fashion and the prospects for virtual economies.

- **Gucci x Roblox: Virtual Luxury Experiences**

Gucci's involvement with Roblox entailed constructing virtual locations and limited-edition digital items that allowed users to explore and purchase virtual fashion pieces. The campaign aimed at Gen Z customers, using the popularity of gaming platforms to bring luxury branding in a new environment. User interaction data revealed a high degree of activity, with virtual things selling out quickly and generating significant attention on social media. Some customers questioned the value proposition of digital-only goods, emphasizing the necessity for clear explanation of the benefits and distinctiveness of virtual fashion fashion.

- **Prada x Adidas: Sustainable Streetwear Synergy**

The Prada x Adidas collaboration aimed to combine premium fashion with streetwear aesthetics, emphasizing sustainability and creativity. Marketing efforts focused on the artistry and eco-friendly materials used in the items, appealing to environmentally conscious customers. The social media sentiment research revealed a largely positive response, with users complimenting the design and ethical considerations. Nonetheless, considerable criticism surfaced regarding the limited availability and high cost of the items, implying a tension between exclusivity and accessibility.

- **Cross-Case Insights**

The examination of co-branding efforts between luxury fashion labels with digital platforms indicates several fundamental aspects that contribute to the success of these alliances.

Brand Alignment: Successful co-branding initiatives show a strong alignment of the collaborating businesses' values and aesthetics, resulting in a unified and authentic product offering. For example, Apple and Hermès' collaboration seamlessly combines Apple's new technology with Hermès' traditional craftsmanship, resulting in a product that appeals to consumers looking for both functionality and luxury. Similarly, Balmain's collaboration with Pokémon combines the fashion house's striking designs with the franchise's lively and adventurous spirit, appealing to a wide audience.

Innovative Engagement: Including interactive and immersive aspects, such as virtual experiences and gamification, increases customer engagement and broadens brand reach. Gucci's foray into the virtual world of Roblox, with projects such as the Gucci Garden and Gucci Town, enables consumers to interact with the brand in an interactive setting, building a stronger relationship with younger audiences.

Consumer Perception: Transparency in messaging and a clear value proposition are essential for developing good consumer impressions, especially when launching new concepts such as digital-only fashion items. Consumers are more inclined to adopt new goods when businesses successfully convey their purpose and benefits, ensuring that digital products are consistent with the brand's identity and consumer expectations.

Accessibility vs. Exclusivity: Balancing the need for exclusivity with the requirement for increased accessibility presents a problem, with pricing and availability influencing consumer attitude. While exclusivity can boost a company's reputation, providing accessible options guarantees that a broader audience can interact with the brand, encouraging inclusivity and broadening the market. This detailed methodological methodology and content analysis shed light on the strategic execution and consumer acceptance of digital co-branding projects in the fashion sector. By studying these high-profile collaborations, the report identifies optimal practices and potential dangers, providing insight for future co-branding efforts in the rapidly changing digital context.

CONCLUSION

This study looked at the nexus of co-branding and digital fashion in the luxury sector, specifically how strategic collaborations and technology advances are transforming consumer involvement and brand experiences. Through qualitative case analyses of collaborations such as Apple x Hermès, Balmain x Pokémon, and Gucci x Roblox, we found significant characteristics that highlight the success and challenges of digital co-branding ventures. One of the key results is the significance of brand alignment in co-branding initiatives. Successful collaborations are distinguished by the seamless integration of brand values, aesthetics, and target audiences. For example, the Apple x Hermès collaboration perfectly combines Apple's technological acumen and Hermès' artisanal artistry, resulting in a device that appeals to consumers seeking both innovation and luxury. Similarly, Balmain's cooperation with Pokémon combines high fashion with pop culture, garnering a diverse audience. Innovative engagement tactics have developed as essential components of successful digital co-branding. The use of interactive and immersive components, such as virtual experiences and gamification, boosts customer engagement and broadens brand reach. Gucci's foray into the virtual world of Roblox, for example, allows customers to interact with the brand in a dynamic and engaging setting, forging stronger ties with younger audiences.

Consumer perception is critical to the success of digital cobranding activities. Transparency in messaging and a clear value proposition are critical for developing good consumer attitudes, especially when launching new concepts such as digital-only fashion items. Brands need to successfully convey the objective and advantages of such services in order to line with customer desires and values. Balancing accessibility and exclusivity is a significant difficulty for digital co-branding. While exclusivity can boost a brand's status, providing more accessible options promotes more consumer engagement. Pricing and availability options must be carefully considered in order to preserve brand equity while extending the consumer base. The incorporation of developing technologies, like as artificial intelligence (AI), augmented reality (AR), and blockchain, is changing the landscape of digital co-branding. AI personalizes consumer experiences through data analytics and predictive modeling, whilst AR provides immersive shopping experiences that bridge the gap between online and offline retail. Blockchain technology makes transactions more transparent and secure, especially when it comes to non-fungible tokens (NFTs) and digital ownership. Environmental and moral concerns are becoming more significant in co-branding efforts. People are expecting increased openness and liability from businesses, resulting in collaborations that promote social and ecological consciousness. For example, using digital product passports (DPPs) improves supply chain accountability and promotes environmentally friendly procedures. The changing online ecosystem needs constant modification and creativity in co-branding tactics. To remain relevant and competitive, brands must be nimble and sensitive to evolving trends and customer behaviors. This involves using data-driven insights that guide decisions and adopting novel channels and technology to successfully communicate with customers.

Finally, the intersection of co-branding and online fashion creates potential and difficulties for premium businesses. Strategic alliances that align brand values and capitalize on advances in technology can boost consumer engagement and generate brand growth. However, accomplishment in this constantly changing industry necessitates a thorough understanding of customer opinions, ethical standards, and the right balance of distinctiveness and availability. Future study should focus on the consequences of digital co-branding in the high-end market, especially as emerging technologies and customer demands emerge.

REFERENCES

Aaker, D. A. (1991). Managing Brand Equity. The Free Press. Barney, J. (1991). Firm Resources and Sustained Competitive Advantage. *Journal of Management.*

Bain & Company. (2024, November 13). *Global luxury spending to land near €1.5 trillion in 2024, remaining relatively flat as consumers prioritize experiences over products amid uncertainty.* https://www.bain.com/about/media-center/press-releases/2024/global-luxury-spending-to-land-near-1.5-trillion-in-2024-remaining-relatively-flat-as-consumers-prioritize-experiences-over-products-amid-uncertainty/Bain

Beverland, M. B. (2006). The 'Real Thing': Branding Authenticity in the Luxury Market. *Journal of Business Research, 59*(2), 251–258. DOI: 10.1016/j.jbusres.2005.04.007

Business of Fashion. (2023, December 13). *The year ahead: Why brand marketing in 2024 may reclaim its cultural relevance.* https://www.businessoffashion.com/articles/marketing-pr/the-state-of-fashion-2024-report-brand-marketing-strategy-community-building/Business of Fashion

Fletcher, K., & Tham, M. (2019). Earth Logic Fashion Action Research Plan. Hillyer, C., & Tikoo, S. (1995). Effect of Co-Branding on Brand Equity. *Journal of Product and Brand Management.*

Keller, K. L. (1993). Conceptualizing, Measuring, and Managing Customer-Based Brand Equity. *Journal of Marketing, 57*(1), 1–22. DOI: 10.1177/002224299305700101

Kim, A. J., & Ko, E. (2012). Do Social Media Marketing Activities Enhance Customer Equity? *Journal of Business Research, 65*(10), 1480–1486. DOI: 10.1016/j.jbusres.2011.10.014

McKinsey & Company. (2025, January 13). *The state of luxury: How to navigate a slowdown.* https://www.mckinsey.com/industries/retail/our-insights/state-of-luxuryMcKinsey & Company

Park, C. W., Jun, S. Y., & Shocker, A. D. (1996). Composite Branding Alliances: An Investigation of Extension and Feedback Effects. *JMR, Journal of Marketing Research, 33*(4), 453–466. DOI: 10.1177/002224379603300407

Rissanen, T., & McQuillan, H. (2016). *Zero Waste Fashion Design.* Bloomsbury Publishing. DOI: 10.5040/9781474222297

Rogers, E. M. (2003). *Diffusion of Innovations. Free Press. Tapscott, D., & Tapscott, A. (2018).* Blockchain Revolution. Portfolio.

Your Majesty Co. (2024). *State of digital in fashion 2024*. https://yourmajesty.co/article/state-of-digital-in-fashion-2024

Chapter 9

Exploring Consumer–Brand Relationships and Organizational Behavior in the Automotive Sector: An Embryonic Study

Abílio Bragança Milheiro
Polytechnic University of Cávado and Ave, Portugal

Bruno Barbosa Sousa
https://orcid.org/0000-0002-8588-2422
Polytechnic University of Cávado and Ave, Portugal

Estela Vilhena
https://orcid.org/0000-0002-3063-5117
Polytechnic Institute of Cávado and Ave, Portugal

Theodore Tarnanidis
https://orcid.org/0000-0002-4836-3906
International Hellenic University, Greece

ABSTRACT

The automotive sector is highly characterized by innovation and the development of competitive strategies, but also by specific consumer behavior and strong relationships of trust with brands. From an exploratory perspective, this chapter aims to present some discussion around the importance that brand attachment may represent in the behavior of luxury car consumers. A brand that has significant commercial value and brand awareness facilitates, in many cases, strong feelings of affection

DOI: 10.4018/979-8-3373-3010-5.ch009

and loyalty in its consumers (and their respective repurchase and recommendation). Emotional variables play a critical role in the decision-making process of luxury car consumers. As a preliminary perspective, the study will present some initial insights for marketing in the (luxury) automotive sector and will present lines of research for future studies.

1. INTRODUCTION

Brand attachment, characterized by its exploratory potential and various practical applications, is the focus of this study, which examines its impact on consumer behavior within the luxury automotive market. Specifically, the chapter investigates the opinions, perceptions, and attitudes of luxury automobile brand customers regarding the effects of brand attachment (Milheiro et al., 2025). With a preliminary approach, the main objectives of this research are to analyze the predisposition of luxury car brand customers, using a luxury car brand as a case study, towards satisfaction, commitment, trust, and loyalty to the brand, and their relationship to brand attachment. Additionally, the study aims to understand the specificities of the decision-making process in choosing luxury car brands based on existing alternatives, analyze consumer profiles and behaviors in specific marketing contexts (specifically, the profiles and behaviors of luxury car brand customers), and identify the practical implications of brand attachment in the luxury segment of the automotive market. Consumer-brand relationships in the automotive sector are pivotal in shaping customer loyalty and brand perception. These relationships are built on trust, satisfaction, and emotional connections that consumers develop with automotive brands (Adhikari& Panda, 2019; Kaufmann et al., 2019; Al-Sarayreh, 2025). As technological advancements and market dynamics evolve, companies must adapt their strategies to meet changing consumer demands (Haris, 2025). This requires a proactive approach to organizational behavior, where continuous improvement and agility are key. By fostering a culture of responsiveness and innovation, automotive brands can strengthen their consumer relationships, differentiate themselves from competitors, and achieve sustainable growth in the market (Milheiro et al., 2024). Co-branding strategies in the automotive sector, particularly for smart luxury cars, involve the collaboration between luxury car manufacturers and technology companies to create innovative and high-value products (Nedergaard & Gyrd-Jones, 2013). These partnerships leverage the strengths of both brands to offer advanced features such as autonomous driving, enhanced connectivity, and state-of-the-art powertrain technologies. For instance, collaborations between car brands and tech giants like Google or Apple can result in vehicles equipped with cutting-edge infotainment systems and AI-driven functionalities, enhancing the overall driving experience. Exploring consumer-brand relationships in the automotive sector involves understanding how emotional connections between consumers and brands influence purchasing decisions and brand loyalty. Research indicates that factors such as brand love, attachment, satisfaction, and loyalty play crucial roles in shaping consumer behavior. For instance, a study focusing on the Portuguese automotive market, particularly brands like Audi, BMW, and Mercedes-Benz, revealed that brand attachment and satisfaction significantly enhance brand love, which in turn fosters brand loyalty and positive word of mouth (Rodrigues et al., 2023).

2. THEORETICAL BACKGROUND

The concept of luxury can be personified as a symbol of distinction and, simultaneously, social exclusion, often associated with superficiality, ostentation, the pursuit of visibility and status, and uncontrolled consumption patterns (McNeil & Riello, 2016; Lourenço et al., 2021). The concept of luxury implies access to a restricted level, suggesting superiority and a higher quality of life, which few can attain due to economic and social exclusivity. However, perceptions of luxury are highly individual, influenced by personal characteristics, experiences, feelings of belonging, self-esteem, and aspirations for status (Nyadzayo, Johnson & Rossi, 2020; Petravičiūtė, Šeinauskienė, Rūtelionė & Krukowski, 2021).

Luxury marketing in the automotive sector plays a crucial role in differentiating high-end vehicles from their mainstream counterparts. By emphasizing exclusivity, superior craftsmanship, and advanced technology, luxury marketing creates a perception of prestige and desirability among consumers (Štrach & Everett, 2006). This perception is vital for attracting affluent customers who seek not only transportation but also a status symbol that reflects their personal success and lifestyle. Effective luxury marketing strategies often include personalized experiences, high-quality content, and strategic partnerships with influencers and celebrities, all of which contribute to building a strong brand image and loyalty (Vinerean & Opreana, 2019). Luxury marketing in the automotive industry drives innovation and technological advancements. As manufacturers strive to meet the high expectations of luxury consumers, they invest heavily in research and development to introduce cutting-edge features and designs. This continuous innovation not only enhances the appeal of luxury vehicles but also sets new standards for the entire automotive industry (Pinkse et al., 2014). Additionally, luxury marketing helps in expanding the market reach by targeting niche segments and emerging markets, thereby increasing overall sales and profitability. In essence, luxury marketing is indispensable for maintaining competitiveness and fostering growth in the automotive sector. Communication and marketing are crucial in the luxury automotive sector, as they play a significant role in shaping brand perception and consumer engagement. Effective communication strategies help convey the unique value propositions of luxury car brands, highlighting their superior craftsmanship, innovative technology, and exclusive features (Mukhtar, Mohan, Chandra & Gotmare, 2023). By clearly articulating these attributes, brands can differentiate themselves from competitors and attract discerning customers who seek not just a vehicle, but a symbol of status and personal achievement (Holt, 2002). Marketing in the luxury automotive sector is essential for building and maintaining strong consumer-brand relationships. Through targeted campaigns, luxury car brands can reach their specific audience, emphasizing the emotional and aspirational aspects of their products (Duarte et al., 2024). Personalized marketing efforts, such as exclusive events, tailored advertisements, and bespoke customer experiences, foster a deeper connection between the brand and its customers. This emotional bond is vital for cultivating brand loyalty, as it encourages repeat purchases and brand advocacy among satisfied customers. Furthermore, communication and marketing are key to navigating the competitive landscape of the luxury automotive market. As consumer preferences and market dynamics evolve, luxury car brands must adapt their strategies to stay relevant and appealing (Truong et al, 2008). By continuously refining their communication and marketing approaches, luxury car brands can enhance their market presence, drive customer loyalty, and achieve sustainable growth in a highly competitive industry. In summary, luxury marketing should be seen as a set of sophisticated strategies, developed with clarity, that align with the desires, habits, trends, and routines of individuals. These strategies aim to predict and create impactful responses that resonate with customers, making them feel that their needs are fully met (Som & Blanckaert, 2015).

The automotive sector exhibits a wide range of characteristics that can be analyzed, as it includes various purposes and variants. This sector caters to a diverse audience in terms of gender, age, and social class, addressing the needs of families and individuals from different backgrounds and social groups. Effective marketing strategies in this sector must account for these specificities to develop successful plans and actions. In a study by Castro et al. (2015), which examined the factors influencing purchasing decisions in the automotive sector, several key elements were identified as fundamental in the buying process, structured around a consumer behavior model. The study concluded that the most valued factors in the automotive purchasing process are related to the level of involvement, brand trust, the quality of the consumer's relationship with the brand, consumer satisfaction, loyalty intentions, and brand recommendations. The current landscape is characterized by the importance of interactions and connectivity, which are highly valued. Consequently, an organization's or brand's image is often recognized by its presence on social media and its digital operations (Huerta-Álvarez, Cambra-Fierro & Fuentes-Blasco, 2020).

Planning and implementing marketing strategies has thus become a complex task, necessitating the adoption of research methodologies and the adaptation of effective tools and instruments in similar contexts. Market segmentation criteria must consider demographic context, social conditions, economic factors, personality traits, living and behavioral standards, as well as responses to production, promotion, and product distribution. In the automotive sector, market segmentation should be planned to integrate these criteria with marketing strategies that focus on meeting customer needs and satisfaction parameters. Developing and executing internal marketing strategies enhances customer interaction and contributes to the growth of the organization or brand, as well as increased profit levels (Milheiro et al., 2024).

Luxury marketing in the automotive sector will face significant challenges in the future, particularly due to shifting consumer preferences and increasing environmental concerns. As consumers become more aware of the environmental impact of vehicles, there will be growing pressure on luxury brands to develop and promote sustainable and eco-friendly vehicles (Ali et al., 2019). This will require substantial investments in research and development of green technologies, such as electric and hybrid vehicles, without compromising the performance and luxury that customers expect. Additionally, the need for transparency and corporate social responsibility will increase, demanding that luxury brands demonstrate their commitment to sustainable and ethical practices (Braidy, Pokharel & ElMekkawy, 2025).

One of the key benefits of co-branding in the luxury automotive market is the ability to appeal to a broader audience while maintaining exclusivity and prestige (Patrick & Monga, 2020). By combining the luxury and craftsmanship of high-end car brands with the innovation and technological prowess of tech companies, these collaborations can create unique selling points that differentiate their products in a competitive market. Successful examples include the partnership between BMW and Louis Vuitton, which resulted in a bespoke luggage set designed to fit perfectly in the BMW i8, blending luxury travel with automotive excellence. Another challenge will be adapting to rapid technological innovations and digitalization. With the growing importance of digital platforms and social media, luxury brands will need to develop sophisticated digital marketing strategies to reach and engage their target audiences (Dahiya, 2018). This includes leveraging artificial intelligence and data analytics to personalize the customer experience and predict consumption trends. Furthermore, competition from new entrants in the market, particularly tech startups, may intensify the need for continuous innovation and differentiation. Maintaining brand exclusivity and authenticity in a rapidly evolving digital environment will be crucial for the success of luxury marketing in the automotive sector.

Studies on consumer behavior indicate that various factors influence individuals, with their ambitions often linked to needs for socialization, recognition, and esteem, as described by Maslow's hierarchy of needs (Cui, Wang, Chen, Wen & Han, 2021;

Teixeira et al., 2021; Milheiro et al., 2025). Luxury products, characterized by exclusive access, higher price points, and the promise of recognition, are associated with greater satisfaction and pleasure. The concept of luxury is subjective and depends on individual characteristics, experiences, and the symbolic value attributed to it (Ko et al., 2019). Minini (2017) identifies several categories of luxury brand customers:

- Absolute Luxurers: High social class individuals with frequent exposure to luxury items, possessing extensive knowledge and experience;
- Megacities: Residents of large metropolitan areas who regularly purchase luxury brand products;
- Social Wearers: Consumers who prioritize luxury products with high quality and durability;
- Experiencers: Individuals with refined tastes and reserved consumption patterns;
- Princes: Young people or young families with high purchasing power, influenced by brand concepts;
- Fashionistas: Predominantly women with high consumption standards, often spending significantly on luxury items;
- Status Seekers: Individuals who manage their purchases based on external appearance, status, and the image they wish to project, valuing accessories and extras.

Overall, most studies and authors describe luxury product and brand customers as those who prioritize the prestige and recognition associated with luxury purchases, driven by a strong sense of exclusivity (Bairrada, Coelho & Coelho, 2018; Byun, Jones & Wooldridge, 2018).

Brand attachment is crucial in the luxury automobile market because it fosters a deep emotional connection between consumers and brands (Sousa & Rocha, 2019). This connection goes beyond the functional benefits of the car, such as performance and safety, and taps into the consumer's identity and self-expression (Reis et al., 2023). When consumers feel emotionally attached to a luxury car brand, they are more likely to perceive the brand as an extension of themselves, which enhances their overall satisfaction and loyalty (Rodrigues et al, 2023). This emotional bond can lead to a stronger preference for the brand, making consumers more likely to choose it over competitors. Moreover, brand attachment significantly influences consumer behavior by increasing commitment and trust. When consumers are attached to a brand, they are more likely to trust its products and services, believing that the brand will consistently meet their expectations. This trust reduces the perceived risk associated with purchasing high-value items like luxury cars. Additionally, committed consumers are more likely to engage in repeat purchases and recommend the brand to others, which can lead to increased market share and brand equity for the luxury car manufacturer. Finally, brand attachment plays a vital role in shaping consumer loyalty (Sousa & Rocha, 2019). Loyal customers are not only repeat buyers but also advocates for the brand, promoting it within their social circles. This word-of-mouth marketing is particularly valuable in the luxury automobile market, where personal recommendations can significantly influence purchasing decisions. By cultivating strong brand attachment, luxury car brands can create a loyal customer base that supports long-term business success and helps the brand maintain a competitive edge in a highly competitive market (Milheiro et al., 2024).

3. CAR MARKET IN SPECIFIC CONTEXTS

The automotive market in Portugal and across Europe has shown significant vitality in recent years, being recognized as a highly competitive sector with substantial economic importance. The Automobile Association of Portugal (ACAP), representing all organizations in the sector, describes it as one of the most dynamic and innovative parts of the national economy. Identifying the key factors influencing the sale of new motor vehicles in Portugal and understanding the demand function is therefore crucial for all stakeholders involved in the sector. The automotive sector is a vital component of Portugal's economy, contributing significantly to its industrial output and employment. The industry encompasses a wide range of activities, including manufacturing, sales, and services, which collectively support economic growth. Major automotive manufacturers and suppliers have established operations in Portugal, creating numerous job opportunities and fostering technological advancements. This sector's robust performance is crucial for maintaining economic stability and driving innovation within the country (Milheiro et al., 2024). In the luxury automotive sector, Kapferer & Bastien (2009) note that the current ambiguity between luxury and premium products does not directly affect customers. Luxury car brands can be categorized as "luxury, premium, and top-of-the-range" (Kapferer & Bastien, 2009, p. 47). The top-of-the-range category represents the highest quality car in a lineup, showcasing the brand's expertise. Premium brands offer a wide range of vehicles to build customer loyalty over their lifetime, from small cars to family cars, minivans for larger families, and top-end models with personalized options (Kapferer & Bastien, 2009, p. 48). Understanding the positioning of premium segment cars is essential, even if this value is intangible, as these cars are designed for daily use in all conditions. This is not the primary expectation for buyers of brands like Aston Martin or Lamborghini (Kapferer & Bastien, 2012, p. 49). Premium brands are chosen for their rational appeal and excellence, such as the Mercedes S or BMW 8 series, which aim to reduce stress, discomfort, risks, and insecurity for both drivers and passengers. These cars are multipurpose, fulfilling the essential function of transporting people (Kapferer & Bastien, 2009). The high price reflects the owner's social status and the superior quality, safety, handling, technology, connectivity, energy efficiency, passenger space, interior comfort, and reduced pollution and noise (Kapferer & Bastien, 2012, p. 50).

The global luxury automobile market is dominated by several prestigious brands known for their innovation, design, and engineering excellence. Mercedes-Benz, a German manufacturer, has been a leader in the luxury segment for decades, offering a range of vehicles that combine performance, comfort, and cutting-edge technology. BMW, another German brand, is renowned for its sporty and elegant vehicles, consistently leading global sales in the luxury car market. Rolls-Royce, now owned by BMW, is synonymous with opulence and craftsmanship, producing bespoke vehicles that cater to the elite. Bugatti, a French brand with German engineering, is famous for its hypercars that push the boundaries of speed and performance. Ferrari, an Italian icon, is celebrated for its high-performance sports cars and its significant presence in motorsports (Aversa et al., 2023). In addition to these established names, other brands such as Porsche, Aston Martin, and Koenigsegg have carved out their niches in the luxury automobile market. Porsche is known for its versatile sports cars that offer a blend of luxury and performance. Aston Martin, a British manufacturer, is revered for its elegant and powerful vehicles, often associated with James Bond films. Koenigsegg, a Swedish brand, is at the forefront of automotive innovation with its futuristic designs and advanced technologies. These brands not only represent the pinnacle of automotive luxury but also continuously push the envelope in terms of engineering, design, and exclusivity, making them highly sought after by car enthusiasts and collectors worldwide.

4. PRELIMINARY RESULTS

The automotive sector, at European and global level, is very much characterized by decisions involving aspects such as competitive differentiation and innovation. Consumers are very observant and pay attention to aspects of safety, quality and comfort. However, brand recognition and the value of the car brand (particularly in the luxury context) are often important in the final consumer's purchasing decision. Some consumers can develop strong feelings of belonging and affection for brands (of high status, good notoriety and positive reputation). Brand attachment develops a relationship of loyalty and commitment among consumers in the brands with greater recognition. In our preliminary study, after an initial reading of the literature review and some secondary data collection, we realized that consumer decision-making in the context of the luxury automotive sector can foster the importance of emotional and affective bonding among consumers (in the luxury segmentation).

The luxury car market is becoming increasingly diverse and intricate, showing significant growth and vast potential. Presently, the automotive sector is a major focus within the luxury market. In addition to traditional features like comfort, convenience, entertainment, and safety, luxury vehicles now offer advanced connectivity, autonomous driving capabilities, and state-of-the-art powertrain electrification technologies. Establishing brand attachment, as highlighted in this study, is one of the most effective ways to build a relationship between consumers and luxury car brands that goes beyond the functional aspects of comfort, convenience, entertainment, and safety. Brand attachment can create emotional connections, including satisfaction, commitment, trust, and loyalty, which are crucial elements that marketing aims to highlight and secure. Luxury car brands cater to consumers' varied needs and interests in comfort, convenience, entertainment, and safety through their products and services. This study underscores the importance of recognizing that, similar to traditional marketing, relationships in this context can be strengthened through marketing efforts. When luxury car brands use marketing strategies that foster brand attachment, consumers are more likely to be satisfied with the products and services, more dedicated to the brand, and more confident in the brand's mission. This can lead to increased loyalty, with consumers more inclined to purchase the same brand in the future and recommend it to others. Therefore, it is crucial for marketers in the luxury car segment, to prioritize relational and emotional brand management, given the numerous positive aspects associated with brand attachment. The luxury car segment can reap significant benefits if its marketing departments consider this issue. Marketers must conduct thorough exploration to identify strengths and weaknesses and develop and implement the most appropriate strategic plans. The importance of marketing in the luxury car market segment is fully justified. The sector should focus on enhancing the relationship between customers and luxury car brands to transform a single sale or service provision into a continuous and lasting experience based on an emotional connection.

This chapter contributes significantly to the study of brand attachment in marketing and strategic management, an area of growing interest among academics. The value of a luxury automobile brand is intrinsically linked to the loyalty of its customers, which is built on trust, emotional connection, and brand commitment. Trust is a fundamental aspect of brand loyalty; when customers trust a luxury car brand, they believe in the reliability, quality, and performance of its vehicles. This trust is often established through consistent positive experiences with the brand, including exceptional customer service and the delivery of high-quality products. As trust grows, customers become more confident in their purchasing decisions, reinforcing their loyalty to the brand. Loyalty to a luxury automobile brand is also deeply rooted in the emotional connection that customers develop with the brand. This emotional bond is formed when the brand resonates with the customer's identity, values, and aspirations. Luxury car brands often evoke feelings of prestige, exclusivity, and personal achievement, which can significantly enhance the emotional attachment customers feel. When customers perceive a brand as an extension of themselves, they are more likely to remain loyal, choosing the brand repeatedly and advocating for it within their social circles. Emotional variables play a critical role in the decision-making process of luxury car consumers (Godey et al., 2012). Unlike standard car buyers, luxury car consumers often seek more than just functionality and performance; they desire an emotional connection with their vehicle. This connection can be influenced by factors such as brand prestige, the aesthetic appeal of the car, and the exclusivity associated with owning a luxury vehicle. These emotional drivers can significantly impact the perceived value of the car, leading consumers to justify higher expenditures for the emotional satisfaction they derive from their purchase (Santos et al., 2022).

Brand loyalty and emotional attachment are also pivotal in the luxury car market (Rehman et al., 2025). Consumers who develop a strong emotional bond with a brand are more likely to remain loyal and make repeat purchases. This loyalty is often cultivated through personalized customer experiences, exceptional service, and the consistent delivery of high-quality products that resonate with the consumer's self-image and aspirations. Emotional attachment to a brand can also lead to positive word-of-mouth referrals, further enhancing the brand's reputation and attracting new customers who seek similar emotional fulfillment. Emotional variables can influence the overall customer satisfaction and long-term relationship with the brand (Walter et al., 2023). Positive emotional experiences, such as the thrill of driving a high-performance vehicle or the pride of ownership, can enhance customer satisfaction and foster a deeper connection with the brand (Fernandes & Moreira, 2019). Conversely, negative emotional experiences, such as poor customer service or unmet expectations, can damage the brand's reputation and lead to customer attrition. Therefore, luxury car manufacturers must prioritize understanding and addressing the emotional needs of their consumers to maintain a competitive edge in the market.

The value of a luxury automobile brand is amplified by the commitment of its loyal customers. Brand commitment reflects the customer's dedication to the brand, often manifesting in repeat purchases and brand advocacy. Loyal customers are not only more likely to buy from the brand again but also to recommend it to others, thereby increasing the brand's market presence and reputation. This commitment is a powerful driver of brand value, as it ensures a steady stream of revenue and helps maintain the brand's competitive edge in the luxury automotive market. By fostering trust, emotional connection, and commitment, luxury car brands can significantly enhance their value and secure long-term customer loyalty.

5. PRELIMINARY CONCLUSIONS AND NEXT STEPS

From an exploratory perspective, this chapter aims to provide preliminary insights that may facilitate the establishment of causal relationships between the study variables in future research. It serves as a work in progress for individuals involved in brand management, automobile marketing, and the luxury automobile market. Preliminary conclusions suggest that an increase in brand attachment is significantly associated with heightened levels of satisfaction, commitment, and trust, which in turn are linked to stronger brand loyalty. Therefore, if luxury car brands implement marketing techniques that enhance brand attachment, it is likely that consumers will be more satisfied with the products and services, more committed to the brand, and more confident in the brand's mission. This could result in increased loyalty, with consumers more inclined to repurchase and recommend the brand. Marketing luxury automobiles presents several unique challenges, primarily due to the high expectations and discerning nature of the target audience. One of the foremost challenges is maintaining the exclusivity and prestige associated with luxury brands. This requires a delicate balance between making the brand accessible enough to attract new customers while ensuring it remains exclusive to retain its allure.

The automotive sector, at European and global level, is very much characterized by decisions involving aspects such as competitive differentiation and innovation. Consumers are very observant and pay attention to aspects of safety, quality and comfort. However, brand recognition and the value of the car brand (particularly in the luxury context) are often important in the final consumer's purchasing decision. Some consumers can develop strong feelings of belonging and affection for brands (of high status, good notoriety and positive reputation). Brand attachment develops a relationship of loyalty and commitment among consumers in the brands with greater recognition. In our preliminary study, after an initial reading of the literature review and some secondary data collection, we realized that consumer decision-making in the context of the luxury automotive sector can foster the importance of emotional and affective bonding among consumers (in the luxury segmentation).

Marketing strategies must emphasize the unique qualities and craftsmanship of luxury vehicles, highlighting their superior performance, innovative technology, and bespoke features that set them apart from non-luxury brands. Another significant challenge is building and maintaining trust with consumers. Luxury car buyers expect exceptional quality and service, and any deviation from these expectations can damage the brand's reputation. Ensuring consistent quality across all touchpoints, from the initial marketing message to after-sales service, is crucial. Brands must also be transparent and authentic in their communications, as today's consumers are more informed and skeptical of marketing tactics. Building trust involves not only delivering on promises but also engaging with customers in meaningful ways that reinforce the brand's values and commitment to excellence. Creating an emotional connection with consumers is also a critical aspect of marketing luxury automobiles. Luxury car brands must go beyond functional benefits and tap into the aspirational and emotional desires of their customers. This involves crafting compelling brand stories that resonate with the target audience's lifestyle and values. Personalized marketing experiences, exclusive events, and tailored communications can help foster a deeper emotional bond between the brand and its customers.

This emotional connection is essential for cultivating brand loyalty and turning customers into brand advocates. Finally, adapting to the rapidly changing digital landscape poses a challenge for luxury car marketing. With the rise of digital and social media, luxury brands must find innovative ways to engage with tech-savvy consumers while maintaining their exclusive image. This includes leveraging digital platforms to provide immersive and interactive experiences, such as virtual showrooms and personalized online consultations. Additionally, luxury brands must navigate the complexities of digital advertising, ensuring their online presence aligns with their high-end image and reaches the right audience without diluting the brand's prestige.

For instance, artificial intelligence is poised to revolutionize the automotive industry by enhancing vehicle safety, efficiency, and user experience. Advanced artificial intelligence systems can process vast amounts of data from sensors and cameras to enable autonomous driving, reducing human error and the likelihood of accidents (Kolasani, 2024. These systems can also optimize fuel consumption and route planning, leading to more efficient and environmentally friendly transportation. AI-driven predictive maintenance can anticipate and address vehicle issues before they become critical, minimizing downtime and repair costs. The integration of artificial intelligence in the automotive sector will also transform the way consumers interact with their vehicles. AI-powered infotainment systems can provide personalized recommendations for music, navigation, and other services based on user preferences and behavior. Voice recognition and natural language processing technologies will allow for seamless communication between drivers and their cars, enhancing convenience and satisfaction. Additionally, AI can facilitate the development of smart cities by enabling connected vehicle networks that improve traffic management and reduce congestion, contributing to a more sustainable urban environment (Khayyam et al., 2019).

In this context, the consumer-brand relationship will be a pivotal factor in the future success of the automotive sector. As the market becomes increasingly competitive, fostering strong, loyal relationships with consumers will be essential for brand differentiation and customer retention (Nnindini & Dankwah, 2022). Brands that can create emotional connections with their customers through personalized experiences, consistent quality, and reliable service will likely see higher levels of customer loyalty and advocacy. These relationships can lead to repeat purchases and positive word-of-mouth, which are crucial for long-term brand sustainability and growth. The evolution of digital technologies will further enhance the consumer-brand relationship in the automotive industry. With the rise of connected vehicles and digital platforms, brands have more opportunities to engage with consumers in real-time, offering tailored services and support (Hamzah & Pontes, 2024). This continuous interaction can help brands better understand consumer preferences and behaviors, allowing for more targeted marketing and product development. Additionally, leveraging data analytics and AI can enable brands to anticipate and meet consumer needs proactively, thereby strengthening the overall relationship and ensuring a more satisfying ownership experience.

Balancing traditional marketing methods with modern digital strategies is key to staying relevant and appealing to a new generation of luxury car buyers. From an interdisciplinary perspective, this study contributes to the fields of management, marketing, and the automotive market.

Future research should empirically test the conceptual model presented here by administering questionnaire surveys to customers and users of luxury car brands and analyzing the data using structural equation modeling with SPSS software. Additionally, focus group methodology should be employed to gain insights from various market stakeholders, including customers, users, salespeople, marketing directors, and entrepreneurs. Finally, it is recommended that the topic of brand attachment continues to be explored in diverse contexts, as understanding brand affection and emotional management can yield significant insights and practical conclusions regarding the emotional relationships and attachments consumers have with brands.

REFERENCES

Adhikari, K., & Panda, R. K. (2019). The role of consumer-brand engagement towards driving brand loyalty: Mediating effect of relationship quality. *Journal of Modelling in Management, 14*(4), 987–1005. DOI: 10.1108/JM2-03-2019-0067

Al-Sarayreh, A. (2025). The Effect of Interactive Marketing on Consumer Engagement in the Automotive Sector. *American Journal of Quantitative Analysis in Business Management and Economics, 1*(1), 33–41.

Ali, A., Xiaoling, G., Ali, A., Sherwani, M., & Muneeb, F. M. (2019). Customer motivations for sustainable consumption: Investigating the drivers of purchase behavior for a green-luxury car. *Business Strategy and the Environment, 28*(5), 833–846. DOI: 10.1002/bse.2284

Aversa, P., Schreiter, K., & Guerrini, F. (2023). The birth of a business icon through cultural branding: Ferrari and the Prancing Horse, 1923–1947. *Enterprise and Society, 24*(1), 28–58. DOI: 10.1017/eso.2021.22

Bairrada, C. M., Coelho, F., & Coelho, A. (2018). Antecedents and outcomes of brand love: Utilitarian and symbolic brand qualities. *European Journal of Marketing, 52*(3/4), 656–682. DOI: 10.1108/EJM-02-2016-0081

Braidy, A., Pokharel, S., & ElMekkawy, T. Y. (2025). Research Perspectives on Innovation in the Automotive Sector. *Sustainability (Basel), 17*(7), 2795. DOI: 10.3390/su17072795

Castro, B. G. A., Torres, C. V., Nascimento, T. G., & Demo, G. (2015). Are Men More Rational Than Women When Buying Cars? A Comparison of Cultural Influences on Product Judgment between Brazil and the United States of America. *Brazilian Business Review, 12*(6), 73–101.

Cui, L., Wang, Y., Chen, W., Wen, W., & Han, M. S. (2021). Predicting determinants of consumers' purchase motivation for electric vehicles: An application of Maslow's hierarchy of needs model. *Energy Policy, 151*, 112167. https://doi.org/https://doi.org/10.1016/j.enpol.2021.112167. DOI: 10.1016/j.enpol.2021.112167

Dahiya, R. (2018). A research paper on digital marketing communication and consumer buying decision process: An empirical study in the Indian passenger car market. *Journal of Global Marketing, 31*(2), 73–95. DOI: 10.1080/08911762.2017.1365991

Duarte, A., Till, B., & Hafeez, K. (2024). Brand Positioning Strategies: Creating and Communicating Unique Value Propositions. *Strategies for Brand Communications and Management*, 86.

Fernandes, T., & Moreira, M. (2019). Consumer brand engagement, satisfaction and brand loyalty: A comparative study between functional and emotional brand relationships. *Journal of Product and Brand Management, 28*(2), 274–286. DOI: 10.1108/JPBM-08-2017-1545

Godey, B., Pederzoli, D., Aiello, G., Donvito, R., Chan, P., Oh, H., Singh, R., Skorobogatykh, I. I., Tsuchiya, J., & Weitz, B. (2012). Brand and country-of-origin effect on consumers' decision to purchase luxury products. *Journal of Business Research, 65*(10), 1461–1470. DOI: 10.1016/j.jbusres.2011.10.012

Hamzah, M. I., & Pontes, N. (2024). What drives car buyers to accept a rejuvenated brand? The mediating effects of value and pricing in a consumer-brand relationship. *Journal of Strategic Marketing, 32*(2), 114–136. DOI: 10.1080/0965254X.2022.2129749

Haris, A. (2025). Consumer Behavior Shifts in Digital Age: Impact on Brand Loyalty. *Advances: Jurnal Ekonomi & Bisnis, 3*(1), 38–51.

Holt, D. B. (2002). Why do brands cause trouble? A dialectical theory of consumer culture and branding. *The Journal of Consumer Research, 29*(1), 70–90. DOI: 10.1086/339922

Huerta-Álvarez, R., Cambra-Fierro, J. J., & Fuentes-Blasco, M. (2020). The interplay between social media communication, brand equity and brand engagement in tourist destinations: An analysis in an emerging economy. *Journal of Destination Marketing & Management, 16*, 100413. https://doi.org/https://doi.org/10.1016/j.jdmm.2020.100413. DOI: 10.1016/j.jdmm.2020.100413

Kapferer, J. N., & Bastien, V. (2009). The specificity of luxury management: Turning marketing upside down. *Journal of Brand Management, 16*(5-6), 311–322. DOI: 10.1057/bm.2008.51

Kaufmann, H. R., Filho, C. G., Silva, C. R. E., & Monteiro, P. R. R. (2019). Consumer-brand relationship development in the automotive market: An integrative model. *International Journal of Automotive Technology and Management, 19*(3-4), 321–340.

Khayyam, H., Javadi, B., Jalili, M., & Jazar, R. N. (2019). Artificial intelligence and internet of things for autonomous vehicles. In *Nonlinear approaches in engineering applications: Automotive applications of engineering problems* (pp. 39–68). Springer International Publishing.

Ko, E., Costello, J. P., & Taylor, C. R. (2019). What is a luxury brand? A new definition and review of the literature. *Journal of Business Research*, *99*, 405–413. https://doi.org/https://doi.org/10.1016/j.jbusres.2017.08.023. DOI: 10.1016/j.jbusres.2017.08.023

Kolasani, S. (2024). Connected cars and autonomous vehicles: personalizing owner/customer experiences and innovation using AI, IoT, Blockchain, and Big Data. *International numeric journal of machine learning and robots, 8*(8), 1-17.

Lourenço, V. L., Catarino, A. W., Fonseca, M. J., & Sousa, B. B. (2021). Consumer-brand relationship and use of the website in virtual communication in the luxury furniture industry. In *Building Consumer-Brand Relationship in Luxury Brand Management* (pp. 158–177). IGI Global Scientific Publishing. DOI: 10.4018/978-1-7998-4369-6.ch008

McNeil, P., & Riello, G. (2016). *Luxury A Rich History*. Oxford University Press.

Milheiro, A., Sousa, B., Milheiro, C., & Vilhena, E. (2025). Understanding Consumers' Feelings of Affectivity in the Luxury Car Segment. In Valeri, M. (Ed.), *Emotional Intelligence and Networking Competencies. Contributions to Management Science*. Springer., DOI: 10.1007/978-3-031-77543-7_6

Milheiro, A. B., Sousa, B., Santos, V., Milheiro, C., & Vilhena, E. (2024). Understanding the Role of Brand Attachment in the Automotive Luxury Brand Segment. *Administrative Sciences*, *14*(6), 119. DOI: 10.3390/admsci14060119

Mukhtar, M. S., Mohan, A. C., Chandra, D., & Gotmare, P. R. (2023). Brand Personality and the Automotive Market: An In-depth Study of Consumer Preferences and Purchase Decisions. *Journal of International Economics*, *14*(2), 1–19.

Nedergaard, N., & Gyrd-Jones, R. (2013). Sustainable brand-based innovation: The role of corporate brands in driving sustainable innovation. *Journal of Brand Management*, *20*(9), 762–778. DOI: 10.1057/bm.2013.16

Nnindini, S. I., & Dankwah, J. B. (2022). Consumer loyalty and brand marketing programs in an emerging economy: Evidence from the automobile industry. *British Journal of Marketing Studies*, *10*(3), 38–58.

Nyadzayo, M. W., Johnson, L. W., & Rossi, M. (2020). Drivers and outcomes of brand engagement in self-concept for luxury fashion brands. *Journal of Fashion Marketing and Management*, *24*(4), 589–609. DOI: 10.1108/JFMM-05-2018-0070

Patrick, V. M., & Monga, A. S. B. (2020). Building and growing luxury brands: Strategies for pursuing growth while maintaining brand coherence. In *Research handbook on luxury branding* (pp. 117–137). Edward Elgar Publishing. DOI: 10.4337/9781786436351.00018

Petravičiūtė, K., Šeinauskienė, B., Rūtelionė, A., & Krukowski, K. (2021). Linking luxury brand perceived value, brand attachment, and purchase intention: The role of consumer vanity. *Sustainability (Basel)*, *13*(12), 6912. DOI: 10.3390/su13126912

Pinkse, J., Bohnsack, R., & Kolk, A. (2014). The role of public and private protection in disruptive innovation: The automotive industry and the emergence of low-emission vehicles. *Journal of Product Innovation Management*, *31*(1), 43–60. DOI: 10.1111/jpim.12079

Rehman, M., Zelin, T., & Hussain, T. (2025). Influence of consumer satisfaction on brand allegiance: An empirical investigation in Pakistan's safety and luxury automobile sector. *Acta Psychologica*, *252*, 104667. DOI: 10.1016/j.actpsy.2024.104667 PMID: 39701004

Reis, R., Catarino, A. P., Sousa, B. B., Malheiro, A., & Santos, V. (2023). Brand attachment and agile marketing: responses to the (post-) pandemic context in the sports fashion context. In *Promoting Organizational Performance Through 5G and Agile Marketing* (pp. 42–54). IGI Global Scientific Publishing.

Rodrigues, D., Sousa, B., Gomes, S., Oliveira, J., & Lopes, E. (2023). Exploring consumer behavior and brand management in the automotive sector: Insights from a digital and territorial perspective. *Administrative Sciences*, *13*(2), 36. DOI: 10.3390/admsci13020036

Santos, V., Ramos, P., Sousa, B., Almeida, N., & Valeri, M. (2022). Factors influencing touristic consumer behavior. *Journal of Organizational Change Management*, *35*(3), 409–429. DOI: 10.1108/JOCM-02-2021-0032

Som, A. & Blanckaert, C. (2015). The Road to Luxury: The Evolution, Markets, and Strategies of Luxury brand management. Singapore: john Wiley & Sons Singapore Pte Ltd

Sousa, B., & Rocha, A. T. (2019). The role of attachment in public management and place marketing contexts: A case study applied to Vila de Montalegre (Portugal). *International Journal of Public Sector Performance Management*, *5*(2), 189–205. DOI: 10.1504/IJPSPM.2019.099094

Štrach, P., & Everett, A. M. (2006). Brand corrosion: Mass-marketing's threat to luxury automobile brands after merger and acquisition. *Journal of Product and Brand Management*, *15*(2), 106–120. DOI: 10.1108/10610420610658947

Teixeira, S. F., Barbosa, B., Cunha, H., & Oliveira, Z. (2021). Exploring the antecedents of organic food purchase intention: An extension of the theory of planning behavior. *Sustainability (Basel)*, *14*(1), 242. DOI: 10.3390/su14010242

Truong, Y., Simmons, G., McColl, R., & Kitchen, P. J. (2008). Status and conspicuousness–are they related? Strategic marketing implications for luxury brands. *Journal of Strategic Marketing*, *16*(3), 189–203. DOI: 10.1080/09652540802117124

Vinerean, S., & Opreana, A. (2019). Social media marketing efforts of luxury brands on Instagram. *Expert Journal of Marketing*, *7*(2), 144–152.

Walter, C. E., Vale, V. T., Au-Yong-Oliveira, M., Veloso, C. M., & Sousa, B. B. (2023). "The Dark Side of the Brand" and Brand Hate: A Review and Future Research Agenda. *Administrative Sciences*, *13*(11), 234. DOI: 10.3390/admsci13110234

KEY TERMS AND DEFINITIONS

Automotive Sector: is the part of a country's economy that includes all the companies and activities involved with the production, distribution, and sale of motor vehicles. It also includes the maintenance of these vehicles.

Brand attachment: is the emotional connection between a consumer and a brand. It's a psychological state that can influence a consumer's loyalty, engagement, and purchasing behavior.

Consumer-brand relationship: is the connection between a consumer and a brand or company. It's based on the consumer's thoughts, feelings, and experiences with the brand.

Luxury Marketing: is a set of strategies used to promote and sell high-end products and services. It's based on traditional marketing, but focuses on luxury goods.

Compilation of References

Aaker, D. A. (1991). Managing Brand Equity. The Free Press. Barney, J. (1991). Firm Resources and Sustained Competitive Advantage. *Journal of Management.*

Abeyatunge, W. G. N. N. (2024). Crafting Tomorrow's Elegance Through the Evolution of Luxury Marketing in a Digital World. *Cross-Cultural Perspectives on Luxury Marketing*, 173.

Achabou, M. A., & Dekhili, S. (2013). Luxury and sustainable development: Is there a match? *Journal of Business Research, 66*(10), 1896–1903. DOI: 10.1016/j.jbusres.2013.02.011

Adhikari, K., & Panda, R. K. (2019). The role of consumer-brand engagement towards driving brand loyalty: Mediating effect of relationship quality. *Journal of Modelling in Management, 14*(4), 987–1005. DOI: 10.1108/JM2-03-2019-0067

Agrawal, K., & Bhat, A. R. R. (2025). Advances in 3D Printing with Eco-Friendly Materials: A Sustainable Approach to Manufacturing. *RSC Sustainability.*

Aguirre, E., Roggeveen, A. L., Grewal, D., & Wetzels, M. (2016). The personalization-privacy paradox: Implications for new media. *Journal of Consumer Marketing, 33*(2), 98–110. DOI: 10.1108/JCM-06-2015-1458

Ahmed, A., & Ahmed, M. (2024). Strategic alliances and partnerships: Expanding brand reach and influence. Strategies for Brand Communications and Management, 257.

Ajitha, S., & Sivakumar, V. (2017). Understanding the effect of personal and social value on attitude and usage behavior of luxury cosmetic brands. *Journal of Retailing and Consumer Services, 39*, 103–113. DOI: 10.1016/j.jretconser.2017.07.009

Akram, U., & Lavuri, R. (2023). Understanding the consumer's luxury webrooming intention: Moderating role of perceived risk and review. *Journal of Consumer Behaviour, 23*(3), 1602–1619. DOI: 10.1002/cb.2295

Al Mahmud, M. A., Hossan, M. Z., Tiwari, A., Khatoon, R., Sharmin, S., Hosain, M. S., & Ferdousmou, J. (2025). Reviewing the Integration of RFID and IoT in Supply Chain Management: Enhancing Efficiency and Visibility. *Journal of Posthumanism*, *5*(3), 409–437. DOI: 10.63332/joph.v5i3.746

Alexander, B., & Contreras, L. O. (2016). Inter-industry creative collaborations incorporating luxury fashion brands. *Journal of Fashion Marketing and Management*, *20*(3), 254–275. DOI: 10.1108/JFMM-09-2015-0075

Ali, A., Xiaoling, G., Ali, A., Sherwani, M., & Muneeb, F. M. (2019). Customer motivations for sustainable consumption: Investigating the drivers of purchase behavior for a green-luxury car. *Business Strategy and the Environment*, *28*(5), 833–846. DOI: 10.1002/bse.2284

Al-Issa, N., Kwiatek, P., & Dens, N. (2024). Masstige buyers: Profile, perceived luxury values and purchase intentions. *International Journal of Consumer Studies*, *48*(1), e13005. DOI: 10.1111/ijcs.13005

Al-Issa, N., & Thanasi, M. (2024). Mapping the Future of Tech-Infused Luxury: A Roadmap and Research Directions. [wiley.com]. *International Journal of Consumer Studies*, *48*(6), e13103. DOI: 10.1111/ijcs.13103

Al-Sarayreh, A. (2025). The Effect of Interactive Marketing on Consumer Engagement in the Automotive Sector. *American Journal of Quantitative Analysis in Business Management and Economics*, *1*(1), 33–41.

Alsharif, A. H., Salleh, N. Z. M., Abdullah, M., Khraiwish, A., & Ashaari, A. (2023). Neuromarketing Tools used in the Marketing Mix: A Systematic literature and Future Research agenda. *SAGE Open*, *13*(1), 21582440231156563. Advance online publication. DOI: 10.1177/21582440231156563

Alwi, S. F. S., Ali, S. M., & Nguyen, B. (2017). The importance of Ethics in branding: Mediating effects of ethical branding on company reputation and brand loyalty. *Business Ethics Quarterly*, *27*(3), 393–422. DOI: 10.1017/beq.2017.20

Anadon, L. D., Chan, G., Harley, A. G., Matus, K., Moon, S., Murthy, S. L., & Clark, W. C. (2016). Making technological innovation work for sustainable development. *Proceedings of the National Academy of Sciences of the United States of America*, *113*(35), 9682–9690. DOI: 10.1073/pnas.1525004113 PMID: 27519800

Anisimova, T. A. (2007). The effects of corporate brand attributes on attitudinal and behavioural consumer loyalty. *Journal of Consumer Marketing*, *24*(7), 395–405. DOI: 10.1108/07363760710834816

Anuar, N. S. K., Abidin, I. H. Z., Samsudin, A., Adzmy, A., & Zulkornain, L. H. (2021). Revisiting the Issues of Co-Branding. In Jurnal Intelek (Vol. 16, Issue 2, p. 11). https://doi.org/DOI: 10.24191/ji.v16i2.391

Arantes, L., Vaz, L., & Costa, S. B. (2025). From Exclusive to Regenerative: The Transformative Power of Luxury Tourism. In Sousa, B. B., Arantes, L., & Bhartiya, S. (Eds.), *Regenerative Tourism for Social Development* (pp. 93–118). IGI Global., DOI: 10.4018/979-8-3373-0235-5.ch005

Arcidiacono, D. (2024). Industry 4.0 and platforms: The new convergence. In Disruptive digitalisation and platforms (pp. 185-199). Routledge.

Arrigo, E. (2018). Social media marketing in luxury brands. *Management Research Review*, *41*(6), 657–679. DOI: 10.1108/MRR-04-2017-0134

Ashfaq, M., Jiang, Y., Yu, S., & Loureiro, S. M. C. (2020). I, Chatbot: Modeling the determinants of users' satisfaction and continuance intention of AI-powered service agents. In Telematics and Informatics (Vol. 54, p. 101473). Elsevier BV. https://doi.org/DOI: 10.1016/j.tele.2020.101473

Athwal, N., Wells, V. K., Carrigan, M., & Henninger, C. E. (2019). Sustainable Luxury Marketing: A synthesis and research agenda. *International Journal of Management Reviews*, *21*(4), 405–426. DOI: 10.1111/ijmr.12195

Atwal, G., & Williams, A. (2009). Luxury brand marketing – The experience is everything! *Journal of Brand Management*, *16*(5–6), 338–346. DOI: 10.1057/bm.2008.48

Aversa, P., Schreiter, K., & Guerrini, F. (2023). The birth of a business icon through cultural branding: Ferrari and the Prancing Horse, 1923–1947. *Enterprise and Society*, *24*(1), 28–58. DOI: 10.1017/eso.2021.22

Babatunde, N. S. O., Odejide, N. O. A., Edunjobi, N. T. E., & Ogundipe, N. D. O. (2024). THE ROLE OF AI IN MARKETING PERSONALIZATION: A THEORETICAL EXPLORATION OF CONSUMER ENGAGEMENT STRATEGIES. *International Journal of Management & Entrepreneurship Research*, *6*(3), 936–949. DOI: 10.51594/ijmer.v6i3.964

Baghi, I., & Gabrielli, V. (2018). Brand prominence in cause-related marketing: Luxury versus non-luxury. *Journal of Product and Brand Management*, *27*(6), 716–731. DOI: 10.1108/JPBM-07-2017-1512

Bagwell, L. S., & Bernheim, B. D. (1996). Veblen effects in a theory of conspicuous consumption. *The American Economic Review*, •••, 349–373.

Bain & Company. (2020). *The luxury goods market: Outlook 2020 and beyond.* https://www.bain.com

Bain & Company. (2024, November 13). *Global luxury spending to land near €1.5 trillion in 2024, remaining relatively flat as consumers prioritize experiences over products amid uncertainty.* https://www.bain.com/about/media-center/press-releases/2024/global-luxury-spending-to-land-near-1.5-trillion-in-2024-remaining-relatively-flat-as-consumers-prioritize-experiences-over-products-amid-uncertainty/Bain

Bairrada, C. M., Coelho, F., & Coelho, A. (2018). Antecedents and outcomes of brand love: Utilitarian and symbolic brand qualities. *European Journal of Marketing, 52*(3/4), 656–682. DOI: 10.1108/EJM-02-2016-0081

Bakir, V., Laffer, A., McStay, A., Miranda, D., & Urquhart, L. (2024). On manipulation by emotional AI: UK adults' views and governance implications. *Frontiers in Sociology, 9*, 1339834. Advance online publication. DOI: 10.3389/fsoc.2024.1339834 PMID: 38912311

Bakpayev, M., Baek, T. H., Van Esch, P., & Yoon, S. (2020). Programmatic creative: AI can think but it cannot feel. [AMJ]. *Australasian Marketing Journal, 30*(1), 90–95. DOI: 10.1016/j.ausmj.2020.04.002

Ballini, E., Battaglia, L., & Cedrola, E. (2025). Blockchain in luxury fashion: a collaborative approach to innovation and trust. In *24° International Marketing Trends Conference-What Managers Want and Real-World Needs* (Vol. 1, pp. 1-9). International Conference Marketing Trends.

Balmain. (2023, December). Balmain x Ant Kai: Balmain continues its trailblazing journey with the Unicorn sneakers. https://int.balmain.com/en/experience/balmain-x-antkai

Bandyopadhyay, A., Septianto, F., & Nallaperuma, K. (2021). Mixed feelings enhance the effectiveness of luxury advertising. [AMJ]. *Australasian Marketing Journal, 30*(1), 28–34. DOI: 10.1177/1839334921998848

Barile, S., Grimaldi, M., Loia, F., & Sirianni, C. A. (2020). Technology, Value Co-Creation and Innovation in service ecosystems: Toward Sustainable Co-Innovation. *Sustainability (Basel), 12*(7), 2759. DOI: 10.3390/su12072759

Barnes, F., & Higgins, D. M. (2017). Brand image, cultural association and marketing: 'New Zealand' butter and lamb exports to Britain, c. 1920–1938. *Business History, 62*(1), 70–97. DOI: 10.1080/00076791.2017.1344223

Barrenechea, F. (2025). Co-branding as a value-creating strategy: The MoonSwatch case study of Omega and Swatch. Available at *SSRN*, 5134152. DOI: 10.2139/ssrn.5134152

Basile, V., Brandão, A., & Ferreira, M. (2024). Does user-generated content influence value co-creation in the context of luxury fashion brand communities? Matching inclusivity and exclusivity. *Italian Journal of Marketing*, *2024*(4), 419–444. DOI: 10.1007/s43039-024-00103-9

Baudier, P., De Boissieu, E., & Duchemin, M. (2022). Source Credibility and Emotions generated by Robot and Human Influencers: The perception of luxury brand representatives. *Technological Forecasting and Social Change*, *187*, 122255. DOI: 10.1016/j.techfore.2022.122255

Baumgarth, C., & Wieker, J. B. (2020). From the classical art to the urban art infusion effect: The effect of street art and graffiti on the consumer evaluation of products. *Creativity and Innovation Management*, *29*(S1), 116–127. DOI: 10.1111/caim.12362

Baxter, P., & Jack, S. (2008). Qualitative case study methodology: Study design and implementation for novice researchers. *The qualitative report*, *13*(4), 544–559.

Bazi, S., Filieri, R., & Gorton, M. (2020). Customers' motivation to engage with luxury brands on social media. *Journal of Business Research*, *112*, 223–235. DOI: 10.1016/j.jbusres.2020.02.032

BBC. (2018, July 19). Burberry burns bags, clothes and perfume worth millions. *BBC News*. https://www.bbc.com/news/business-44885983

Becker-Leifhold, C., & Iran, S. (2018). Collaborative fashion consumption–drivers, barriers and future pathways. *Journal of Fashion Marketing and Management*, *22*(2), 189–208. DOI: 10.1108/JFMM-10-2017-0109

Berzowska, J. (2005). Electronic textiles: Wearable computers, reactive fashion, and soft computation. *Textile (Oxford)*, *3*(1), 58–75. DOI: 10.2752/147597505778052639

Beunardeau, P. (2024, August 11). The future of luxury: How Web3 is elevating high-end brands. LinkedIn. https://www.linkedin.com/pulse/future-luxury-how-web3-elevating-high-end-brands-pierre-beunardeau-pmjoe

Beverland, M. B. (2006). The 'Real Thing': Branding Authenticity in the Luxury Market. *Journal of Business Research*, *59*(2), 251–258. DOI: 10.1016/j.jbusres.2005.04.007

Beyond Market Insights. (2024, October). Smart fabrics market size, share, and trends analysis report by product (passive, active, very smart), by functionality (sensing, energy harvesting, luminescence & aesthetics), by end-use, by region, and segment forecasts, 2024–2032 (Report No. 6300). https://beyondmarketinsights.com/report/smart-fabrics-market/

Bhandari, M., Rodgers, S., & Pan, P. (2020). Brand feedback to negative eWOM messages: Effects of stability and controllability of problem causes on brand attitudes and purchase intentions. *Telematics and Informatics*, *58*, 101522. DOI: 10.1016/j.tele.2020.101522

Bhardwaj, P., Priya, P. S., & Sharma, S. (2025). Exploring the Influence of Circular Economy Paradigm on Convergence of Sustainability and Luxury Apparel: A Systematic Review. *Circular Economy and Sustainability*, *5*(3), 2295–2316. Advance online publication. DOI: 10.1007/s43615-024-00489-w

Bian, X., & Haque, S. (2020). Counterfeit versus original patronage: Do emotional brand attachment, brand involvement, and past experience matter? In Journal of Brand Management (Vol. 27, Issue 4, p. 438). Palgrave Macmillan. https://doi.org/ DOI: 10.1057/s41262-020-00189-4

Bilancia, A., Costa, F., & Staudacher, A. P. (2025). Achieving sustainability and circular economy in the luxury fashion industry through lean practices: A systematic literature review. *Computers & Industrial Engineering*, *111107*, 111107. Advance online publication. DOI: 10.1016/j.cie.2025.111107

Bi, M. (2024). Narrative brand design in the field of emotional consumption. *Highlights in Art and Design*, *6*(3), 22–25. DOI: 10.54097/yy2n8f93

Bishnoi, S. K., & Singh, S. (2021). A study on consumer buying behaviour for fashion and luxury brands under emotional influence. *Research Journal of Textile and Apparel*, *26*(4), 405–418. DOI: 10.1108/RJTA-03-2021-0026

Blazquez, M., Alexander, B., & Fung, K. (2020). Exploring Millennial's perceptions towards luxury fashion wearable technology. *Journal of Fashion Marketing and Management*, *24*(3), 343–359. DOI: 10.1108/JFMM-09-2019-0200

Bleier, A., Goldfarb, A., & Tucker, C. E. (2020). Consumer privacy and the future of Data-Based Innovation and marketing. SSRN *Electronic Journal*. https://doi.org/ DOI: 10.2139/ssrn.3570156

Böhm, N., Watkowski, L. M. L., & Buck, C. (2024). Establishing platform-based ecosystems: How automotive manufacturers adapt their value creation to a digital end. In *Handbook on digital platforms and business ecosystems in manufacturing* (pp. 147–163). Edward Elgar Publishing. DOI: 10.4337/9781035301003.00016

Bohn, D. (2019, September 30). Google's Project Jacquard is available on new Levi's jackets. The Verge. https://www.theverge.com/2019/9/30/20888909/google-levi-project-jacquard-available-trucker-sherpa-jackets-price-design

Boka, T. F., Niu, Z., & Neupane, R. B. (2024). A survey of sequential recommendation systems: Techniques, evaluation, and future directions. *Information Systems*, *125*, 102427. DOI: 10.1016/j.is.2024.102427

Bonifield, C., & Cole, C. (2006). Affective responses to service failure: Anger, regret, and retaliatory versus conciliatory responses. *Marketing Letters*, *18*(1–2), 85–99. DOI: 10.1007/s11002-006-9006-6

Braidy, A., Pokharel, S., & ElMekkawy, T. Y. (2025). Research Perspectives on Innovation in the Automotive Sector. *Sustainability (Basel)*, *17*(7), 2795. DOI: 10.3390/su17072795

Brooks, A., Fletcher, K., Francis, R. A., Rigby, E. D., & Roberts, T. (2017). Fashion, sustainability, and the anthropocene. *Utopian Studies*, *28*(3), 482–504. DOI: 10.5325/utopianstudies.28.3.0482

Brynjolfsson, E., & McAfee, A. (2014). *The second machine age: Work, progress, and prosperity in a time of brilliant technologies*. W. W. Norton & Company.

Budnitsky, S., & Jia, L. (2018). Branding Internet sovereignty: Digital media and the Chinese–Russian cyberalliance. *European Journal of Cultural Studies*, *21*(5), 594–613. DOI: 10.1177/1367549417751151

Burberry (2025, April 23). Burberry and The RealReal join forces to make fashion circular. https://www.burberryplc.com/news/sustainability/2019/burberry-and-the-realreal-join-forces-to-make-fashion-circular

Burger, J. M. (1986). Increasing compliance by improving the deal: The that's-not-all technique. *Journal of Personality and Social Psychology*, *51*(2), 277–283. DOI: 10.1037/0022-3514.51.2.277

Business of Fashion. (2023, December 13). *The year ahead: Why brand marketing in 2024 may reclaim its cultural relevance*. https://www.businessoffashion.com/articles/marketing-pr/the-state-of-fashion-2024-report-brand-marketing-strategy-community-building/Business of Fashion

Büyüközkan, G., & Görener, A. (2015). Evaluation of product development partners using an integrated AHP-VIKOR model. In Kybernetes (Vol. 44, Issue 2, p. 220). Emerald Publishing Limited. https://doi.org/DOI: 10.1108/K-01-2014-0019

Cabigiosu, A. (2020). An Overview of the Luxury Fashion Industry. In *Palgrave advances in luxury* (p. 9). Springer International Publishing., DOI: 10.1007/978-3-030-48810-9_2

Cabigiosu, A. (2020). *Digitalization in the luxury fashion industry*. Springer International Publishing. DOI: 10.1007/978-3-030-48810-9

Cain, L. N., Mistry, T. G., Douglas, S., Rahman, I., & Moreo, A. (2024). Luxury hotel technology trends: A multigenerational analysis. *International Journal of Contemporary Hospitality Management, 36*(12), 4031–4054. DOI: 10.1108/IJCHM-10-2023-1597

Campbell, C., Plangger, K., Sands, S., & Kietzmann, J. (2021). Preparing for an era of Deepfakes and AI-Generated Ads: A framework for understanding responses to manipulated advertising. *Journal of Advertising, 51*(1), 22–38. DOI: 10.1080/00913367.2021.1909515

Caniato, F., Caridi, M., Crippa, L., & Moretto, A. (2012). Environmental sustainability in fashion supply chains: An exploratory case based research. *International Journal of Production Economics, 135*(2), 659–670. DOI: 10.1016/j.ijpe.2011.06.001

Caponigri, F. (2023). Splitting the atom of heritage to respond to luxury's challenges and crises. *Luxury Studies the in Pursuit of Luxury Journal, 2*(1), 37–57. DOI: 10.1386/ipol_00019_1

Cardoso, C., Casteleiro-Pitrez, J., & Lobo, H. (2025). Embracing the Luxury Paradox: New Luxury, Digitalization, and Digital Experiences in the Metaverse. In Raposo, D., Neves, J., Silva, R., Correia Castilho, L., & Dias, R. (Eds.), *Advances in Design, Music and Arts III* (Vol. 49, pp. 449–466). Springer Nature Switzerland., DOI: 10.1007/978-3-031-73705-3_34

Carlotto, F. (2024). *Luxury brand and art collaborations: Postmodern consumer culture*. Taylor & Francis. DOI: 10.4324/9781003274094

Carnevale, M., Yucel-Aybat, O., & Kachersky, L. (2017). Meaningful stories and attitudes toward the brand: The moderating role of consumers' implicit mindsets. *Journal of Consumer Behaviour, 17*(1). Advance online publication. DOI: 10.1002/cb.1687

Carrington, M. J., Neville, B. A., & Whitwell, G. J. (2010). Why ethical consumers don't walk their talk: Towards a framework for understanding the gap between the ethical purchase intentions and actual buying behaviour of ethically minded consumers. *Journal of Business Ethics, 97*(1), 139–158. DOI: 10.1007/s10551-010-0501-6

Carroll, A. B. (1991). The pyramid of corporate social responsibility: Toward the moral management of organizational stakeholders. *Business Horizons, 34*(4), 39–48. DOI: 10.1016/0007-6813(91)90005-G

Castillo-Villar, F. R., & Villasante-Arellano, A. J. (2020). Applying the multisensory sculpture technique to explore the role of brand usage on multisensory brand experiences. *Journal of Retailing and Consumer Services, 57*, 102185. DOI: 10.1016/j.jretconser.2020.102185

Castro, B. G. A., Torres, C. V., Nascimento, T. G., & Demo, G. (2015). Are Men More Rational Than Women When Buying Cars? A Comparison of Cultural Influences on Product Judgment between Brazil and the United States of America. *Brazilian Business Review, 12*(6), 73–101.

Catoire, P. (2024). *The feedback effect of co-branding collaborations on perceived brand luxury and attitude toward luxury fashion brands: Repercussions of the "fast-fashion co-branding" trend*. Louvain School of Management.

Chakraborty, D., Polisetty, A. G. S., Rana, N. P., & Khorana, S. (2024). Unlocking the potential of AI: Enhancing consumer engagement in the beauty and cosmetic product purchases. *Journal of Retailing and Consumer Services, 79*, 103842. DOI: 10.1016/j.jretconser.2024.103842

Chakraborty, S., & Biswas, M. C. (2019). *Fused deposition modeling 3D printing technology in textile and fashion industry: Materials and innovation. Modern Concepts in Material Science*. Iris Publishers.

Chakraborty, U., & Biswal, S. K. (2024). Diffusion of innovation in direct-to-avatar (D2A): A study of immersive communication and branding in the metaverse. *Computers in Human Behavior, 158*, 108318. DOI: 10.1016/j.chb.2024.108318

Chala, V. (2018). Surrealism in the design of clothes: Elsa Schiaparelli and Salvador Dali (Doctoral dissertation).

Chan, E. Y., & Northey, G. (2021). Luxury goods in online retail: How high/low positioning influences consumer processing fluency and preference. *Journal of Business Research, 132*, 136–145. DOI: 10.1016/j.jbusres.2021.04.031

Chantecaille. (2025, April 15). *Space for Giants*. https://chantecaille.com/pages/space-for-giants

Chapman, A., & Dilmperi, A. (2022). Luxury brand value co-creation with online brand communities in the service encounter. *Journal of Business Research, 144*, 902–921. DOI: 10.1016/j.jbusres.2022.01.068

Chauhan, S. (2024). The growing energy demand of data centers: Impacts of AI and cloud computing. *International Journal for Multidisciplinary Research, 6*(4), 26591. Advance online publication. DOI: 10.36948/ijfmr.2024.v06i04.26591

Chaurasia, M., Kumar, A., & Panda, R. K. (2024). Two decades of research on "masstige" marketing: A systematic literature review and future research agenda. *International Journal of Consumer Studies, 48*(5), e13077. DOI: 10.1111/ijcs.13077

Cheah, I., Shimul, A. S., & Parker, J. (2021). Guest editorial: Luxury Branding – Strategy, Innovation and Sustainability. [AMJ]. *Australasian Marketing Journal, 29*(4), 275–276. DOI: 10.1177/18393349211052617

Chen, C., Guo, L., Zhou, M., Tsaur, W., Sun, H., Zhan, W., Deng, Y., & Li, C. (2022). Blockchain-Based Anti-Counterfeiting Management System for traceable luxury products. *Sustainability (Basel), 14*(19), 12814. DOI: 10.3390/su141912814

Cheng, R., Zhu, J., Zhu, X., & Zhang, C. (2021). Marketing Strategies in Times of Crisis Exploring More Digital Strategies in Luxury Industry. In *Advances in economics, business and management research/Advances in Economics, Business and Management Research*. Atlantis Press., DOI: 10.2991/assehr.k.211209.365

Chen, L. (2024). Using marketing mix theory to analyze the successful reasons for Heytea's co-branding strategy. Advances in Economics. *Management and Political Sciences, 110*, 209–216.

Chen, X., & Li, Z. (2022). Investigating the consumer evaluation of the co-branding of luxury brands. *International Journal of Research in Business & Social Science, 11*(4), 1–14. DOI: 10.20525/ijrbs.v11i4.1808

Chen, Y., & Choi, J. (2025). The Effectiveness of Collaboration: The Features of Co-Branded Products Affecting Brand Equity and Consumer Sentiment. *The Journal of Society for E-Business Studies, 30*(1), 1–22. DOI: 10.7838/jsebs.2025.30.1.001

Chernev, A., Hamilton, R., & Gal, D. (2011). Competing for consumer identity: Limits to Self-Expression and the perils of lifestyle branding. *Journal of Marketing, 75*(3), 66–82. DOI: 10.1509/jmkg.75.3.66

Cherubino, P., Martinez-Levy, A. C., Caratù, M., Cartocci, G., Di Flumeri, G., Modica, E., Rossi, D., Mancini, M., & Trettel, A. (2019). Consumer Behaviour through the Eyes of Neurophysiological Measures: State-of-the-Art and Future Trends. *Computational Intelligence and Neuroscience, 2019*, 1–41. DOI: 10.1155/2019/1976847 PMID: 31641346

Cheshire, P., & Sheppard, S. (1989). British planning policy and access to housing: Some empirical estimates. *Urban Studies (Edinburgh, Scotland), 26*(5), 469–485. DOI: 10.1080/00420988920080541

Cheung, T. W., & Li, L. (2018). Sustainable development of smart textiles: A review of 'self-functioning' abilities which makes textiles alive. *Journal of textile engineering & fashion technology, 4*(2), 151-156.

Chiang, C. (2023). A systematic literature network analysis of green information technology for sustainability: Toward smart and sustainable livelihoods. *Technological Forecasting and Social Change, 199*, 123053. DOI: 10.1016/j.techfore.2023.123053

Chiang, C., & Jang, S. S. (2007). The effects of perceived price and brand image on value and purchase intention: Leisure travelers' attitudes toward online hotel booking. *Journal of Hospitality & Leisure Marketing, 15*(3), 49–69. DOI: 10.1300/J150v15n03_04

Childs, M., & Jin, B. E. (2020). Brand and retailer co-branding. *Journal of Fashion Marketing and Management, 24*(1), 49–65. DOI: 10.1108/JFMM-03-2019-0061

Chloé. (2021). Chloé becomes the first luxury fashion Maison to receive B Corp certification. https://www.chloe.com

Choi, C. (2024). The impact of perceived brand prestige on brand engagement: the mediating roles of brand trust and brand commitment. *Korea International Trade Research Institute, 20*(2), 351–368. https://doi.org/DOI: 10.16980/jitc.20.2.202404.351

Choi, E., Ko, E., & Kim, A. J. (2016). Explaining and predicting purchase intentions following luxury-fashion brand value co-creation encounters. *Journal of Business Research, 69*(12), 5827–5832. DOI: 10.1016/j.jbusres.2016.04.180

Choi, J., & Kim, S. (2016). Is the smartwatch an IT product or a fashion product? A study on factors affecting the intention to use smartwatches. *Computers in Human Behavior, 63*, 777–786. DOI: 10.1016/j.chb.2016.06.007

Choudhary, N., & Vijaya Kittu, M. (2021). Revenge Travel or the Road Less Traveled? Opting for a suitable option in Pandemic Times. *International Journal of Advance & Innovative Research, 8*(3). Advance online publication. DOI: 10.17605/OSF.IO/83BCG

Christodoulides, G. (2009). Branding in the post-internet era. *Marketing Theory, 9*(1), 141–144. DOI: 10.1177/1470593108100071

Clero, L. M. (2023). *Deadstock in the Fashion Industry: Is the On-Demand Business Model a Viable and Sustainable Solution?* (Master's thesis, Universidade Catolica Portuguesa (Portugal)).

Cointelegraph. (2022, July 19). Fashion meets NFTs: Original Cristóbal Balenciaga designs released in NFT drop slated for July 21. https://cointelegraph.com/press-releases/fashion-meets-nfts-original-cristobal-balenciaga-designs-released-in-nft-drop-slated-for-july-21

Comert, O. (2020). *Blockchain revolution: how the technology behind bitcoin and other cryptocurrencies is changing the world.* Hein Online.

Correia, A., & Kozak, M. (2021). Luxury product and brand purchasing behavior. https://doi.org/, https://www.tandfonline.com/doi/full/10.1080/21639159.2020.1808823DOI: 10.1080/21639159.2020.1808823

Crane, A., Matten, D., Glozer, S., & Spence, L. J. (2019). *Business ethics: Managing corporate citizenship and sustainability in the age of globalization.* Oxford university press. DOI: 10.1093/hebz/9780198810070.001.0001

Cranny-Francis, A. (2013). Smart Textiles: Giving a New Sense to the Notion That Clothing Makes the (Hu) man. In *Technology and Touch: The Biopolitics of Emerging Technologies* (pp. 151–171). Palgrave Macmillan UK. DOI: 10.1057/9781137268310_8

Creevey, D., Coughlan, J., & O'Connor, C. (2021). Social media and luxury: A systematic literature review. *International Journal of Management Reviews, 24*(1), 99–129. DOI: 10.1111/ijmr.12271

Csaba, F. F., & Skjold, E. (2018). Fur and sustainability: Oxymoron or key to 'deeper'luxury? *Journal of Design. Business & Society, 4*(2), 131–149.

Cui, L., Wang, Y., Chen, W., Wen, W., & Han, M. S. (2021). Predicting determinants of consumers' purchase motivation for electric vehicles: An application of Maslow's hierarchy of needs model. *Energy Policy, 151*, 112167. https://doi.org/https://doi.org/10.1016/j.enpol.2021.112167. DOI: 10.1016/j.enpol.2021.112167

Cunha, M.Jr, Forehand, M. R., & Angle, J. W. (2014). Riding Coattails: When Co-Branding Helps versus Hurts Less-Known Brands. *The Journal of Consumer Research, 41*(5), 1284–1300. DOI: 10.1086/679119

Cutrona, V., De Paoli, F., Košmerlj, A., Nikolov, N., Palmonari, M., Perales, F., & Roman, D. (2019). Semantically-Enabled optimization of digital marketing campaigns. In *Lecture notes in computer science* (pp. 345–362). https://doi.org/DOI: 10.1007/978-3-030-30796-7_22

Dahiya, R. (2018). A research paper on digital marketing communication and consumer buying decision process: An empirical study in the Indian passenger car market. *Journal of Global Marketing*, *31*(2), 73–95. DOI: 10.1080/08911762.2017.1365991

Dai, D., An, Z., Pan, Q., & Yang, L. (2024). Harnessing NFC to generate standard optical barcodes for NFC-Missing smartphones. *IEEE Transactions on Mobile Computing*, *23*(12), 12952–12968. DOI: 10.1109/TMC.2024.3420410

Dallabona, A. (2024). *Luxury fashion marketing and branding: A strategic approach.* Taylor & Francis. DOI: 10.4324/9781003264811

Danziger, P. (2015). Five luxe trends for 2015. Unity Marketing. https://unitymarketingonline.com/wp-content/uploads/Five-Key-Luxury-Market-Trends-FINAL-NEW.pdf

Danziger, P. N. (2025, April 11). *As luxury brands brace for tariffs, affluent consumers hit pause.* Forbes. https://www.forbes.com/sites/pamdanziger/2025/04/11/as-luxury-brands-brace-for-tariffs-affluent-consumers-hit-pause/

Das, A. K. (2016). On partially included intuitionistic fuzzy rough relations. *Afrika Matematika*, *27*(5-6), 993–1001. DOI: 10.1007/s13370-016-0395-2

Das, A. K. (2018). Weighted fuzzy soft multiset and decision-making. *International Journal of Machine Learning and Cybernetics*, *9*(5), 787–794. DOI: 10.1007/s13042-016-0607-y

Das, A. K., & Granados, C. (2021). An advanced approach to fuzzy soft group decision-making using weighted average ratings. *SN Computer Science*, *2*(6), 471. DOI: 10.1007/s42979-021-00873-5

Das, A. K., & Granados, C. (2022a). A new fuzzy parameterized intuitionistic fuzzy soft multiset theory and group decision-making. *Journal of Current Science and Technology*, *12*(3), 547–567. DOI: 10.14456/jcst.2022.42

Das, A. K., & Granados, C. (2022b). FP-intuitionistic multi fuzzy N-soft set and its induced FP-Hesitant N soft set in decision-making. *Decision Making: Applications in Management and Engineering*, *5*(1), 67–89. DOI: 10.31181/dmame181221045d

Das, A. K., & Granados, C. (2022c). IFP-intuitionistic multi fuzzy N-soft set and its induced IFP-hesitant N-soft set in decision-making. *Journal of Ambient Intelligence and Humanized Computing*, *14*(8), 10143–10152. DOI: 10.1007/s12652-021-03677-w

Das, A. K., & Granados, C. (2023). Preference intuitionistic fuzzy rough relation and its theoretical approach. *Journal of the Indian Mathematical Society*, *90*(3-4), 199–212.

Das, A. K., Granados, C., & Bhattacharya, J. (2022). Some new operations on fuzzy soft sets and their applications in decision-making. *Songklanakarin Journal of Science and Technology*, *44*(2), 440–449. DOI: 10.14456/sjst-psu.2022.61

Das, A. K., Gupta, N., & Granados, C. (2023). Weighted hesitant bipolar-valued fuzzy soft set in decision-making. *Songklanakarin Journal of Science and Technology*, *45*(6), 681–690. https://sjst.psu.ac.th/journal/45-6/10.pdf

Das, A. K., Gupta, N., Mahmood, T., Tripathy, B. C., Das, R., & Das, S. (2024). An innovative fuzzy multi-criteria decision making model for analyzing anthropogenic influences on urban river water quality. *Iran Journal of Computer Science*. Advance online publication. DOI: 10.1007/s42044-024-00211-x

Das, A. K., Gupta, N., Mahmood, T., Tripathy, B. C., Das, R., & Das, S. (2025a). An efficient water quality evaluation model using weighted hesitant fuzzy soft sets for water pollution rating. In Kumar, A., Kumar, P., Rathee, S., & Kumar, B. (Eds.), *Mechatronics: Concepts, Tools, Applications, and New Trends* (1st ed., p. 17). CRC Press., DOI: 10.1201/9781003494478-10

Das, A. K., Gupta, N., Patra, S., Tripathy, B. C., Das, R., & Datta, R. (2025b). An effective fuzzy soft decision-making model for rating water pollution: A case study of Manu, the longest river in Tripura, India, Vietnam. *Journal of Science and Technology*.

Das, M., Habib, M., Saha, V., & Jebarajakirthy, C. (2021). Bandwagon vs snob luxuries: Targeting consumers based on uniqueness dominance. *Journal of Retailing and Consumer Services*, *61*, 102582. DOI: 10.1016/j.jretconser.2021.102582

Davenport, T. H. (2023). Hyper-Personalization for Customer Engagement with Artificial Intelligence. *Management and Business Review.*, *3*(1–2), 29–36. DOI: 10.1177/26941058202303010006

De Boissieu, E., Kondrateva, G., Baudier, P., & Ammi, C. (2021). The use of blockchain in the luxury industry: Supply chains and the traceability of goods. *Journal of Enterprise Information Management*, *34*(5), 1318–1338. DOI: 10.1108/JEIM-11-2020-0471

De Kerviler, G., & Rodriguez, C. M. (2019). Luxury brand experiences and relationship quality for Millennials: The role of self-expansion. *Journal of Business Research*, *102*, 250–262. DOI: 10.1016/j.jbusres.2019.01.046

Delmas, M. A., & Burbano, V. C. (2011). The drivers of greenwashing. *California Management Review*, *54*(1), 64–87. DOI: 10.1525/cmr.2011.54.1.64

Deloitte. (2023). *Global Powers of Luxury Goods 2023*. https://www.deloitte.com/global/en/Industries/consumer/analysis/gx-cb-global-powers-of-luxury-goods.html

Desai, D. (2021). Hyper-Personalization. In *Advances in marketing, customer relationship management, and e-services book series* (pp. 40–53). https://doi.org/DOI: 10.4018/978-1-7998-7959-6.ch003

Dey, P. P., Pramanik, S., & Giri, B. C. (2016). TOPSIS for solving multi-attribute decision-making problems under bi-polar neutrosophic environment. In Smarandache, F., & Pramanik, S. (Eds.), *New trends in neutrosophic theory and applications* (pp. 65–77). Pons Editions.

Dhaoui, C., & Webster, C. M. (2020). Brand and consumer engagement behaviors on Facebook brand pages: Let's have a (positive) conversation. *International Journal of Research in Marketing*, *38*(1), 155–175. DOI: 10.1016/j.ijresmar.2020.06.005

Di Leo, A., Sfodera, F., Cucari, N., Mattia, G., & Dezi, L. (2023). Sustainability reporting practices: An explorative analysis of luxury fashion brands. *Management Decision*, *61*(5), 1274–1297. DOI: 10.1108/MD-02-2022-0142

Diallo, M. F., Mouelhi, N. B. D., Gadekar, M., & Schill, M. (2020). CSR actions, brand value, and willingness to pay a premium price for luxury brands: Does Long-Term orientation matter? *Journal of Business Ethics*, *169*(2), 241–260. DOI: 10.1007/s10551-020-04486-5

Diawara, M. (1995). Dial Ouaga for AFRICAN Cinema. *Nka Journal of Contemporary African Art*, *1995*(3), 46–49. DOI: 10.1215/10757163-3-1-46

Dogaru, I., Furnham, A., & McClelland, A. (2024). Understanding how the presence of music in advertisements influences consumer behaviour. *Acta Psychologica*, *248*, 104333. DOI: 10.1016/j.actpsy.2024.104333 PMID: 39032271

Dong, Y., & Huanug, X. (2024). The rejuvenation of luxury brands: How to attract young consumers in China? Advances in Economics. *Management and Political Sciences*, *70*, 287–293.

Drakou, A., Papachristou, E., Tarnanidis, T., & Bilalis, N. (2024). Transforming fashion education: The impact of ai-driven 3d design on learning, creativity, and sustainability Proceedings of the 13th Hellenic Conference on Artificial Intelligence, https://doi.org/DOI: 10.1145/3688671.3688794

Duarte, A., Till, B., & Hafeez, K. (2024). Brand Positioning Strategies: Creating and Communicating Unique Value Propositions. *Strategies for Brand Communications and Management*, 86.

Dubois, D., Jung, S., & Ordabayeva, N. (2020). The psychology of luxury consumption. *Current Opinion in Psychology*, *39*, 82–87. DOI: 10.1016/j.copsyc.2020.07.011 PMID: 32861090

Dutta, S., & Bansal, P. (2024). *Evolution of sustainable wearables: integrating cutting-edge techniques for future textile innovation*. Research Journal of Textile and Apparel.

Eisenhardt, K. M. (1989). Building theories from case study research. *Academy of Management Review*, *14*(4), 532–550. DOI: 10.2307/258557

El Periódico. (2024, April 2). El cronógrafo IWC de Lewis Hamilton recauda más de 230.000 euros para la Fundación Laureus. https://gentleman.elperiodico.com/estilo/el-cronografo-iwc-de-lewis-hamilton-recauda-mas-de-230-0000-euros-para-la-fundacion-laureus/

Elkhwesky, Z., Manzani, Y. E., & Salem, I. E. (2022). Driving hospitality and tourism to foster sustainable innovation: A systematic review of COVID-19-related studies and practical implications in the digital era. *Tourism and Hospitality Research*, *24*(1), 115–133. DOI: 10.1177/14673584221126792 PMID: 40479339

El-Sayed, A., & El-Fanagely, B. (2022). The importance of digital lightening fashion technique in.

Endres, T., Eitel, A., Renninger, K. A., Vössing, C., & Renkl, A. (2024). Why narrative frames matter for instructional videos: A value-evoking narrative frame is essential to foster sustained learning with emotional design videos. *Learning and Instruction*, *94*, 101962. DOI: 10.1016/j.learninstruc.2024.101962

Escobedo, P., Bhattacharjee, M., Nikbakhtnasrabadi, F., & Dahiya, R. (2020). Smart bandage with wireless strain and temperature sensors and batteryless NFC tag. *IEEE Internet of Things Journal*, *8*(6), 5093–5100. DOI: 10.1109/JIOT.2020.3048282

European Commission. (2022). *EU Strategy for Sustainable and Circular Textiles.* https://ec.europa.eu

Fakhravar, H., & Tahami, H. (2022). International Co-Branding and Firms Finance Performance. In arXiv (Cornell University). Cornell University. https://doi.org/DOI: 10.48550/arXiv.2202

Faludi, J., Cline-Thomas, N., & Agrawala, S. (2017). *3D printing and its environmental implications. The Next Production Revolution Implications for Governments and Business, Organization for Economic Cooperation and Development.* OECD.

Fernandes, T., & Moreira, M. (2019). Consumer brand engagement, satisfaction and brand loyalty: A comparative study between functional and emotional brand relationships. *Journal of Product and Brand Management, 28*(2), 274–286. DOI: 10.1108/JPBM-08-2017-1545

Ferraz, L., Nobre, H., & Barbosa, B. (2021). When luxury vinous-concept hotel meets premium wine brands: An exploratory study on co-branding. In *Handbook of research on applied data science and artificial intelligence in business and industry* (pp. 366–380). IGI Global. DOI: 10.4018/978-1-7998-6985-6.ch017

Fionda, A. M., & Moore, C. M. (2009). The anatomy of the luxury fashion brand. *Journal of Brand Management, 16*(5–6), 347–363. DOI: 10.1057/bm.2008.45

Fletcher, K.-A., & Stephen, C. (2025). Luxury Brands in Dynamic Times. In Gbadamosi, A. (Ed.), *Brands, Branding, and Consumerism* (pp. 311–344). Springer Nature Switzerland., DOI: 10.1007/978-3-031-80859-3_8

Fletcher, K., & Tham, M. (2019). Earth Logic Fashion Action Research Plan. Hillyer, C., & Tikoo, S. (1995). Effect of Co-Branding on Brand Equity. *Journal of Product and Brand Management.*

Forbes (2017). La moda lucha por los océanos: Stella McCartney se alía con Parley for the Oceans. https://forbes.com.mx/forbes-life/stella-mccartney-parley-for-the-oceans/

Forbes (2019). What Is Really Driving The RealReal's New Partnership With Burberry. https://www.forbes.com/sites/pamdanziger/2019/10/07/circular-fashion-is-not-whats-most-important-about-burberry-partnering-with-the-realreal/

Frew, B., Brinkman, G., Denholm, P., Narwade, V., Stephen, G., Bloom, A., & Lau, J. (2021). Impact of operating reserve rules on electricity prices with high penetrations of renewable energy. *Energy Policy, 156*, 112443. DOI: 10.1016/j.enpol.2021.112443

Fuchs, C., & Diamantopoulos, A. (2010). Evaluating the effectiveness of brand-positioning strategies from a consumer perspective. *European Journal of Marketing*, *44*(11/12), 1763–1786. DOI: 10.1108/03090561011079873

Füller, J. (2010). Refining Virtual Co-Creation from a Consumer Perspective. *California Management Review*, *52*(2), 98–122. DOI: 10.1525/cmr.2010.52.2.98

Gao, E., & Yang, X. (2022). The Effectiveness of High-tech Product on Customer Satisfaction towards Luxury Hotel. *BCP Business & Management*, *20*, 858–864. DOI: 10.54691/bcpbm.v20i.1073

Giglio, S., Pantano, E., Bilotta, E., & Melewar, T. (2019). Branding luxury hotels: Evidence from the analysis of consumers' "big" visual data on TripAdvisor. *Journal of Business Research*, *119*, 495–501. DOI: 10.1016/j.jbusres.2019.10.053

Gil, M. Á. C. (2019). Standard and precordial leads obtained with an Apple watch. *Annals of Internal Medicine*, *172*(6), 436–437. DOI: 10.7326/M19-2018 PMID: 31766051

Gkiouzepas, L., & Tarnanidis, T. (2024). Revisiting Visual Communication: Conceptualizing How Extremity of Claims Work in the Context of Advertising Visuals. In Tarnanidis, T., & Sklavounos, N. (Eds.), *New Trends in Marketing and Consumer Science* (pp. 337–359). IGI Global., DOI: 10.4018/979-8-3693-2754-8.ch018

Godey, B., Manthiou, A., Pederzoli, D., Rokka, J., Aiello, G., Donvito, R., & Singh, R. (2016). Social media marketing efforts of luxury brands: Influence on brand equity and consumer behavior. *Journal of Business Research*, *69*(12), 5833–5841. DOI: 10.1016/j.jbusres.2016.04.181

Godey, B., Pederzoli, D., Aiello, G., Donvito, R., Chan, P., Oh, H., Singh, R., Skorobogatykh, I. I., Tsuchiya, J., & Weitz, B. (2012). Brand and country-of-origin effect on consumers' decision to purchase luxury products. *Journal of Business Research*, *65*(10), 1461–1470. DOI: 10.1016/j.jbusres.2011.10.012

Gonçalves, A. R., Pinto, D. C., Rita, P., & Pires, T. (2023). Artificial intelligence and its ethical implications for marketing. *Emerging Science Journal*, *7*(2), 313–327. DOI: 10.28991/ESJ-2023-07-02-01

Górska-Warsewicz, H. (2024). Relationship between entrepreneurial orientation, innovative co-branding partnership, and business performance. Journal of Entrepreneurship. *Management and Innovation*, *20*(2), 139–159. DOI: 10.7341/20242022

Górska-Warsewicz, H. (2024). *Trust and brand management: The role of brand heritage*. Taylor & Francis. DOI: 10.4324/9781032707754

Grandhi, B., Patwa, N., & Saleem, K. (2020). Data-driven marketing for growth and profitability. *EuroMed Journal of Business*, *16*(4), 381–398. DOI: 10.1108/EMJB-09-2018-0054

Grappe, C. G., Lombart, C., Louis, D., & Durif, F. (2021). "Not tested on animals": How consumers react to cruelty-free cosmetics proposed by manufacturers and retailers? *International Journal of Retail & Distribution Management*, *49*(11), 1532–1553. DOI: 10.1108/IJRDM-12-2020-0489

Green, A. (2024). Private Art Museums in Paris: How France's Luxury Titans Shape Urban Branding Through Iconic Architecture. *Luxury*, *11*(1–3), 69–91. DOI: 10.1080/20511817.2025.2466345

Grewal, D., Krishnan, R., Baker, J., & Borin, N. (1998). The effect of store name, brand name and price discounts on consumers' evaluations and purchase intentions. *Journal of Retailing*, *74*(3), 331–352. DOI: 10.1016/S0022-4359(99)80099-2

Guan, Y. (2023). Emotional branding Strategies and effects in the U.S market. *Communications in Humanities Research*, *4*(1), 524–530. DOI: 10.54254/2753-7064/4/20220838

Guercini, S., Ranfagni, S., & Runfola, A. (2020). E-commerce internationalization for top luxury fashion brands: some emerging strategic issues. In Journal of Management Development (Vol. 39, Issue 4, p. 423). Emerald Publishing Limited. https://doi.org/DOI: 10.1108/JMD-10-2019-0434

Gungunawat, A., Khandelwal, N., & Gupta, N. (2024). AI-Powered Personalization in Digital Marketing: Transforming consumer engagement and strategy. *RESEARCH REVIEW International Journal of Multidisciplinary*, *9*(11), 183–191. DOI: 10.31305/rrijm.2024.v09.n11.026

Gupta, V., Hushain, J., & Mathur, A. (2024). The future of luxury brand management: A study on the impact of new technology and relationship marketing. In *AI in business: Opportunities and limitations* (Vol. 1, pp. 57–68). Springer Nature Switzerland. DOI: 10.1007/978-3-031-48479-7_6

Halwani, L. (2020). The Online Experience of Luxury Consumers: Insight into Motives and Reservations. In International Journal of Business and Management (Vol. 15, Issue 11, p. 157). Canadian Center of Science and Education. https://doi.org/DOI: 10.5539/ijbm.v15n11p157

Hamzah, M. I., & Pontes, N. (2022). What drives car buyers to accept a rejuvenated brand? the mediating effects of value and pricing in a consumer-brand relationship. *Journal of Strategic Marketing*, *32*(2), 114–136. DOI: 10.1080/0965254X.2022.2129749

Han, S., & Kim, K. (2020). Role of consumption values in the luxury brand experience: Moderating effects of category and the generation gap. *Journal of Retailing and Consumer Services*, *57*, 102249. DOI: 10.1016/j.jretconser.2020.102249

Han, Y. J., Nunes, J. C., & Drèze, X. (2010). Signaling Status with Luxury Goods: The Role of Brand Prominence. *Journal of Marketing*, *74*(4), 15–30. DOI: 10.1509/jmkg.74.4.015

Haris, A. (2025). Consumer Behavior Shifts in Digital Age: Impact on Brand Loyalty. *Advances: Jurnal Ekonomi & Bisnis*, *3*(1), 38–51.

Harmon-Kizer, T. R., Kumar, A., Ortinau, D., & Stock, J. (2013). When multiple identities compete: The role of centrality in self-brand connections. *Journal of Consumer Behaviour*, *12*(6), 483–495. DOI: 10.1002/cb.1449

Hassan, S. Q. U., Ilyas, M., & Rehman, C. A. (2016). Corporate branding, emotional attachment and brand loyalty : The case of luxury fashion branding. *Deleted Journal*, *6*(2), 43–57. DOI: 10.12816/0031561

Hassan, S., Husić-Mehmedović, M., & Duverger, P. (2015). Retaining the allure of luxury brands during an economic downturn. *Journal of Fashion Marketing and Management*, *19*(4), 416–429. DOI: 10.1108/JFMM-03-2015-0030

Hayes, J. L., Brinson, N. H., Bott, G. J., & Moeller, C. M. (2021). The influence of Consumer–Brand Relationship on the personalized Advertising privacy calculus in social media. *Journal of Interactive Marketing*, *55*(1), 16–30. DOI: 10.1016/j.intmar.2021.01.001

Heath, J., & Potter, A. (2005). *The rebel sell: Why the culture can't be jammed.* HarperCollins.

Heath, T. B., & Chatterjee, S. (1995). Asymmetric Decoy Effects on Lower-Quality versus Higher-Quality Brands: Meta-Analytic and Experimental Evidence. *The Journal of Consumer Research*, *22*(3), 268. DOI: 10.1086/209449

Helmig, B., Huber, J. A., & Leeflang, P. S. (2008). Co-branding: The state of the art. *Schmalenbach Business Review*, *60*(4), 359–377. DOI: 10.1007/BF03396775

Henninger, C. E., Alevizou, P. J., & Oates, C. J. (2016). What is sustainable fashion? *Journal of Fashion Marketing and Management*, *20*(4), 400–416. DOI: 10.1108/JFMM-07-2015-0052

Here are the provided references formatted in APA 7th edition style:Abedsoltan, H., Taleizadeh, A. A., & Sarker, R. B. (2022). Optimal production of remanufactured products with collaboration and co-branding of luxury brands. *Computers & Industrial Engineering*, *172*, 108533. DOI: 10.1016/j.cie.2022.108533

Herinckx, J., & Ghislain, R. (2022). The use of blockchain to fight counterfeiting in the second-hand luxury fashion market. *cit. a*, 48.

Hermann, E. (2021). Leveraging Artificial Intelligence in Marketing for Social Good—An Ethical Perspective. *Journal of Business Ethics*, *179*(1), 43–61. DOI: 10.1007/s10551-021-04843-y PMID: 34054170

Heyes, A. (2021). What is luxury hospitality? A need to move towards a scientific understanding. In Research in Hospitality Management (Vol. 11, Issue 2, p. 67). Taylor & Francis. https://doi.org/DOI: 10.1080/22243534.2021.1917919

Hildebrandt, J., Thrän, D., & Bezama, A. (2021). The circularity of potential bio-textile production routes: Comparing life cycle impacts of bio-based materials used within the manufacturing of selected leather substitutes. *Journal of Cleaner Production*, *287*, 125470. DOI: 10.1016/j.jclepro.2020.125470

Hirschmiller, S. (2023, June 7). Unpacking the Web3 strategy behind Louis Vuitton's €39k digital Treasure Trunk NFTs. Forbes. https://www.forbes.com/sites/stephaniehirschmiller/2023/06/07/unpacking-the-strategy-behind-louis-vuittons-39k-digital-treasure-trunk-nfts/

Hoang, D., Kousi, S., & Martinez, L. F. (2022). Online customer engagement in the post-pandemic scenario: A hybrid thematic analysis of the luxury fashion industry. *Electronic Commerce Research*, *23*(3), 1401–1428. DOI: 10.1007/s10660-022-09635-8

Hoffmann, J., Ramirez, R., & Lecamp, L. (2018). Right on time - Socioecological strategy and implications of turbulence in the Swiss watchmaking field. *Technological Forecasting and Social Change*, *137*, 101–117. DOI: 10.1016/j.techfore.2018.06.047

Ho, H., Chiu, C. L., Liu, S., Lee, L. L., Lado, N., & Cesaronic, F. (2019). Understanding the role of attitude components in co-branding: A comparison of Spanish and Taiwanese consumers. *Journal of International Consumer Marketing*, *31*(3), 203–224. DOI: 10.1080/08961530.2018.1531363

Ho, H., Lado, N., & Rivera-Torres, P. (2017). Detangling consumer attitudes to better explain co-branding success. *Journal of Product and Brand Management, 26*(7), 704–721. DOI: 10.1108/JPBM-11-2015-1039

Holmqvist, J., Berger, C., De Keyser, A., & Verleye, K. (2025). Luxury in the Circular Economy: An Engagement Journey Perspective. *Journal of Consumer Behaviour*, cb.2460. DOI: 10.1002/cb.2460

Holmqvist, J., Ruiz, C. D., & Peñaloza, L. (2019). Moments of luxury: Hedonic escapism as a luxury experience. *Journal of Business Research, 116*, 503–513. DOI: 10.1016/j.jbusres.2019.10.015

Holmqvist, J., Wirtz, J., & Fritze, M. P. (2020). Luxury in the digital age: A multi-actor service encounter perspective. In *Journal of Business Research* (Vol. 121, p. 747). Elsevier BV., DOI: 10.1016/j.jbusres.2020.05.038

Holt, D. B. (2002). Why do brands cause trouble? A dialectical theory of consumer culture and branding. *The Journal of Consumer Research, 29*(1), 70–90. DOI: 10.1086/339922

Ho, S. C., Yeh, Y. Y., & Lin, T. M. Y. (2021). Effects of luxury perceptions on purchase intention of High-Tech products. *Journal of International Consumer Marketing, 34*(2), 168–183. DOI: 10.1080/08961530.2021.1937769

Hoyer, W. D., Kroschke, M., Schmitt, B., Kraume, K., & Shankar, V. (2020). Transforming the customer experience through new technologies. *Journal of Interactive Marketing, 51*(1), 57–71. DOI: 10.1016/j.intmar.2020.04.001

Hsiao, K. (2017). What drives smartwatch adoption intention? Comparing Apple and non-Apple watches. *Library Hi Tech, 35*(1), 186–206. DOI: 10.1108/LHT-09-2016-0105

Hsieh, S. H., & Chang, A. (2016). The psychological mechanism of brand co-creation engagement. *Journal of Interactive Marketing, 33*(1), 13–26. DOI: 10.1016/j.intmar.2015.10.001

Huang, R. (2023; Vol. 47). Marketing Strategy and Operation Mode of Luxury Goods Lead to Brand Premium. In BCP Business & Management., DOI: 10.54691/bcpbm.v47i.5168

Huang, R., Ha, S., & Kim, S. (2018). Narrative persuasion in social media: An empirical study of luxury brand advertising. *Journal of Research in Interactive Marketing, 12*(3), 274–292. DOI: 10.1108/JRIM-07-2017-0059

Huang, X., Yang, F., Zheng, J., Feng, C., & Zhang, L. (2023). Personalized human resource management via HR analytics and artificial intelligence: Theory and implications. *Asia Pacific Management Review, 28*(4), 598–610. DOI: 10.1016/j.apmrv.2023.04.004

Huerta-Álvarez, R., Cambra-Fierro, J. J., & Fuentes-Blasco, M. (2020). The interplay between social media communication, brand equity and brand engagement in tourist destinations: An analysis in an emerging economy. *Journal of Destination Marketing & Management, 16*, 100413. https://doi.org/https://doi.org/10.1016/j.jdmm.2020.100413. DOI: 10.1016/j.jdmm.2020.100413

Huh, J., Kim, H., & Lee, G. (2023). "Oh, happy day!" Examining the role of AI-powered voice assistants as a positive technology in the formation of brand loyalty. *Journal of Research in Interactive Marketing, 17*(5), 794–812. DOI: 10.1108/JRIM-10-2022-0328

Huld, A. (2025, April 11). Trump raises tariffs on China to 145% – Overview and trade implications. China Briefing. https://www.china-briefing.com/news/trump-raises-tariffs-on-china-to-125-overview-and-trade-implications/

Hunt, S. D. (2018). The ethics of branding, customer-brand relationships, brand-equity strategy, and branding as a societal institution. *Journal of Business Research, 95*, 408–416. DOI: 10.1016/j.jbusres.2018.07.044

Husain, R. V. S., Kamath, R., & Moorthy, J. (2025). Navigating the Sustainability–Luxury Paradox: Bibliometric Insights and Research Directions. *Journal of Consumer Behaviour*, cb.2469. DOI: 10.1002/cb.2469

Husain, R., Paul, J., & Koles, B. (2022). The role of brand experience, brand resonance and brand trust in luxury consumption. *Journal of Retailing and Consumer Services, 66*, 102895. DOI: 10.1016/j.jretconser.2021.102895

Hypebeast. (2008, March 4). Hermès x Bugatti Veyron FBG. https://hypebeast.com/2008/3/hermes-x-bugatti-veyron-fbg

Indrakumari, R., Kumar, R., Balusamy, B., & Asirvadam, V. S. (2021). Convergence of Blockchain, AI, and IoT. In *CRC Press eBooks*. Informa., DOI: 10.1201/9781003081180

Ismyrlis, V., Tarnanidis, T., & Moschidis, E. (Eds.). (2020). *Knowledge Management Practices in the Public Sector*. IGI Global., DOI: 10.4018/978-1-7998-1940-0

IWC. (2025, April 20). *Laureus Sport for Good*. https://www.iwc.com/es/es/company/partnerships/laureus-sport.html

Jain, S., & Rathi, R. (2023). Do consumer values and perceived readiness impact secondhand luxury consumption? A goal-framing theory approach. In Journal of Product & Brand Management (Vol. 32, Issue 7, p. 973). Emerald Publishing Limited. https://doi.org/DOI: 10.1108/JPBM-10-2021-3703

Jain, S., Mishra, S., & Mukhopadhyay, S. (2021). Critical success factors for luxury fashion brands in emerging markets: Insights from a qualitative study. In Journal of Global Fashion Marketing (Vol. 12, Issue 1, p. 47). Taylor & Francis. https://doi.org/DOI: 10.1080/20932685.2020.1845765

Jain, N. C. (2024). Predictive analytics for personalized health interventions using wearable data. *Deleted Journal*, *12*(3), 188–198. DOI: 10.36676/dira.v12.i3.66

Jang, J., & Kang, J. (2025b, January 18). From New to Pre-Loved: The Impact of Blockchain-enabled NFT Authentication on Warranting Value and Assurance in Luxury Markets. *Making Waves Toward A Sustainable and Equitable Future*. Making Waves Toward A Sustainable and Equitable Future. DOI: 10.31274/itaa.18818

Jang, J., & Kang, J. (2025a). Blockchain meets luxury: The role of NFT authentication in luxury retail platforms. *Journal of Retailing and Consumer Services*, *84*, 104262. DOI: 10.1016/j.jretconser.2025.104262

Javornik, A., Duffy, K., Rokka, J., Scholz, J., Nobbs, K., Motala, A., & Goldenberg, A. (2021). Strategic approaches to augmented reality deployment by luxury brands.[). *Journal of Business Research*, *136*, 284–292. gla.ac.uk. DOI: 10.1016/j.jbusres.2021.07.040

Jeddi, Z., & Bohr, A. (2020). Remote patient monitoring using artificial intelligence. In *Elsevier eBooks* (p. 203). Elsevier BV., DOI: 10.1016/B978-0-12-818438-7.00009-5

Jia, W., & Ran, H. (2021, December). The analysis of brand co-branding strategy. In 2021 3rd International Conference on Economic Management and Cultural Industry (ICEMCI 2021) (pp. 1567-1574). Atlantis Press. DOI: 10.2991/assehr.k.211209.254

Jiang, Y. (2023, July). The Research on the Current Situation and Trends of Social Media Marketing. In 2023 2nd International Conference on Social Sciences and Humanities and Arts (SSHA 2023) (pp. 348-355). Atlantis Press. https://doi.org/, atlantis-press.comDOI: 10.2991/978-2-38476-062-6_44

Jiang, X., Guangkuo, G., & Xuezheng, Y. (2022). Evolutionary game analysis on live streaming commerce considering brand awareness and anchor influence. *Kybernetes*, *52*(12), 6467–6493. DOI: 10.1108/K-04-2022-0593

Jihua, S., & Lin, S. (2024). Research on Data-driven Marketing Strategy Optimization. *Journal of Business and Marketing.*, *1*(2), 157–161. DOI: 10.62517/jbm.202409224

Jitta, A. S., Manda, V. K., & Tarnanidis, T. (2025). Geospatial Technologies for Smart Cities. In Darwish, D., & Chemingui, H. (Eds.), *Recent Trends in Geospatial AI* (pp. 103–136). IGI Global Scientific Publishing., DOI: 10.4018/979-8-3693-8054-3.ch004

Ji, Y., Sun, Z., & Xia, Y. (2023). Leveraging Digital Transformation: Analysis of LVMH's Impact on Economic, Social, and Environmental Performance. *Advances in Economics. Management and Political Sciences*, *33*, 189–193.

Johnson, D. S., Muzellec, L., Sihi, D., & Zahay, D. (2019). The marketing organization's journey to become data-driven. *Journal of Research in Interactive Marketing*, *13*(2), 162–178. DOI: 10.1108/JRIM-12-2018-0157

Joshi, I. (2024). Neuromarketing in Fashion Advertising: Enhancing consumer engagement through sensory and neuroscientific insights. *International Journal of Social Science and Economic Research*, *09*(11), 5637–5642. DOI: 10.46609/IJSSER.2024.v09i11.043

Joy, A., Sherry, J. F.Jr, Venkatesh, A., Wang, J., & Chan, R. (2012). Fast fashion, sustainability, and the ethical appeal of luxury brands. *Fashion Theory*, *16*(3), 273–295. DOI: 10.2752/175174112X13340749707123

Joy, A., Zhu, Y., Peña, C., & Brouard, M. (2022). Digital future of luxury brands: Metaverse, digital fashion, and non-fungible tokens. *Strategic Change*, *31*(3), 337–343. DOI: 10.1002/jsc.2502

Jun, S., & Park, D. (2017). Visualization of brand positioning based on consumer web search information. *Internet Research*, *27*(2), 381–407. DOI: 10.1108/IntR-02-2016-0037

Jun, Y., & Lee, H. (2021). A sound brand identity design: The interplay between sound symbolism and typography on brand attitude and memory. *Journal of Retailing and Consumer Services*, *64*, 102724. DOI: 10.1016/j.jretconser.2021.102724

Kajla, T., Raj, S., Kansra, P., Gupta, S. L., & Singh, N. (2023). Neuromarketing and consumer behavior: A bibliometric analysis. *Journal of Consumer Behaviour*, *23*(2), 959–975. DOI: 10.1002/cb.2256

Kamalakannan, R., Ramesh, C., Shunmugasundaram, M., Sivakumar, P., & Mohamed, A. (2020). Evaluvation and selection of suppliers using TOPSIS. In *Materials Today Proceedings* (Vol. 33, p. 2771). Elsevier BV., DOI: 10.1016/j.matpr.2020.02.105

Kampakaki, E., Papachristou, E., Tarnanidis, T., & Bilalis, N. (2022). What are the most important digital qualifications of the employees in the clothing and fashion companies? In *Proceedings of the 12th Hellenic Conference on Artificial Intelligence* (pp. 1-7). https://doi.org/DOI: 10.1145/3549737.3549804

Kania, R., Sukoyo, K. S., & Wibisono, N. (2021). Effect of co-branding strategy elements on consumer attitude and purchase intention: A study in Indonesia. [JMI]. *Journal of Marketing Innovation*, *1*(01), 30–50. DOI: 10.35313/jmi.v1i01.17

Kapferer, J. N., & Bastien, V. (2009). The specificity of luxury management: Turning marketing upside down. *Journal of Brand Management*, *16*(5-6), 311–322. DOI: 10.1057/bm.2008.51

Kapferer, J. N., & Bastien, V. (2012). *The luxury strategy: Break the rules of marketing to build luxury brands*. Kogan Page Publishers.

Kapferer, J. N., & Michaut-Denizeau, A. (2017). Is luxury compatible with sustainability? Luxury consumers' viewpoint. In *Advances in Luxury Brand Management* (pp. 123–156). Springer. DOI: 10.1007/978-3-319-51127-6_7

Kapferer, J. N., & Michaut-Denizeau, A. (2019). Are millennials really more sensitive to sustainable luxury? A cross-generational international comparison of sustainability consciousness when buying luxury. *Journal of Brand Management*, *26*, 123–138.

Kapferer, J., & Valette-Florence, P. (2017). The impact of brand penetration and awareness on luxury brand desirability. *Journal of Business Research*, *83*, 38–50. DOI: 10.1016/j.jbusres.2017.09.025

Karangi, S. W., & Lowe, B. (2021). Haptics and brands: The effect of touch on product evaluation of branded products. *Journal of Consumer Behaviour*, *20*(6), 1480–1496. DOI: 10.1002/cb.1959

Karypidis, M., Tarnanidis, T., & Papachristou, E. (2024). Treating Textile Effluents for Sustainable Fashion and Green Marketing. In *Driving Green Marketing in Fashion and Retail* (pp. 166–179). IGI Global. DOI: 10.4018/979-8-3693-3049-4.ch010

Kaufmann, H. R., Filho, C. G., Silva, C. R. E., & Monteiro, P. R. R. (2019). Consumer-brand relationship development in the automotive market: An integrative model. *International Journal of Automotive Technology and Management*, *19*(3-4), 321–340.

Kauppinen-Räisänen, H., Mühlbacher, H., & Taishoff, M. (2020). Exploring consumers' subjective shopping experiences in directly operated luxury brand stores. *Journal of Retailing and Consumer Services*, *57*, 102251. DOI: 10.1016/j.jretconser.2020.102251

Kaur, K., & Soch, H. (2025). From prominence to preference: Social and emotional value perception as mediator between the brand prominence and purchase intentions in the luxury market. *Journal of Fashion Marketing and Management, 29*(4), 639–661. Advance online publication. DOI: 10.1108/JFMM-05-2024-0164

Kaur, U., & Siddhey, P. K. (2024). The Influence of Digital Transformation on the Retail Value Chain. In *Reshaping Marketing Science in Wholesaling and Retailing* (pp. 334–353). IGI Global. DOI: 10.4018/979-8-3693-6145-0.ch014

Keller, K. L. (2003). Strategic branding management: Building measuring, and managing brand equity.

Keller, K. L. (1993). Conceptualizing, Measuring, and Managing Customer-Based Brand Equity. *Journal of Marketing, 57*(1), 1–22. DOI: 10.1177/002224299305700101

Keller, K. L. (2009). Managing the growth tradeoff: Challenges and opportunities in luxury branding. *Journal of Brand Management, 16*(5–6), 290–301. DOI: 10.1057/bm.2008.47

Kenyans.co.ke. (2025, April 10). *Chantecaille Foundation: Details of make-up company that just donated Land Cruisers worth Ksh 200M to KWS.* https://www.kenyans.co.ke/news/108862-chantecaille-foundation-details-make-company-just-donated-landcruisers-worth-ksh-200m

Kering. (2021). *Sustainability progress report.* https://www.kering.com

Khaliq, L. N., & Manda, V. K. (2023). Customer Experience in the Web 3.0 Era: The Meeting of Blockchain and the Metaverse. In Majeed, M., Ofori, K. S., Amoako, G. K., Alolo, A.-R., & Awini, G. (Eds.), (pp. 258–274). Advances in Marketing, Customer Relationship Management, and E-Services. IGI Global., DOI: 10.4018/978-1-6684-7649-9.ch015

Khayyam, H., Javadi, B., Jalili, M., & Jazar, R. N. (2019). Artificial intelligence and internet of things for autonomous vehicles. In *Nonlinear approaches in engineering applications: Automotive applications of engineering problems* (pp. 39–68). Springer International Publishing.

Kim, A. J., & Ko, E. (2012). Do Social Media Marketing Activities Enhance Customer Equity? *Journal of Business Research, 65*(10), 1480–1486. DOI: 10.1016/j.jbusres.2011.10.014

Kim, J. C., Park, B., & Dubois, D. (2018). How consumers' political ideology and Status-Maintenance goals interact to shape their desire for luxury goods. *Journal of Marketing, 82*(6), 132–149. DOI: 10.1177/0022242918799699

Kim, J., Lloyd, S., & Cervellon, M. (2015). Narrative-transportation storylines in luxury brand advertising: Motivating consumer engagement. *Journal of Business Research*, *69*(1), 304–313. DOI: 10.1016/j.jbusres.2015.08.002

Kim, K., Ko, E., Lee, M., Mattila, P., & Kim, K. H. (2014). Fashion collaboration effects on consumer response and customer equity in global luxury and SPA brand marketing. *Journal of Global Scholars of Marketing Science*, *24*(3), 350–364. DOI: 10.1080/21639159.2014.913376

Kim, Y. (2018). Power moderates the impact of desire for exclusivity on luxury experiential consumption. *Psychology and Marketing*, *35*(4), 283–293. DOI: 10.1002/mar.21086

Kim, Y., & Sullivan, P. (2019). Emotional branding speaks to consumers' heart: The case of fashion brands. *Fashion and Textiles*, *6*(1), 2. Advance online publication. DOI: 10.1186/s40691-018-0164-y

King, M. F. (2011). *Fashion, the body and technology: Tracing early 20th century techno-utopian ideas, aesthetics and impulses in 21st century wearable technology* (Doctoral dissertation, Queensland University of Technology).

Kłeczek, R., & Hajdas, M. (2025). The appreciation task for luxury brand–artist collaborations. *Journal of Brand Management*, *32*(3), 203–226. Advance online publication. DOI: 10.1057/s41262-025-00378-z

Knoeferle, K. M., Knoeferle, P., Velasco, C., & Spence, C. (2016). Multisensory brand search: How the meaning of sounds guides consumers' visual attention. *Journal of Experimental Psychology. Applied*, *22*(2), 196–210. DOI: 10.1037/xap0000084 PMID: 27295466

Ko, E., Costello, J. P., & Taylor, C. R. (2019). What is a luxury brand? A new definition and review of the literature. *Journal of Business Research*, *99*, 405–413. https://doi.org/https://doi.org/10.1016/j.jbusres.2017.08.023. DOI: 10.1016/j.jbusres.2017.08.023

Ko, E., Phau, I., & Aiello, G. (2016). Luxury brand strategies and customer experiences: Contributions to theory and practice. *Journal of Business Research*, *69*(12), 5749–5752. DOI: 10.1016/j.jbusres.2016.04.170

Kolasani, S. (2024). Connected cars and autonomous vehicles: personalizing owner/customer experiences and innovation using AI, IoT, Blockchain, and Big Data. *International numeric journal of machine learning and robots, 8*(8), 1-17.

Kongsri, N., & Jaroenwanit, P. (2024). Key aspects of personal brand identity in social media commerce: Impact on successful personal branding. *International Journal of Data and Network Science*, *8*(4), 2663–2676. DOI: 10.5267/j.ijdns.2024.4.012

Kordzadeh, N., & Ghasemaghaei, M. (2021). Algorithmic bias: Review, synthesis, and future research directions. *European Journal of Information Systems*, *31*(3), 388–409. DOI: 10.1080/0960085X.2021.1927212

Kowalczyk, C. M., & Mitchell, N. A. (2021). Understanding the antecedents to luxury brand consumer behavior. *Journal of Product and Brand Management*, *31*(3), 438–453. DOI: 10.1108/JPBM-09-2020-3126

Kozlowski, A., Searcy, C., & Bardecki, M. (2015). Corporate sustainability reporting in the apparel industry: An analysis of indicators disclosed. *International Journal of Productivity and Performance Management*, *64*(3), 377–397. DOI: 10.1108/IJPPM-10-2014-0152

Kressmann, F., Sirgy, M. J., Herrmann, A., Huber, F., Huber, S., & Lee, D. (2006). Direct and indirect effects of self-image congruence on brand loyalty. *Journal of Business Research*, *59*(9), 955–964. DOI: 10.1016/j.jbusres.2006.06.001

Kumagai, K., & Nagasawa, S. (2019). Psychological switching mechanism of consumers' luxury and non-luxury brand attitude formation: The effect of store location prestige and self-congruity. *Heliyon*, *5*(5), e01581. DOI: 10.1016/j.heliyon.2019.e01581 PMID: 31193173

Kumar, D., & Suthar, N. (2024). Ethical and legal challenges of AI in marketing: An exploration of solutions. *Journal of Information Communication and Ethics in Society*, *22*(1), 124–144. DOI: 10.1108/JICES-05-2023-0068

Kumar, J. (2024). Psychological mechanisms behind access-based luxury brand consumption: Empirical investigation from the lens of new ownership paradigm. *Journal of Business Research*, *175*, 114539. DOI: 10.1016/j.jbusres.2024.114539

Kumar, J., & Nayak, J. K. (2019). Consumer psychological motivations to customer brand engagement: A case of brand community. *Journal of Consumer Marketing*, *36*(1), 168–177. DOI: 10.1108/JCM-01-2018-2519

Kumar, P. (2023). Luxury consumption amidst the COVID-19 pandemic. [HTML]. *Marketing Intelligence & Planning*, *41*(1), 62–82. Advance online publication. DOI: 10.1108/MIP-02-2022-0081

Kumar, V., & Kaushal, V. (2021). Perceived brand authenticity and social exclusion as drivers of psychological brand ownership. *Journal of Retailing and Consumer Services*, *61*, 102579. DOI: 10.1016/j.jretconser.2021.102579

Kumar, V., Rajan, B., Venkatesan, R., & Lecinski, J. (2019). Understanding the role of artificial intelligence in personalized engagement marketing. *California Management Review*, *61*(4), 135–155. DOI: 10.1177/0008125619859317

Kunz, J., May, S., & Schmidt, H. J. (2020). Sustainable luxury: Current status and perspectives for future research. *BuR -. Business Research*, *13*(2), 541–601. DOI: 10.1007/s40685-020-00111-3

Kur, A. (2016). Brand symbols, the consumer, and the internet. SSRN *Electronic Journal*. https://doi.org/DOI: 10.2139/ssrn.2721740

Kurtz, J. (2020). On foreign investor 'Privilege' and the limits of the law: A reply to Ivar Alvik. *European Journal of International Law*, *31*(1), 313–319. DOI: 10.1093/ejil/chaa020

Kusumasondjaja, S. (2019). Exploring the role of visual aesthetics and presentation modality in luxury fashion brand communication on Instagram. *Journal of Fashion Marketing and Management*, *24*(1), 15–31. DOI: 10.1108/JFMM-02-2019-0019

Kuusela, M. (2024). Reimagining luxury: The potential of digital product passports in accelerating circular business models In high-end fashion.

Land, K. M. (2023). *Improving CAD Designs with Autodesk Fusion 360: A project-based guide to modelling effective parametric designs*. Packt Publishing Ltd.

Lankton, N., McKnight, D. H., & Tripp, J. (2015). Technology, Humanness, and Trust: Rethinking trust in technology. *Journal of the Association for Information Systems*, *16*(10), 880–918. DOI: 10.17705/1jais.00411

Laureus (2025). *Sporting legends launch IWC watch design to raise funds for Laureus*. https://www.laureus.com/news/sporting-legends-launch-iwc-watch-design-to-raise-funds-for-laureus

Leban, M., Seo, Y., & Voyer, B. G. (2019). Transformational effects of social media lurking practices on luxury consumption. *Journal of Business Research*, *116*, 514–521. DOI: 10.1016/j.jbusres.2019.09.010

Ledger Insights. (2021, July 15). *Luxury brand Dolce & Gabbana to launch NFT collection*. https://www.ledgerinsights.com/luxury-brand-dolce-gabbana-to-launch-nft-collection-blockchain/

Ledger Insights. (2022, January 18). Gucci to launch NFT collection with Superplastic brand. https://www.ledgerinsights.com/gucci-to-launch-nft-collection-with-superplastic-brand/

Lee, J. E., Hur, S., & Watkins, B. (2018). Visual communication of luxury fashion brands on social media: Effects of visual complexity and brand familiarity. *Journal of Brand Management*, *25*(5), 449–462. DOI: 10.1057/s41262-018-0092-6

Le, H. T. P. M., Park, J., Nguyen, T. T., & Yun, J. (2024). Exploring the relationship between social media use and anti-luxury behavior: Implications for interactive marketing strategies. [HTML]. *Journal of Research in Interactive Marketing*, *18*(4), 549–569. DOI: 10.1108/JRIM-04-2023-0125

Leonhard, H., Nagl, M., & Schaefers, W. (2025). Virtual land in the Metaverse? Exploring the dynamic correlation with physical real estate. *Journal of European Real Estate Research*, *18*(1), 121–151. DOI: 10.1108/JERER-04-2024-0025

Leuthesser, L., Kohli, C., & Suri, R. (2003). 2+2=5? A framework for using co-branding to leverage a brand. *Journal of Brand Management*, *11*(1), 35–47. DOI: 10.1057/palgrave.bm.2540146

Li, J. (2018). *The Lighting Technology of Fashion Designs in the Contemporary Social Context* (Master's thesis, State University of New York at Stony Brook).

Li, A., Sun, J., Li, P., Pu, Y., Li, H., & Chen, Y. (2021). Hermes. *Proceedings of the 28th Annual International Conference on Mobile Computing and Networking*, 420–437. https://doi.org/DOI: 10.1145/3447993.3483278

Liang, Y., Xu, Q., & Jin, L. (2021). The effect of smart and connected products on consumer brand choice concentration. *Journal of Business Research*, *135*, 163–172. DOI: 10.1016/j.jbusres.2021.06.039

Li, J., & Kang, J. (2025). Human-centred AI vs machine-centred AI: Navigating AI ethical crises in luxury retail. *International Journal of Retail & Distribution Management*, *53*(7), 639–653. Advance online publication. DOI: 10.1108/IJRDM-11-2024-0600

Lin, S. (2023, June). The influential factors on consumer purchase intention: A case study on MINISO. In Proceedings of the 6th International Conference on Economic Management and Green Development (pp. 1265-1277). Springer Nature Singapore. DOI: 10.1007/978-981-19-7826-5_121

Lin, J. (2023). Analysis of luxury brand marketing strategies based on consumer psychology. *Advances in Economics Management and Political Sciences*, *24*(1), 240–245. DOI: 10.54254/2754-1169/24/20230443

Lin, Y. (2022). Augmented Reality in customer interaction for branding of high-end vehicles. *BCP Business & Management*, *21*, 330–339. DOI: 10.54691/bcpbm.v21i.1258

Liu, Y. (2023). Digital Transformation in the Luxury Industry: Strategies, Challenges, And Opportunities. In Highlights in Business Economics and Management (Vol. 23, p. 650). DOI: 10.54097/xgftec52

Liu, X. (2024). Research on the marketing strategy and effect of sportswear brand co-branding with entertainment celebrities. Advances in Economics. *Management and Political Sciences*, *84*, 212–217.

Liu, X., Shin, H., & Burns, A. C. (2019). Examining the impact of luxury brand's social media marketing on customer engagement: Using big data analytics and natural language processing. *Journal of Business Research*, *125*, 815–826. DOI: 10.1016/j.jbusres.2019.04.042

Li, Y., Liu, J., & Ren, J. (2019). Social recommendation model based on user interaction in complex social networks. *PLoS One*, *14*(7), e0218957. DOI: 10.1371/journal.pone.0218957 PMID: 31291288

Loh, L., & Tan, S. (2020). Impact of sustainability Reporting on brand Value: An examination of 100 leading brands in Singapore. *Sustainability (Basel)*, *12*(18), 7392. DOI: 10.3390/su12187392

Loureiro, S. M. C., Bilro, R. G., & Japutra, A. (2019). The effect of consumer-generated media stimuli on emotions and consumer brand engagement. *Journal of Product and Brand Management*, *29*(3), 387–408. DOI: 10.1108/JPBM-11-2018-2120

Lourenço, V. L., Catarino, A. W., Fonseca, M. J., & Sousa, B. B. (2021). Consumer-brand relationship and use of the website in virtual communication in the luxury furniture industry. In *Building Consumer-Brand Relationship in Luxury Brand Management* (pp. 158–177). IGI Global Scientific Publishing. DOI: 10.4018/978-1-7998-4369-6.ch008

Lowe, M. L., & Haws, K. L. (2016). Sounds big: The effects of acoustic pitch on product perceptions. *JMR, Journal of Marketing Research*, *54*(2), 331–346. DOI: 10.1509/jmr.14.0300

Lu, C. (2023). Co-branding marketing strategy and optimization of Perfect Diary from the perspective of brand equity. *Academic Journal of Business & Management*, *5*(25), 64–69.

Lu, C., Zhang, Y., Chen, Z., Sang, Z., & Liu, Z. (2025). How do marketing practices impact price premium of co-branded sportswear? The moderating effect of product category fit and partner brand name. *International Journal of Sports Marketing & Sponsorship*. Advance online publication. DOI: 10.1108/IJSMS-12-2023-0248

LVMH. (2020). [*Our sustainability program.* https://www.lvmh.com]. *Life (Chicago, Ill.), 360*, •••.

Machado, P. M., Lawson-Tovey, S., Strangfeld, A., Mateus, E., Hyrich, K., Gossec, L., Carmona, L., Rodrigues, A., Raffeiner, B., Duarte, C., Hachulla, E., Veillard, E., Strakova, E., Burmester, G. R., Yardimci, G. K., Gómez-Puerta, J. A., Zepa, J., Kearsley-Fleet, L., Trefond, L., & Mariette, X. (2022). OA01 Safety of vaccination against SARS-CoV-2 in people with rheumatic and musculoskeletal diseases: Results from the EULAR Coronavirus Vaccine (COVAX) physician-reported registry. *Lara D. Rheumatology (Oxford, England), 61*(Supplement_1), keac132. Advance online publication. DOI: 10.1093/rheumatology/keac132

Mahmoud, A. B., Kumar, V., Fuxman, L., & Mohr, I. (2023). Climate change risks, sustainability and luxury branding: Friend or a foe. *Industrial Marketing Management, 115*, 57–68. DOI: 10.1016/j.indmarman.2023.09.008

Majeed, M. U., Aftab, H., Arslan, A., & Shakeel, Z. (2024). Determining online consumer's luxury purchase intention: The influence of antecedent factors and the moderating role of brand awareness, perceived risk, and web atmospherics. *PLoS One, 19*(2), e0295514. DOI: 10.1371/journal.pone.0295514 PMID: 38394125

Malarvizhi, C. A., Al Mamun, A., Jayashree, S., Naznen, F., & Abir, T. (2022). Modelling the significance of social media marketing activities, brand equity and loyalty to predict consumers' willingness to pay premium price for portable tech gadgets. Heliyon, 8(8). https://doi.org/ cell.comDOI: 10.1016/j.heliyon.2022.e10145

Maldonado-Canca, L., Casado-Molina, A., Cabrera-Sánchez, J., & Bermúdez-González, G. (2024). Beyond the post: An SLR of enterprise artificial intelligence in social media. *Social Network Analysis and Mining, 14*(1), 219. Advance online publication. DOI: 10.1007/s13278-024-01382-y

Mallipeddi, R., Kumar, S., Sriskandarajah, C., & Zhu, Y. (2018). A framework for Analyzing influencer marketing in Social Networks: Selection and Scheduling of Influencers. SSRN *Electronic Journal.* https://doi.org/DOI: 10.2139/ssrn.3255198

Mallon, S. (2024). Materialities of 'Mushroom-Leather': A Critical Appraisal of the Material Dimensions and Sustainability of Fungus-Based Alternatives to Leather for use in Fashion and Textiles. *Ethnologia Fennica, 51*(1), 81–110. DOI: 10.23991/ef.v51i1.141682

Manda, V. K., Tarnanidis, T., & Majeed, M. (2025a). Advanced topics in robotics. In D. Darwish (Ed.), Advances in Robots Technologies and Implementations (pp. 96-112). Deep Science Publishing. https://doi.org/DOI: 10.70593/978-81-983916-1-2_5

Manda, V. K., Bikkina, A., & Tarnanidis, T. (2025b). Future trends in Virtual Reality (VR). In *Virtual Reality Technologies and Real Life Applications* (pp. 83–96). Deep Science Publishing., DOI: 10.70593/978-81-982935-1-0_5

Manda, V. K., Manda, V. K., & Katneni, V. (2022). Blockchain for the asset management industry. *World Review of Science, Technology and Sustainable Development*, *19*(1-2), 170–185. DOI: 10.1504/WRSTSD.2023.127266

Manda, V. K., Sagi, S., & Yadav, A. (2024). Blockchain in Advertising and Marketing: Revolutionizing the Industry Through Transparency and Trust. In Tarnanidis, T. K., & Sklavounos, N. (Eds.), (pp. 89–112). Advances in Marketing, Customer Relationship Management, and E-Services. IGI Global., DOI: 10.4018/979-8-3693-2754-8.ch005

Mandler, T., Johnen, M., & Gräve, J. (2019). Can't help falling in love? How brand luxury generates positive consumer affect in social media. *Journal of Business Research*, *120*, 330–342. DOI: 10.1016/j.jbusres.2019.10.010

Margariti, K., Boutsouki, C., & Hatzithomas, L. (2019). The dilemma of luxury brand extensions. *Journal of Global Fashion Marketing*, *10*(4), 305–323. DOI: 10.1080/20932685.2019.1649167

Mars, M. M. (2023). Looking behind the label: Firm-centric brand storytelling and the development of local consumption spaces. *Local Development & Society*, *4*(2), 370–392. DOI: 10.1080/26883597.2023.2175713

Ma, Y. (2023). Analysis of the Hybridity Translation Strategy of Moment in Peking from the Perspective of Post-colonialism. *Communications in Humanities Research*, *6*(1), 304–310. DOI: 10.54254/2753-7064/6/20230284

Mbah, G. O. (2024). Data privacy in the era of AI: Navigating regulatory landscapes for global businesses.

McCormick, H., & Ram, P. (2022). 'Take a stand': The importance of social sustainability and its effect on Generation Z consumption of luxury fashion brands. In Gardetti, M., & Torres, A. J. (Eds.), *Sustainable luxury: An international perspective* (pp. 219–239). Springer International Publishing. DOI: 10.1007/978-3-031-06928-4_11

McKinsey & Company. (2020). *The state offashion 2020*. https://www.mckinsey.com

McKinsey & Company. (2025, January 13). *The state of luxury: How to navigate a slowdown*. https://www.mckinsey.com/industries/retail/our-insights/state-of-luxuryMcKinsey & Company

McNeil, P., & Riello, G. (2016). *Luxury A Rich History*. Oxford University Press.

Melé, D. (2003). The challenge of humanistic management. *Journal of Business Ethics*, *44*(1), 77–88. DOI: 10.1023/A:1023298710412

Meyerding, S. G., & Mehlhose, C. M. (2018). Can neuromarketing add value to the traditional marketing research? An exemplary experiment with functional near-infrared spectroscopy (fNIRS). *Journal of Business Research*, *107*, 172–185. DOI: 10.1016/j.jbusres.2018.10.052

Michel, G., & Willing, R. (2020). *The art of successful brand collaborations: Partnerships with artists, designers, museums, territories, sports, celebrities, science, good cause… and more*. Routledge. DOI: 10.4324/9781351014472

Migicovsky, Z., Gardner, K. M., Richards, C., Chao, C. T., Schwaninger, H. R., Fazio, G., Zhong, G., & Myles, S. (2021). Genomic consequences of apple improvement. *Horticulture Research*, *8*(1), 9. Advance online publication. DOI: 10.1038/s41438-020-00441-7 PMID: 33384408

Milanesi, M., Guercini, S., & Runfola, A. (2022). Let's play! Gamification as a marketing tool to deliver a digital luxury experience. In Electronic Commerce Research (Vol. 23, Issue 4, p. 2135). Springer Science+Business Media. https://doi.org/DOI: 10.1007/s10660-021-09529-1

Milheiro, A. B., Sousa, B., Santos, V., Milheiro, C., & Vilhena, E. (2024). Understanding the Role of Brand Attachment in the Automotive Luxury Brand Segment. *Administrative Sciences*, *14*(6), 119. DOI: 10.3390/admsci14060119

Milheiro, A., Sousa, B., Milheiro, C., & Vilhena, E. (2025). Understanding Consumers' Feelings of Affectivity in the Luxury Car Segment. In Valeri, M. (Ed.), *Emotional Intelligence and Networking Competencies. Contributions to Management Science*. Springer., DOI: 10.1007/978-3-031-77543-7_6

Minkkinen, T. (2024). The role of co-branding in brand loyalty. Unpublished manuscript.

Mirzajani, H., Istif, E., Abbasiasl, T., Mirlou, F., Özkahraman, E. E., Hasanreisoglu, M., & Beker, L. (2022). Femtosecond laser ablation assisted NFC antenna fabrication for smart contact lenses. *Advanced Materials Technologies*, *7*(10), 2101629. Advance online publication. DOI: 10.1002/admt.202101629

Modhoriye, P., Yadav, P., & Jadhav, S. (2023). AI Transformation in Business: Unveiling the dual effects of advancement and challenges. *INTERANTIONAL JOURNAL OF SCIENTIFIC RESEARCH IN ENGINEERING AND MANAGEMENT*, *07*(11), 1–11. DOI: 10.55041/IJSREM27359

Mohtar, M., & Taha, A. Z. (2024). Case 12: From scraps to style—The art of Hermès' upcycling journey with "Petit H". In *Marketing case studies in emerging markets: Contemporary multi-industry issues and best-practices* (pp. 163–176). Springer Nature Switzerland.

Moon, H., & Sprott, D. E. (2015). SELECTING THE RIGHT PARTNERS FOR THE RIGHT REASONS: THE IMPORTANCE OF FIT IN INGREDIENT BRANDING STRATEGY IN LUXURY BRAND. *Global Fashion Management Conference, 2*(4), 199–200. https://doi.org/DOI: 10.15444/GFMC2015.02.04.04

Moreau, C. P., Prandelli, E., Schreier, M., & Hieke, S. (2020). Customization in luxury brands: Can Valentino get personal? *JMR, Journal of Marketing Research, 57*(5), 937–947. DOI: 10.1177/0022243720943191

Mrad, M., Farah, M. F., & Haddad, S. (2018). From Karl Lagerfeld to Erdem: A series of collaborations between designer luxury brands and fast-fashion brands. *Journal of Brand Management, 26*(5), 567–582. DOI: 10.1057/s41262-018-00146-2

Mukherjee, A., & Das, A. K. (2015c). Interval valued intuitionistic fuzzy soft multi set theoretic approach to decision making problems, 2015 International Conference on Computer, Communication and Control (IC4), Indore, India, 2015, pp. 1-5, https://doi.org/DOI: 10.1109/IC4.2015.7375640

Mukherjee, A., & Das, A. K. (2016a). Application of fuzzy soft multi sets in decision-making problems. In: Nagar, A., Mohapatra, D., Chaki, N. (eds) Proceedings of 3rd International Conference on Advanced Computing, Networking and Informatics. Smart Innovation, Systems and Technologies, vol. 43. Springer, New Delhi. https://doi.org/DOI: 10.1007/978-81-322-2538-6_3

Mukherjee, A., & Das, A. K. (2023). Essentials of fuzzy soft multisets: Theory and applications, Springer Nature Singapore Pte Ltd., 1-151. ISBN: 978-981-19-2759-1, https://doi.org/DOI: 10.1007/978-981-19-2760-7

Mukherjee, A., & Das, A. K. (2015a). Application of interval valued intuitionistic fuzzy soft set in investment decision making, *2015 Fifth International Conference on Advances in Computing and Communications (ICACC)*, Kochi, India, 2015, pp. 61-64, https://doi.org/DOI: 10.1109/ICACC.2015.37

Mukherjee, A., & Das, A. K. (2015b). Relations on intuitionistic fuzzy soft multi sets. In Kim, K. (Ed.), *Information Science and Applications. Lecture Notes in Electrical Engineering* (Vol. 339). Springer., DOI: 10.1007/978-3-662-46578-3_71

Mukherjee, A., & Das, A. K. (2016b). Topological structure formed by intuitionistic fuzzy rough relations. *Journal of the Indian Mathematical Society, 83*(1-2), 135–144.

Mukherjee, A., & Das, A. K. (2022). Einstein operations on fuzzy soft multisets and decision making. *Boletim da Sociedade Paranaense de Matematica*, *40*, 1–10. DOI: 10.5269/bspm.32546

Mukherjee, A., Das, A. K., Gupta, N., Patra, S., Mahmood, T., Das, S., & Das, R. (2025). *A novel generalized interval-valued neutrosophic rough soft set framework for enhanced decision-making: Application in water quality assessment*. Transactions on Fuzzy Sets and System.

Mukhtar, M. S., Mohan, A. C., Chandra, D., & Gotmare, P. R. (2023). Brand Personality and the Automotive Market: An In-depth Study of Consumer Preferences and Purchase Decisions. *Journal of International Economics*, *14*(2), 1–19.

Muret, D. (2024, March 21). *La percepción del lujo atraviesa un profundo cambio impulsada por la sostenibilidad y la tecnología*. FashionNetwork.com. https://es.fashionnetwork.com/news/La-percepcion-del-lujo-atraviesa-un-profundo-cambio-impulsada-por-la-sostenibilidad-y-la-tecnologia,1648338.html

Murtas, G., Pedeliento, G., Mangiò, F., & Andreini, D. (2022). Co-branding strategies in luxury fashion: The Off-White case. *Journal of Strategic Marketing*, •••, 1–20. DOI: 10.1080/0965254X.2022.2153258

Nadeem, W., Tan, T. M., Tajvidi, M., & Hajli, N. (2021). How do experiences enhance brand relationship performance and value co-creation in social commerce? The role of consumer engagement and self brand-connection. *Technological Forecasting and Social Change*, *171*, 120952. DOI: 10.1016/j.techfore.2021.120952

Nawres, D., Nedra, B., Yousaf, A., & Mishra, A. (2023). The role of augmented reality in shaping purchase intentions and WOM for luxury products. *Journal of Business Research*, *171*, 114368. DOI: 10.1016/j.jbusres.2023.114368

Nedergaard, N., & Gyrd-Jones, R. (2013). Sustainable brand-based innovation: The role of corporate brands in driving sustainable innovation. *Journal of Brand Management*, *20*(9), 762–778. DOI: 10.1057/bm.2013.16

Nejadrezaei, N., Henchion, M., & O'Neill, E. (2024). Consumers Perspectives on Biobased Products: A Scoping Review. *Cleaner Production Letters*, 100073.

Nieroda, M. E., Mrad, M., & Solomon, M. R. (2018). How do consumers think about hybrid products? Computer wearables have an identity problem. *Journal of Business Research*, *89*, 159–170. DOI: 10.1016/j.jbusres.2018.04.024

Niinimäki, K. (2018). *Sustainable fashion in a circular economy*. Aalto University.

Nnindini, S. I., & Dankwah, J. B. (2022). Consumer loyalty and brand marketing programs in an emerging economy: Evidence from the automobile industry. *British Journal of Marketing Studies, 10*(3), 38–58.

Norori, N., Hu, Q., Aellen, F. M., Faraci, F. D., & Tzovara, A. (2021). Addressing bias in big data and AI for health care: A call for open science. *Patterns (New York, N.Y.), 2*(10), 100347. DOI: 10.1016/j.patter.2021.100347 PMID: 34693373

Nurfitriansyah, H. R., & Dirgantari, P. D. (2021). Factors That Affect the Election of Higher Education in Holistic Marketing Point of View. In Advances in economics, business and management research/Advances in Economics, Business and Management Research. Atlantis Press. https://doi.org/DOI: 10.2991/aebmr.k.210831.065

Nurrosidah, A., Djatmika, E. T., & Dewi, T. S. (2025). Exploring the Role of Perceived Value and Customer Attitude in Co-Branding: Impact on Purchase Intention for Somethinc Cosmetics X Bittersweet by Najla. *Journal of Applied Business. Taxation and Economics Research, 4*(3), 445–458. DOI: 10.54408/jabter.v4i3.365

Nyadzayo, M. W., Johnson, L. W., & Rossi, M. (2020). Drivers and outcomes of brand engagement in self-concept for luxury fashion brands. *Journal of Fashion Marketing and Management, 24*(4), 589–609. DOI: 10.1108/JFMM-05-2018-0070

Nyika, J., Mwema, F. M., Mahamood, R. M., Akinlabi, E. T., & Jen, T. C. (2022). Advances in 3D printing materials processing-environmental impacts and alleviation measures. *Advances in Materials and Processing Technologies, 8*(sup3), 1275-1285.

Ochkovskaya, M. S., Mkhitaryan, S. V., & Indichekova, E. P. (2021). Influence of collaborations with visual art on luxury brand identity: Pilot study. *Vestnik of Saint Petersburg University Management, 20*(1), 3–29. DOI: 10.21638/11701/spbu08.2021.101

Oeppen, J., & Jamal, A. (2014). Collaborating for success: Managerial perspectives on co-branding strategies in the fashion industry. *Journal of Marketing Management, 30*(9-10), 925–948. DOI: 10.1080/0267257X.2014.934905

Oğuz, A. (2024). Consumer behavior in the era of AI-Driven marketing. *Human-Computer Interaction, 8*(1), 147. DOI: 10.62802/h9frxh42

Ojadi, J. O., Odionu, C., Onukwulu, E., & Owulade, O. (2024). Big Data Analytics and AI for Optimizing Supply Chain Sustainability and Reducing Greenhouse Gas Emissions in Logistics and Transportation. *International Journal of Multidisciplinary Research and Growth Evaluation, 5*(1), 1536–1548. DOI: 10.54660/.IJMRGE.2024.5.1.1536-1548

Okonkwo, U. (2009). Sustaining the luxury brand on the Internet. *Journal of Brand Management*, *16*(5–6), 302–310. DOI: 10.1057/bm.2009.2

Oliveira, M. I. L. L., & Fernandes, T. (2020). Luxury brands and social media: drivers and outcomes of consumer engagement on Instagram. In Journal of Strategic Marketing (Vol. 30, Issue 4, p. 389). Taylor & Francis. https://doi.org/DOI: 10.1080/0965254X.2020.1777459

Osburg, V. S., Davies, I., Yoganathan, V., & McLeay, F. (2021). Perspectives, opportunities and tensions in ethical and sustainable luxury: Introduction to the thematic symposium. *Journal of Business Ethics*, *169*(2), 201–210. DOI: 10.1007/s10551-020-04487-4

Ottoni, F. M. J., Thomé, K. M., & Porto, R. B. (2024). Advances in B2B Co-Branding: A Systematic Literature Review. [HTML]. *Journal of Business-To-Business Marketing*, •••, 1–17. DOI: 10.1080/1051712X.2024.2380683

Ottoni, F. M. J., Thomé, K. M., & Porto, R. B. (2024). Advances in B2B co-branding: A systematic literature review. *Journal of Business-To-Business Marketing*, •••, 1–17. DOI: 10.1080/1051712X.2024.2197098

Pangarkar, A., & Shukla, P. (2023). Conspicuous and inconspicuous consumption of luxury goods in a digital world: Insights, implications, and future research directions. *International Journal of Advertising*, *42*(7), 1226–1238. DOI: 10.1080/02650487.2023.2246260

Pantano, E., Passavanti, R., Priporas, C., & Verteramo, S. (2018). To what extent luxury retailing can be smart? *Journal of Retailing and Consumer Services*, *43*, 94–100. DOI: 10.1016/j.jretconser.2018.03.012

Parashar, J., Jain, A., & Ali, S. (2023). Artificial intelligence impact on human fitness: Exploring emerging trends. *International Journal for Research in Applied Science and Engineering Technology*, *11*(10), 287–293. DOI: 10.22214/ijraset.2023.55971

Parguel, B., Delécolle, T., & Valette-Florence, P. (2015). How price display influences consumer luxury perceptions. *Journal of Business Research*, *69*(1), 341–348. DOI: 10.1016/j.jbusres.2015.08.006

Park, C. W., Jun, S. Y., & Shocker, A. D. (1996). Composite Branding Alliances: An Investigation of Extension and Feedback Effects. *JMR, Journal of Marketing Research*, *33*(4), 453–466. DOI: 10.1177/002224379603300407

Park, J., & Ahn, S. (2024). Traditional vs. AI-generated brand personalities: Impact on brand preference and purchase intention. *Journal of Retailing and Consumer Services*, *81*, 104009. DOI: 10.1016/j.jretconser.2024.104009

Parley. (2025). *Where creators, thinkers and leaders collaborate for the oceans.* https://parley.tv/

Paska, M., & Budnik, K. (2023). Trust as a participant in the adaptation of technological changes. *Scientific Papers of Silesian University of Technology Organization and Management Series, 2023*(170), 383–391. DOI: 10.29119/1641-3466.2023.170.24

Patrick, V. M., & Monga, A. S. B. (2020). Building and growing luxury brands: Strategies for pursuing growth while maintaining brand coherence. In *Research handbook on luxury branding* (pp. 117–137). Edward Elgar Publishing. DOI: 10.4337/9781786436351.00018

Pavashe, A. S., Kadam, P. D., Zirange, V. B., & Katkar, R. D. (2023). The impact of artificial intelligence on employment and workforce trends in the Post-Pandemic era. *International Journal for Research in Applied Science and Engineering Technology, 11*(11), 154–157. DOI: 10.22214/ijraset.2023.56418

Paydas Turan, C. (2021). Success drivers of co-branding: A meta-analysis. International Journal of Consumer Studies. DOI: , wiley.comDOI: 10.1111/ijcs.12682

Peasley, M. C., & Woodroof, P. J. (2024). Managing conflicting stakeholder interests: A co-branding perspective for how managers can mitigate negative shareholder responses to cause-related marketing. *Journal of Nonprofit & Public Sector Marketing*, •••, 1–27. DOI: 10.1080/10495142.2024.2408556

Peck, J., & Shu, S. B. (2009). The effect of mere touch on perceived ownership. *The Journal of Consumer Research, 36*(3), 434–447. DOI: 10.1086/598614

Pennington, C. R., Monk, R. L., Qureshi, A. W., Kulkarni, R., Li, W., Li, J., & Heim, D. (2022). Dark Nudges: Branding magnifies the decoy effect in alcohol purchasing decisions. *Journal of Studies on Alcohol and Drugs, 83*(6), 857–866. DOI: 10.15288/jsad.21-00398 PMID: 36484583

Pentina, I., Guilloux, V., & Micu, A. C. (2018). Exploring social media engagement behaviors in the context of luxury brands. *Journal of Advertising, 47*(1), 55–69. DOI: 10.1080/00913367.2017.1405756

Peragallo, N. A. (2025). *Empowering Place: Unlocking Value for Investors by Integrating Indigenous Values in Luxury Hospitality* [Program in Real Estate Development., Massachusetts Institute of Technology]. https://hdl.handle.net/1721.1/158878

Perry, P., & Towers, N. (2013). Conceptual framework development: CSR implementation in fashion supply chains. *International Journal of Physical Distribution & Logistics Management, 43*(5/6), 478–501. DOI: 10.1108/IJPDLM-03-2012-0107

PETA. (2020). *Burberry and Chanel ban fur*. https://www.peta.org

PETA. (2025). *Beauty without bunnies*. https://crueltyfree.peta.org/

Petravičiūtė, K., Šeinauskienė, B., Rūtelionė, A., & Krukowski, K. (2021). Linking luxury brand perceived value, brand attachment, and purchase intention: The role of consumer vanity. *Sustainability (Basel), 13*(12), 6912. DOI: 10.3390/su13126912

Petrova, S., & Pruzhansky, V. (2011). The Economics of Luxury Goods: Utility based on exclusivity. SSRN *Electronic Journal*. https://doi.org/DOI: 10.2139/ssrn.1930361

Phau, I., & Prendergast, G. (2000). Consuming luxury brands: The relevance of the 'Rarity Principle.'. *Journal of Brand Management, 8*(2), 122–138. DOI: 10.1057/palgrave.bm.2540013

Philiastides, M. G., & Ratcliff, R. (2013). Influence of branding on Preference-Based Decision making. *Psychological Science, 24*(7), 1208–1215. DOI: 10.1177/0956797612470701 PMID: 23696199

Pinello, C., Picone, P. M., & Mocciaro Li Destri, A. (2022). Co-branding research: Where we are and where we could go from here. *European Journal of Marketing, 56*(2), 584–621. DOI: 10.1108/EJM-02-2021-0073

Pinkse, J., Bohnsack, R., & Kolk, A. (2014). The role of public and private protection in disruptive innovation: The automotive industry and the emergence of low-emission vehicles. *Journal of Product Innovation Management, 31*(1), 43–60. DOI: 10.1111/jpim.12079

Pino, G., Amatulli, C., Peluso, A. M., Nataraajan, R., & Guido, G. (2017). Brand prominence and social status in luxury consumption: A comparison of emerging and mature markets. *Journal of Retailing and Consumer Services, 46*, 163–172. DOI: 10.1016/j.jretconser.2017.11.006

Platt, M., Sedlmeir, J., Platt, D., Xu, J., Tasca, P., Vadgama, N., & Ibañez, J. I. (2021, December). The energy footprint of blockchain consensus mechanisms beyond proof-of-work. In *2021 IEEE 21st International Conference on Software Quality, Reliability and Security Companion (QRS-C)* (pp. 1135-1144). IEEE. DOI: 10.1109/QRS-C55045.2021.00168

Porbundarwala, T., Lalji, S., Patil, H., & Dongardive, J. (2023). Impact of AI in personalised health monitoring using wearable devices. *International Journal for Multidisciplinary Research, 5*(5), 7144. Advance online publication. DOI: 10.36948/ijfmr.2023.v05i05.7144

Pound, J. (2025, April 9). China slaps 84% retaliatory tariffs on U.S. goods in response to Trump. CNBC. https://www.cnbc.com/2025/04/09/china-slaps-retaliatory-tariffs-of-84percent-on-us-goods-in-response-to-trump.html

Pozharliev, R., Verbeke, W. J., Van Strien, J. W., & Bagozzi, R. P. (2015). Merely Being with you Increases My Attention to Luxury Products: Using EEG to Understand Consumers' Emotional Experience with Luxury Branded Products. *JMR, Journal of Marketing Research*, *52*(4), 546–558. DOI: 10.1509/jmr.13.0560

Prada Group. (2020). *Re-Nylon: A sustainable future*. https://www.pradagroup.com

Prahalad, C. K., & Ramaswamy, V. (2005). *The future of competition: Co-creating unique value with customers* (Nachdr.). Harvard Business School Press.

Pramanik, S., Dalapati, S., Alam, S., & Roy, T. K. (2018). VIKOR-based MAGDM strategy under bipolar neutrosophic set environment. *Neutrosophic Sets and Systems*, *19*, 57–69.

Pramanik, S., Das, S., Das, R., & Tripathy, B. C. (2023). Neutrosophic BWM-TOPSIS strategy under SVNS environment. *Neutrosophic Sets and Systems*, *56*, 178–189.

Prandelli, E., Wang, Y., & Weijo, H. (2024). Luxury branding and the creator Economy: Emerging challenges and future avenues. *International Journal of Research in Marketing*, *41*(3), 455–467. DOI: 10.1016/j.ijresmar.2024.07.004

Prentice, C., & Loureiro, S. M. C. (2018). Consumer-based approach to customer engagement – The case of luxury brands. *Journal of Retailing and Consumer Services*, *43*, 325–332. DOI: 10.1016/j.jretconser.2018.05.003

Prestini, S., & Sebastiani, R. (2021). Embracing consumer ambivalence in the luxury shopping experience. [HTML]. *Journal of Consumer Behaviour*, *20*(5), 1243–1268. Advance online publication. DOI: 10.1002/cb.1933

Quiles-Soler, C., Martínez-Sala, A. M., & Monserrat-Gauchi, J. (2023). Fashion industry's environmental policy: Social media and corporate website as vehicles for communicating corporate social responsibility. *Corporate Social Responsibility and Environmental Management*, *30*(1), 180–191. DOI: 10.1002/csr.2347

Rachbini, W., Soeharso, S. Y., Nugroho, A. S., Evi, T., Rahmawati, E., & Suyanto, N. (2023). The Impact of Brand Society and Narrative Branding on Brand Image and its Implications on Consumer Purchase Behavior: A Case Study of Luxury Fashion Brands in Jakarta. *Journal of Law and Sustainable Development*, *11*(7), e1387. DOI: 10.55908/sdgs.v11i7.1387

Raffaelli, R. (2018). Technology Reemergence: Creating new value for old technologies in Swiss mechanical watchmaking, 1970–2008. *Administrative Science Quarterly*, *64*(3), 576–618. DOI: 10.1177/0001839218778505

Rahmadani, D., Fauzi, H., Lubis, R. A., & Ariyanti, M. (2022). Study of Neuromarketing in Consumer Behavior due to Product Logos Color Changes Effect. [CEPAT]. *Journal of Computer Engineering Progress Application and Technology*, *1*(02), 41. DOI: 10.25124/cepat.v1i02.5245

Rahman, M. S., Bag, S., Hossain, M. A., Fattah, F. M. A., Gani, M. O., & Rana, N. P. (2023). The new wave of AI-powered luxury brands online shopping experience: The role of digital multisensory cues and customers' engagement. *Journal of Retailing and Consumer Services*, *72*, 103273. DOI: 10.1016/j.jretconser.2023.103273

Ramaswamy, V., & Ozcan, K. (2014). *The Co-Creation Paradigm*. Stanford University Press., DOI: 10.1515/9780804790758

Ramaswamy, V., & Ozcan, K. (2018). Offerings as Digitalized Interactive Platforms: A Conceptual framework and implications. *Journal of Marketing*, *82*(4), 19–31. DOI: 10.1509/jm.15.0365

Ramirez, M. (2024, May 16). Ralph Lauren Corporation: AI use cases 2024. PitchGrade. https://pitchgrade.com/companies/ralph-lauren-corporation-ai-use-cases

Ranfagni, S., Faraoni, M., Zollo, L., & Vannucci, V. (2021). Combining online market research methods for investigating brand alignment: The case of Nespresso. *British Food Journal*, *123*(13), 37–58. DOI: 10.1108/BFJ-06-2020-0462

Rao, A. R., & Ruekert, R. W. (1994). Brand alliances as signals of product quality. *Sloan Management Review*, *36*(1), 87–87.

Rathi, R., Garg, R., Kataria, A., & Chhikara, R. (2022). Evolution of luxury marketing landscape: a bibliometric analysis and future directions. In Journal of Brand Management (Vol. 29, Issue 3, p. 241). Palgrave Macmillan. https://doi.org/DOI: 10.1057/s41262-022-00273-x

Rehman, M., Zelin, T., & Hussain, T. (2025). Influence of consumer satisfaction on brand allegiance: An empirical investigation in Pakistan's safety and luxury automobile sector. *Acta Psychologica*, *252*, 104667. DOI: 10.1016/j.actpsy.2024.104667 PMID: 39701004

Reimann, M., Castaño, R., Zaichkowsky, J., & Bechara, A. (2011). How we relate to brands: Psychological and neurophysiological insights into consumer–brand relationships. *Journal of Consumer Psychology*, *22*(1), 128–142. DOI: 10.1016/j.jcps.2011.11.003

Reinikainen, H., Tan, T. M., Luoma-aho, V., & Salo, J. (2021). Making and breaking relationships on social media: the impacts of brand and influencer betrayals. In *Technological Forecasting and Social Change* (Vol. 171, p. 120990). Elsevier BV., DOI: 10.1016/j.techfore.2021.120990

Reis, R. (2022). Losing the inflation anchor. *Brookings Papers on Economic Activity, 2021*(2), 307–379. DOI: 10.1353/eca.2022.0004

Reis, R., Catarino, A. P., Sousa, B. B., Malheiro, A., & Santos, V. (2023). Brand attachment and agile marketing: responses to the (post-) pandemic context in the sports fashion context. In *Promoting Organizational Performance Through 5G and Agile Marketing* (pp. 42–54). IGI Global Scientific Publishing.

Remy, N., Catena, M., & Durand-Servoingt, B. (2023). Digital inside: Get wired for the ultimate luxury experience. https://www.mckinsey.com/capabilities/growth-marketing-and-sales/our-insights/digital-inside-get-wired-for-the-ultimate-luxury-experience

Reuters. (2025, May 15). Luxury sector faces more gloom as Bain cuts sales forecast. ET BrandEquity. https://brandequity.economictimes.indiatimes.com/news/marketing/luxury-sector-faces-more-gloom-as-bain-cuts-sales-forecast/121184224

Reyes-Menendez, A., Palos-Sanchez, P., Saura, J. R., & Santos, C. R. (2021). Revisiting the impact of perceived social value on consumer behavior toward luxury brands. *European Management Journal, 40*(2), 224–233. DOI: 10.1016/j.emj.2021.06.006

Rissanen, T., & McQuillan, H. (2016). *Zero Waste Fashion Design*. Bloomsbury Publishing. DOI: 10.5040/9781474222297

Rodrigues, C. (2019). Multisensory brand experiences and brand love. In *IGI Global eBooks* (pp. 615–635). https://doi.org/DOI: 10.4018/978-1-5225-9282-2.ch029

Rodrigues, D., Sousa, B., Gomes, S., Oliveira, J., & Lopes, E. (2023). Exploring consumer behavior and brand management in the automotive sector: Insights from a digital and territorial perspective. *Administrative Sciences, 13*(2), 36. DOI: 10.3390/admsci13020036

Rogers, E. M. (2003). *Diffusion of Innovations. Free Press. Tapscott, D., & Tapscott, A. (2018).* Blockchain Revolution. Portfolio.

Romero, P. M., Lim, R. E., & Park, H. (2024). Branded virtual worlds in fashion gaming: Examining the role of brand narratives and social presence in consumer engagement. *Computers in Human Behavior, 108505*. Advance online publication. DOI: 10.1016/j.chb.2024.108505

Rosendo-Rios, V., & Shukla, P. (2022). The effects of masstige on loss of scarcity and behavioral intentions for traditional luxury consumers. *Journal of Business Research, 156*, 113490. DOI: 10.1016/j.jbusres.2022.113490

Royo-Vela, M., & Sánchez, M. P. (2022). Downward price-based luxury brand line extension: Effects on premium luxury buyer's perception and consequences on buying intention and brand loyalty. *European Research on Management and Business Economics, 28*(3), 100198. DOI: 10.1016/j.iedeen.2022.100198

Roy, S. K., Singh, G., Hope, M., Nguyen, B., & Harrigan, P. (2019). The rise of smart consumers: Role of smart servicescape and smart consumer experience co-creation. *Journal of Marketing Management, 35*(15–16), 1480–1513. DOI: 10.1080/0267257X.2019.1680569

Ruiz, C. D., & Cruz, A. G. B. (2023). Unconventional luxury brand collaborations: A new form of luxury consumption among young adults in China. *International Marketing Review, 40*(7), 1–21. DOI: 10.1108/IMR-04-2022-0099

Ruiz, L. E., Pinho, A. C., & Resende, D. N. (2022). 3D printing as a disruptive technology for the circular economy of plastic components of end-of-life vehicles: A systematic review. *Sustainability (Basel), 14*(20), 13256. DOI: 10.3390/su142013256

Ryu, K., Lehto, X. Y., Gordon, S. E., & Fu, X. (2018). Effect of a brand story structure on narrative transportation and perceived brand image of luxury hotels. *Tourism Management, 71*, 348–363. DOI: 10.1016/j.tourman.2018.10.021

Sahne, S. S. K., & Daronkola, H. K. (2025). Enhancing loyalty in luxury fashion with AI: the mediation role of customer relationships. In *Asia-Pacific Journal of Business Administration*. Emerald Publishing Limited., DOI: 10.1108/APJBA-08-2024-0456

Sanli, M. (2024). Digital transformations in the hotel industry (Doctoral dissertation, Long Island University, CW Post Center).

Santos, V., Costa, C., Au-Yong, M., & Sousa, M. (2024). Innovations in luxury tourism: Developing a novel theoretical framework. In Sustainable Tourism: Frameworks, Practices, and Innovative Solutions (pp. 59-76). Springer International Publishing.

Santos, V., Sousa, B., & Simões, J. (2024). Metaverse and NFT Market as AI Disruptive Technology for Digital Luxury Brands. In Business Development via AI and Digitalization: Volume 2 (pp. 1025-1036). Cham: Springer Nature Switzerland. https://doi.org/, researchgate.netDOI: 10.1007/978-3-031-62106-2_77

Santos, J., & Boote, J. (2003). A theoretical exploration and model of consumer expectations, post-purchase affective states and affective behaviour. *Journal of Consumer Behaviour, 3*(2), 142–156. DOI: 10.1002/cb.129

Santos, V., Ramos, P., Sousa, B., Almeida, N., & Valeri, M. (2022). Factors influencing touristic consumer behavior. *Journal of Organizational Change Management, 35*(3), 409–429. DOI: 10.1108/JOCM-02-2021-0032

Sanz-Lopez, F., Gallego-Losada, R., Montero-Navarro, A., & García-Abajo, E. (2024). IS THE DIGITALISATION THE FUTURE OF THE LUXURY INDUSTRY? *Heliyon, 10*(21), e40029. DOI: 10.1016/j.heliyon.2024.e40029 PMID: 39553661

Sarkar, S., & Banerjee, S. (2019). Brand co-creation through triadic stakeholder participation. *European Business Review, 31*(5), 585–609. DOI: 10.1108/EBR-04-2018-0079

Sass, F., Bergen, H., & Keller, M. M. (2024). *Diamonds & disruption: Evaluating brand identity fit in "Inside the Box" vs. "Outside the Box" co-branding strategies—A case study of Tiffany & Co.* LBMG Strategic Brand Management-Masters Paper Series.

Sathakathulla, A. A. (2025). The Influence of Credit Cards on Consumer Buying Behavior among Expats in the UAE. *International Journal of Advances in Engineering and Management, 7*(1), 347–361. DOI: 10.35629/5252-0701347361

Saunders, S. (2022, June 9). Stratasys 3D printed fashion collection debuts at Milan Design Week. 3DPrint.com. https://3dprint.com/291745/collaborative-stratasys-3d-printed-fashion-collection-debuts-at-milan-design-week/

Saura, J. R., Palacios-Marqués, D., & Ribeiro-Soriano, D. (2021). Digital marketing in SMEs via data-driven strategies: Reviewing the current state of research. *Journal of Small Business Management, 61*(3), 1278–1313. DOI: 10.1080/00472778.2021.1955127

Schaefer, M., & Rotte, M. (2007). Favorite brands as cultural objects modulate reward circuit. *Neuroreport, 18*(2), 141–145. DOI: 10.1097/WNR.0b013e328010ac84 PMID: 17301679

Schmitt, R., Rossi, A., Bensoussan, A., Schmitt, R., Rossi, A., & Bensoussan, A. (2022). Digital Transformation of the Omni-personal Customer Journey. Omni-personal Luxury: How to Transform your Luxury Business for the Digital Age, 63-139. https://doi.org/ [HTML]DOI: 10.1007/978-3-030-85769-1_5

Schmitt, B. (2011). The consumer psychology of brands. *Journal of Consumer Psychology, 22*(1), 7–17. DOI: 10.1016/j.jcps.2011.09.005

Scholz, J., & Duffy, K. (2018). We ARe at home: How augmented reality reshapes mobile marketing and consumer-brand relationships. *Journal of Retailing and Consumer Services, 44*, 11–23. DOI: 10.1016/j.jretconser.2018.05.004

Schuck, K. A., Derqui, B., Grimaldi, D., & Mehn, A. (2025). Sustainability and Luxury in Hospitality: Achieving Compatibility for a Better Future. *International Journal of Hospitality & Tourism Administration*, •••, 1–24. DOI: 10.1080/15256480.2025.2476405

Schuetz, S., Kuai, L., Lacity, M. C., & Steelman, Z. (2024). A qualitative systematic review of trust in technology. *Journal of Information Technology*. Advance online publication. DOI: 10.1177/02683962241254392

Semwal, R., Tripathi, N., Tyagi, P., & Singh, A. (2024). Synergising Brands: Licensing and Collaboration Strategies in Tourism. In Albattat, A., Azman, N., Valeri, M., Talib, Z. M., & Tyagi, P. K. (Eds.), *Advances in Hospitality, Tourism, and the Services Industry* (pp. 163–180). IGI Global., DOI: 10.4018/979-8-3693-4171-1.ch007

Sénéchal, S., Georges, L., & Pernin, J. L. (2013). Alliances between corporate and fair trade brands: Examining the antecedents of overall evaluation of the co-branded product. *Journal of Business Ethics*, *124*(3), 365–381. DOI: 10.1007/s10551-013-1875-z

Septianto, F., Seo, Y., Sung, B., & Zhao, F. (2020). Authenticity and exclusivity appeals in luxury advertising: The role of promotion and prevention pride. *European Journal of Marketing*, *54*(6), 1305–1323. DOI: 10.1108/EJM-10-2018-0690

Shabi, S. B. M., Bhuvaneshwari, M., Kanimozhi, S., & Pavithra, S. (2025, February). Artificial Intelligence, Intelligent Manufacturing, and Sustainable Futures in Next-Gen Textiles: A Comprehensive Approach. In *2025 3rd International Conference on Intelligent Data Communication Technologies and Internet of Things (IDCIoT)* (pp. 2144-2149). IEE

Shah, A. M., & Ali, M. (2023). Social media marketing activities and luxury fashion brands in the post-pandemic world. *Asia Pacific Journal of Marketing and Logistics*, *36*(9), 2104–2120. DOI: 10.1108/APJML-10-2022-0863

Shah, D., & Murthi, B. (2020). Marketing in a data-driven digital world: Implications for the role and scope of marketing. *Journal of Business Research*, *125*, 772–779. DOI: 10.1016/j.jbusres.2020.06.062

Shahid, S., Husain, R., Islam, J. U., & Hollebeek, L. D. (2025). Brand love and customer brand engagement for masstige: A cross-cultural perspective. *Journal of Product and Brand Management*, *34*(4), 545–561. DOI: 10.1108/JPBM-07-2024-5370

Shahid, S., Iqbal, M., Saeed, H., Hira, S., Batool, A., & Goyal, A. (2024). Abstract 4136358: Diagnostic accuracy of Apple Watch Electrocardiogram for Atrial Fibrillation: A Systematic Review and Meta-Analysis. *Circulation*, *150*(Suppl_1). Advance online publication. DOI: 10.1161/circ.150.suppl_1.4136358

Shahid, S., & Paul, J. (2021). Intrinsic motivation of luxury consumers in an emerging market. *Journal of Retailing and Consumer Services*, *61*, 102531. DOI: 10.1016/j.jretconser.2021.102531

Shahzad, M. F., Xu, S., An, X., & Javed, I. (2024). Assessing the impact of AI-chatbot service quality on user e-brand loyalty through chatbot user trust, experience and electronic word of mouth. *Journal of Retailing and Consumer Services*, *79*, 103867. DOI: 10.1016/j.jretconser.2024.103867

Shaikh, S., & Gummerum, M. (2025). A luxury consumption perspective on motivations for brand value co-creation in emerging markets. *International Journal of Emerging Markets*. Advance online publication. DOI: 10.1108/IJOEM-10-2023-1637

Shamim, M. A., Dwivedi, P., Aggarwal, P., Dwivedi, R., Anil, A., Saravanan, A., Tiwari, K., Dodiya, R. P., Yadav, I., Sukhija, S., Varthya, S. B., Shukla, R., Singh, S., & Singh, K. (2024). 802-P: A New Dawn beyond Basal Insulin? Dose-Response Meta-analysis for Efficacy and Safety of Tirzepatide Compared with Basal Insulin in Uncontrolled Diabetes Mellitus. *Diabetes*, *73*(Supplement_1), 802-P. Advance online publication. DOI: 10.2337/db24-802-P

Shams, S. M. R., Brown, D. M., & Hardcastle, K. (2025a). Brand Management, Marketing Communication, and Promotion Management for People, Planet, and Profit. In S. M. R. Shams, D. M. Brown, & K. Hardcastle, *Sustainable Marketing* (pp. 227–298). Springer Nature Switzerland. DOI: 10.1007/978-3-031-79130-7_4

Shams, S. M. R., Brown, D. M., & Hardcastle, K. (2025b). Product and Service Development for People, Planet, and Profit. In S. M. R. Shams, D. M. Brown, & K. Hardcastle, *Sustainable Marketing* (pp. 103–167). Springer Nature Switzerland. DOI: 10.1007/978-3-031-79130-7_2

Shan, J., Lu, H., & Cui, A. P. (2022). 1+1>2? Is co-branding an effective way to improve brand masstige? *Journal of Business Research*, *144*, 556–571. DOI: 10.1016/j.jbusres.2022.01.058

Shankar, A., & Jain, S. (2020). Factors affecting luxury consumers' webrooming intention: A moderated-mediation approach. *Journal of Retailing and Consumer Services*, *58*, 102306. DOI: 10.1016/j.jretconser.2020.102306

Shankar, S. (2024). Enhancing energy efficiency in AI-Powered Data Centers: Challenges and solutions. *International Journal of High School Research*, *6*(11), 93–98. DOI: 10.36838/v6i11.15

Shankar, V. (2024). Managing the Twin Faces of AI: A commentary on "Is AI changing the world for better or worse?". *Journal of Macromarketing*, *44*(4), 892–899. Advance online publication. DOI: 10.1177/02761467241286483

Sharma, P., Pradhan, S., & Srivastava, A. (2021). Understanding the luxury purchase intentions of young consumers: A qualitative analysis. *Asia-Pacific Journal of Business Administration*, *13*(4), 452–476. DOI: 10.1108/APJBA-01-2021-0019

Shee Weng, L. (2024). Digital Product Passports: Transforming Industries Through Transparency, Circularity, and Compliance. *Digital Product Passports: Transforming Industries Through Transparency, Circularity, and Compliance (November 10, 2024).*

Shen, B., Choi, T., & Chow, P. (2017). Brand loyalties in designer luxury and fast fashion co-branding alliances. *Journal of Business Research*, *81*, 173–180. DOI: 10.1016/j.jbusres.2017.06.017

Shin, H. H., & Jeong, M. (2022). Redefining luxury service with technology implementation: The impact of technology on guest satisfaction and loyalty in a luxury hotel. *International Journal of Contemporary Hospitality Management*, *34*(4), 1491–1514. DOI: 10.1108/IJCHM-06-2021-0798

Shi, X., Li, F., & Chumnumpan, P. (2020). The use of product scarcity in marketing. *European Journal of Marketing*, *54*(2), 380–418. DOI: 10.1108/EJM-04-2018-0285

Siepmann, C., & Kowalczuk, P. (2021). Understanding continued smartwatch usage: the role of emotional as well as health and fitness factors. In Electronic Markets (Vol. 31, Issue 4, p. 795). Springer Science+Business Media. https://doi.org/DOI: 10.1007/s12525-021-00458-3

Silverstein, M. J., & Fiske, N. (2003). Luxury for the masses. *Harvard Business Review*, *81*(4), 48–57. PMID: 12687919

Singh, B., & Kaunert, C. (2024). Redefining the Nodes of Metaverse and Luxury Hospitality: Unveiling AI and Tech-Driven Solutions in Hospitality and Tourism Industry. In Impact of AI and Tech-Driven Solutions in Hospitality and Tourism (pp. 175-202). IGI Global. DOI: [HTML]DOI: 10.4018/979-8-3693-6755-1.ch010

Singh, J., & Shukla, P. (2024). *Brand management: Principles and applications for effective branding*. Kogan Page Publishers.

Singh, P. (2024). UNVEILING ZARA'S LUXURY PERSONA: A COMPREHENSIVE ANALYSIS OF SOCIAL MEDIA STRATEGIES AND CONSUMER BEHAVIOUR. *Deleted Journal, 147–151*, 147–151. Advance online publication. DOI: 10.62823/7.1(II).6462

Slaton, K., & Pookulangara, S. (2025). Exploring Consumer Use of Digital Product Passports for Secondary Luxury Consumption: A Conceptual Study. *Advances in Consumer Research. Association for Consumer Research (U. S.)*, *2*(1). https://acr-journal.com/article/exploring-consumer-use-of-digital-product-passports-for-secondary-luxury-consumption-a-conceptual-study-902/

Smarandache, F. A. (1998). unifying field in logics: Neutrosophic logic. Neutrosophy, neutrosophic set, neutrosophic probability and statistics. Rehoboth: American Research Press.

Smarandache, F. (2005). Neutrosophic set: A generalization of the intuitionistic fuzzy sets. *International Journal of Pure and Applied Mathematics*, *24*, 287–297.

Smelik, A. M. (2017). *Cybercouture: the fashionable technology of Pauline van Dongen.* Iris van Herpen and Bart Hess.

Solem, B. A. (2016). Influences of customer participation and customer brand engagement on brand loyalty. *Journal of Consumer Marketing*, *33*(5), 332–342. DOI: 10.1108/JCM-04-2015-1390

Som, A. & Blanckaert, C. (2015). The Road to Luxury: The Evolution, Markets, and Strategies of Luxury brand management. Singapore: john Wiley & Sons Singapore Pte Ltd

Song, G., Gazi, M. I., Waaje, A., Roshid, M. M., Karim, R., Rahaman, M. A., Min, Z., & Senathirajah, A. R. B. S. (2025). The Neuromarketing: Bridging neuroscience and marketing for enhanced consumer engagement. *IEEE Access : Practical Innovations, Open Solutions*, *1*, 40331–40353. Advance online publication. DOI: 10.1109/ACCESS.2025.3545742

Song, X., & Bonanni, C. (2024). AI-Driven Business model: How AI-Powered Try-On technology is refining the luxury shopping experience and customer satisfaction. *Journal of Theoretical and Applied Electronic Commerce Research, 19*(4), 3067–3087. DOI: 10.3390/jtaer19040148

Soni, N., & Kumar, S. (2024). What drives new luxury consumption? Application of schema congruity theory and heuristic systematic framework. *Asia Pacific Journal of Marketing and Logistics*, *36*(9), 2213–2233. DOI: 10.1108/APJML-04-2023-0319

Sousa, B., & Rocha, A. T. (2019). The role of attachment in public management and place marketing contexts: A case study applied to Vila de Montalegre (Portugal). *International Journal of Public Sector Performance Management, 5*(2), 189–205. DOI: 10.1504/IJPSPM.2019.099094

Spangler, A. R. (2023). *Integrating Fashion and Technology in Wearable Devices: A Study on Co-Branding Effects* (Doctoral dissertation, University of Pennsylvania).

Sprott, D. E., & Liu, R. L. (2016). Research trends on branding in consumer psychology. *Current Opinion in Psychology, 10*, 124–128. DOI: 10.1016/j.copsyc.2016.01.002

Stake, R. E. (1995). *The art of case study research.* SAGE Publications.

Steffl, J., Ganassali, S., & Emes, J. (2024). Hybrid product branding strategies for brand value creation: Combining and comparing green product innovations, limited editions, and co-branding. *Journal of Product and Brand Management, 33*(8), 1073–1087. DOI: 10.1108/JPBM-04-2023-4472

Stella McCartney. (2025). *Social responsibility & human rights.* https://www.stellamccartney.com/es/en/sustainability/social-sustainability.html

Štrach, P., & Everett, A. M. (2006). Brand corrosion: Mass-marketing's threat to luxury automobile brands after merger and acquisition. *Journal of Product and Brand Management, 15*(2), 106–120. DOI: 10.1108/10610420610658947

Sun, C., Cryder, C., & Rick, S. (2023). A co-branding conundrum: Consumers underuse co-branded credit cards outside of their featured brands. Available at *SSRN*, 4494708. https://doi.org/DOI: 10.2139/ssrn.4494708

Sung, B., Hatton-Jones, S., Teah, M., Cheah, I., & Phau, I. (2020). Shelf-based scarcity as a cue of luxuriousness: An application of psychophysiology. *European Journal of Marketing, 55*(2), 497–516. DOI: 10.1108/EJM-06-2018-0418

Sun, X. (2023). An investigation into the impact of luxury co-branding on the purchasing behavior of Chinese millennial consumers. *Art and Society, 2*(6), 81–107. DOI: 10.56397/AS.2023.12.11

Sun, Y., Wang, R., Bai, H., Whyatt, G., & Ohandi, T. T. (2024). Perceived values of luxury fashion brands—Evidence from Indonesian Millennials. *SN Business & Economics, 4*(12), 169. DOI: 10.1007/s43546-024-00755-x

Surià, M. H. (2024). *Architecture and Design of Luxury Fashion Brands' Flagships and Boutiques: A Comparison Between Western and Eastern Culture* [Master in Fashion Management Thesis, ISEM Fashion Business Schoo, Universidad de Navarra]. https://dadun.unav.edu/entities/publication/2a3eec15-f55b-4b64-9b38-98b24bd983ac

Sutar, S. (2025, April 16). Case study: The impact of tariff wars on luxury fashion brands and their pricing models. LinkedIn. https://www.linkedin.com/pulse/case-study-impact-tariff-wars-luxury-fashion-brands-pricing-sutar-mvaxc

Suzuki, S., & Kanno, S. (2022). The role of brand coolness in the masstige co-branding of luxury and mass brands. *Journal of Business Research, 149*, 240–249. DOI: 10.1016/j.jbusres.2022.04.061

Tai, F. M., & Chuang, S. H. (2014). Corporate social responsibility. *Ibusiness, 6*(3), 117–130. DOI: 10.4236/ib.2014.63013

Tajeddini, K., & Mueller, S. L. (2011). Corporate entrepreneurship in Switzerland: Evidence from a case study of Swiss watch manufacturers. *The International Entrepreneurship and Management Journal, 8*(3), 355–372. DOI: 10.1007/s11365-011-0179-y

Tang, D. (2021). WHAT IS DIGITAL TRANSFORMATION? In EDPACS (Vol. 64, Issue 1, p. 9). Taylor & Francis. https://doi.org/DOI: 10.1080/07366981.2020.1847813

Tang, Y. (2023). Analysis of the Current Chinese Consumer for Luxury Goods - Generation Z. *Lecture Notes in Education Psychology and Public Media, 4*(1), 531–537. DOI: 10.54254/2753-7048/4/2022171

Tapscott, D., & Tapscott, A. (2017). *Blockchain revolution: How the technology behind bitcoin is changing money, business, and the world*. Penguin.

Tarnanidis, T. (2022). Measuring Sustainable Entrepreneurship in the Food Sector. [IJSECSR]. *International Journal of Sustainable Entrepreneurship and Corporate Social Responsibility, 7*(1), 1–8. https://doi.org/DOI:10.4018/IJSECSR.314157. DOI: 10.4018/IJSECSR.314157

Tarnanidis, T. (2023). Understanding reference points to make purchase decisions: Overview, phases, and time-features. *Marketing Science & Inspirations, 18*(4), 2–13. DOI: 10.46286/msi.2023.18.4.1

Tarnanidis, T. (2024a). Exploring the Impact of Mobile Marketing Strategies on Consumer Behavior: A Comprehensive Analysis. *International Journal of Information, Business and Management, 16*(2), 1–17.

Tarnanidis, T. (Ed.). (2024). *Consumer experience and decision-making in the metaverse*. IGI Global. DOI: 10.4018/979-8-3693-4167-4

Tarnanidis, T. K., Papachristou, E., Karypidis, M., & Ismyrlis, V. (2024a). *Driving Green Marketing in Fashion and Retail. Driving Green Marketing in Fashion and Retail*. IGI Global., DOI: 10.4018/979-8-3693-3049-4

Tarnanidis, T. K.Theodore K. Tarnanidis. (2024). Fashion consumers' response to the COVID-19 pandemic and today's future. *International Journal of Science and Research Archive, 12*(02), 648–655. DOI: 10.30574/ijsra.2024.12.2.1286

Tarnanidis, T., & Gkiouzepas, L. (2024). Leveraging Marketing Strategies to Help Consumers Make Decisions Using Metaverse Technologies. In Tarnanidis, T. (Ed.), *Consumer Experience and Decision-Making in the Metaverse* (pp. 158–171). IGI Global., DOI: 10.4018/979-8-3693-4167-4.ch009

Tarnanidis, T., Karypidis, M., Manda, V. K., Bilali, S., & Papachristou, E. (2025). Discovering the Patterns of Slow Fashion Consumption in the Rich Market. In *Sustainable Practices in the Fashion and Retail Industry* (p. 20). IGI Global., https://www.igi-global.com/gateway/chapter/372015 DOI: 10.4018/979-8-3693-9959-0.ch007

Tarnanidis, T., & Manaf, A. H. A. (2024). Research notes on the future of marketing and consumer insights. In *New trends in marketing and consumer science* (pp. 324–336). IGI Global. DOI: 10.4018/979-8-3693-2754-8.ch017

Tarnanidis, T., Papachristou, E., Karypidis, M., & Ismyrlis, V. (Eds.). (2023a). *Social Media and Online Consumer Decision Making in the Fashion Industry*. IGI Global., DOI: 10.4018/978-1-6684-8753-2

Tarnanidis, T., Papachristou, E., Karypidis, M., & Manda, V. K. (2024). Luxury Fashion Consumption and Branding in the Affluent Market. In Tarnanidis, T. K. (Ed.), (pp. 161–172). Advances in Marketing, Customer Relationship Management, and E-Services. IGI Global., DOI: 10.4018/979-8-3693-7535-8.ch008

Tarnanidis, T., Papachristou, E., Karypidis, M., & Manda, V. K. (2025). Luxury fashion consumption and branding in the affluent market. In *Cross-Cultural Perspectives on Luxury Marketing* (pp. 161–172). IGI Global.

Tarnanidis, T., Papachristou, E., Karypidis, M., & Manda, V. K. (2025). Luxury fashion consumption and branding in the affluent market. In Tarnanidis, T., & Gardetti, M. (Eds.), *Cross-cultural perspectives on luxury marketing* (pp. 161–172). IGI Global.

Tarnanidis, T., Papathanasiou, J., Ismyrlis, V., & Kittu Manda, V. (Eds.). (2025c). *Multiple-Criteria Decision-Making (MCDM) Techniques and Statistics in Marketing*. IGI Global., DOI: 10.4018/979-8-3693-9122-8

Tarnanidis, T., Papathanasiou, J., Mareschal, B., & Vlachopoulou, M. (2025f). MCDM Review in marketing and managerial decisions: Practical implications and Future research. *Management Science Letters, 15*(1), 45–54. Advance online publication. DOI: 10.5267/j.msl.2024.3.004

Tarnanidis, T., Papathanasiou, J., Osei-Frimpong, K., & Owusu-Frimpong, N. (2024b). How young consumers are influenced by the valence of positive and negative frames: A cross-cultural perspective. *International Journal of Information and Decision Sciences, 16*(2), 109–133. DOI: 10.1504/IJIDS.2024.139825

Tarnanidis, T., Papathanasiou, J., & Subeniotis, D. (2023e). How can we teach young consumers to become entrepreneurs? Application of specific instruments. *World Review of Entrepreneurship, Management and Sustainable Development, 19*(6), 431–453. DOI: 10.1504/WREMSD.2023.133735

Tarnanidis, T., Papathanasiou, J., Vlachopoulou, M., & Mareschal, B. (2023d). Review and the Use of PROMETHEE Methods in Marketing (Problems). In Tarnanidis, T., Vlachopoulou, M., & Papathanasiou, J. (Eds.), *Influences of Social Media on Consumer Decision-Making Processes in the Food and Grocery Industry* (pp. 196–212). IGI Global., DOI: 10.4018/978-1-6684-8868-3.ch009

Tarnanidis, T., & Tarnanidis, J. (2024). Analyzing the SERVQUAL method for senior high school education's gaps and factors. *European Journal of Educational Management, 7*(4), 227–243. DOI: 10.12973/eujem.7.4.227

Tarnanidis, T., Vlachopoulou, M., & Papathanasiou, J. (Eds.). (2023c). *Influences of social media on consumer decision-making processes in the food and grocery industry*. IGI Global., DOI: 10.4018/978-1-6684-8868-3

Tarquini, A., Mühlbacher, H., & Kreuzer, M. (2022). The experience of luxury craftsmanship–a strategic asset for luxury experience management. [HTML]. *Journal of Marketing Management, 38*(13-14), 1307–1338. DOI: 10.1080/0267257X.2022.2064899

Tasci, A. D., & Guillet, B. D. (2011). It affects, it affects not: A quasi-experiment on the transfer effect of co-branding on consumer-based brand equity of hospitality products. *International Journal of Hospitality Management, 30*(4), 774–782. DOI: 10.1016/j.ijhm.2010.12.009

Teixeira, S. F., Barbosa, B., Cunha, H., & Oliveira, Z. (2021). Exploring the antecedents of organic food purchase intention: An extension of the theory of planning behavior. *Sustainability (Basel), 14*(1), 242. DOI: 10.3390/su14010242

Terenteva, N., & Vovchenko, D. (2023). FEATURES OF THE SPACE OF NEUROMARKETING SENSORY TECHNOLOGIES. *Aktual'ni Problemi Ekonomìki, 1*(268), 23–30. DOI: 10.32752/1993-6788-2023-1-268-23-30

Thani, S., & Sharma, A. (2021). The Psychology behind Luxury Consumption. In *Journal of Pharmaceutical Research International* (p. 341). Sciencedomain International., DOI: 10.9734/jpri/2021/v33i57B34065

Ting, L. (2023). Factors that Enhance Consumer Purchase Intention of Luxury Brand Collaboration Products. *Korea Industrial Technology Convergence Society, 28*(2), 115–129. https://doi.org/DOI: 10.29279/jitr.2023.28.2.115

Tran, T., Taylor, D. G., & Wen, C. (2022). Value co-creation through branded apps: Enhancing perceived quality and brand loyalty. *Journal of Research in Interactive Marketing, 17*(4), 562–580. DOI: 10.1108/JRIM-04-2022-0128

Truong, Y., Simmons, G., McColl, R., & Kitchen, P. J. (2008). Status and conspicuousness–are they related? Strategic marketing implications for luxury brands. *Journal of Strategic Marketing, 16*(3), 189–203. DOI: 10.1080/09652540802117124

Tsaples, G., & Tarnanidis, T. (2020). A System Dynamics Model and Interface for the Simulation and Analysis of Milk Supply Chains. In I. Management Association (Ed.), Supply Chain and Logistics Management: Concepts, Methodologies, Tools, and Applications (pp. 108-135). IGI Global Scientific Publishing. https://doi.org/ DOI: 10.4018/978-1-7998-0945-6.ch006

Tschorn, A. (2015, August 25). Ralph Lauren's biometric-sensing smart shirt hits retail Thursday. Los Angeles Times. https://www.latimes.com/fashion/alltherage/la-ar-ralph-lauren-s-biometric-sensing-smart-shirt-hits-retail-aug-27-20150825-story.html

Tsicalos, Z. (2023). The Fashion Industry, the Environment & Sustainability: A Call for an International Treaty and What This Means for the Future of the Fashion Industry Notes and Student Works. *Journal of International Business and Law, 23*(1), 7.

Twarogal, P., & Puczyk, A. (2025, July 4). AI in fashion retail industry: Why retailers must break free from old habits. Neontri. https://neontri.com/blog/ai-fashion-retail/

Tynan, C., McKechnie, S., & Chhuon, C. (2010). Co-creating value for luxury brands. *Journal of Business Research, 63*(11), 1156–1163. DOI: 10.1016/j.jbusres.2009.10.012

Ukanwa, K. (2024). Algorithmic bias: Social science research integration through the 3-D Dependable AI Framework. *Current Opinion in Psychology, 58*, 101836. DOI: 10.1016/j.copsyc.2024.101836 PMID: 38981371

Umarova, N., & Jurayeva, M. (2024). NEURO-LINGUISTIC PROGRAMMING ELEMENTS IN ADVERTISING: CRAFTING PERSUASIVE DISCOURSE. *CURRENT RESEARCH JOURNAL OF PHILOLOGICAL SCIENCES*, *5*(10), 42–49. DOI: 10.37547/philological-crjps-05-10-08

Vacca, A., Di Sorbo, A., Visaggio, C. A., & Canfora, G. (2021). A systematic literature review of blockchain and smart contract development: Techniques, tools, and open challenges. *Journal of Systems and Software*, *174*, 110891. DOI: 10.1016/j.jss.2020.110891

Vanhamme, J., Lindgreen, A., & Sarial-Abi, G. (2021). Luxury Ethical Consumers: Who Are They? In Journal of Business Ethics (Vol. 183, Issue 3, p. 805). Springer Science+Business Media. https://doi.org/DOI: 10.1007/s10551-021-04981-3

Varan, N. (2020). *The effect of brand equity on intrinsic and extrinsic motivations of consumers for luxury consumption tendency* (Master's thesis).

Vargo, S. L., Maglio, P. P., & Akaka, M. A. (2008). On value and value co-creation: A service systems and service logic perspective. *European Management Journal*, *26*(3), 145–152. DOI: 10.1016/j.emj.2008.04.003

Varsha, P. S., Akter, S., Kumar, A., Gochhait, S., & Patagundi, B. (2021). The impact of artificial intelligence on branding: A bibliometric analysis (1982-2019). [JGIM]. *Journal of Global Information Management*, *29*(4), 221–246. DOI: 10.4018/JGIM.20210701.oa10

Venturini, A. (2023). *Understanding the Consumer-Brand Relationship in the Hyperconnected Environment* [PhD program in Communication, Markets and Society, IULM]. https://apeiron.iulm.it/bitstream/10808/64689/2/venturini_brands_over_heels.pdf

Vinerean, S., & Opreana, A. (2019). Social media marketing efforts of luxury brands on Instagram. *Expert Journal of Marketing*, *7*(2), 144–152.

Vogue Business. (2019, November 5). *Burberry's partnership with The RealReal signifies a real shift.* https://www.voguebusiness.com/companies/burberrys-partnership-realreal-secondhand

Walter, C. E., Vale, V. T., Au-Yong-Oliveira, M., Veloso, C. M., & Sousa, B. B. (2023). "The Dark Side of the Brand" and Brand Hate: A Review and Future Research Agenda. *Administrative Sciences*, *13*(11), 234. DOI: 10.3390/admsci13110234

Wang, Y. (2022). A conceptual framework of contemporary luxury consumption. International Journal of Research in Marketing. https://doi.org/ sciencedirect.com-DOI: 10.1016/j.ijresmar.2021.10.010

Wang, Z., Yuan, R., Luo, J., & Liu, M. J. (2022). Redefining "masstige" luxury consumption in the post-COVID era. Journal of Business Research. https://doi.org/, nih.govDOI: 10.1016/j.jbusres.2022.01.057

Wang, R. (2023). Analysis of marketing strategies of Louis Vuitton based on the marketing theory of 4Ps. Advances in Economics. *Management and Political Sciences*, *18*, 140–145.

Wang, R. W. Y., Chang, Y., & Chuang, S. (2016). EEG Spectral Dynamics of Video Commercials: Impact of the narrative on the branding product preference. *Scientific Reports*, *6*(1), 36487. Advance online publication. DOI: 10.1038/srep36487 PMID: 27819348

Wang, X., Sung, B., & Phau, I. (2021). Examining the influences of perceived exclusivity and perceived rarity on consumers' perception of luxury. *Journal of Fashion Marketing and Management*, *26*(2), 365–382. DOI: 10.1108/JFMM-12-2020-0254

Wang, X., Yin, H., & Ma, D. (2023). Understanding consumer reference price behavior in co-branding: From the perspective of a dynamic supply chain. *Electronic Commerce Research*. Advance online publication. DOI: 10.1007/s10660-023-09759-5

Wang, Y., & Griskevicius, V. (2013). Conspicuous consumption, relationships, and rivals: Women's luxury products as signals to other women. *The Journal of Consumer Research*, *40*(5), 834–854. DOI: 10.1086/673256

Wang, Z. (2024). Heytea: Revolutionizing the Beverage Industry With Innovation and Customer Engagement. In Elhaoussine, Y., Appelgryn, H., & Wang, L. (Eds.), (pp. 19–42). Advances in Business Strategy and Competitive Advantage. IGI Global., DOI: 10.4018/979-8-3693-2921-4.ch002

Weatherill, S. (2017). The several internal markets. *Yearbook of European Law*, *36*, 125–178. DOI: 10.1093/yel/yex007

Wei, Y. (2022). *A study of the luxury brand marketing strategy and its influence on consumers*. Advances in Economics, Business and Management Research/Advances in Economics, Business and Management Research., DOI: 10.2991/aebmr.k.220603.046

Wei, Y., & Li, F. (2020). Omnichannel supply chain operations for luxury products with conspicuous consumers. In *Transportation Research Part E Logistics and Transportation Review* (Vol. 137, p. 101918). Elsevier BV., DOI: 10.1016/j.tre.2020.101918

Wiedmann, K., Labenz, F., Haase, J., & Hennigs, N. (2017). The power of experiential marketing: Exploring the causal relationships among multisensory marketing, brand experience, customer perceived value and brand strength. *Journal of Brand Management*, *25*(2), 101–118. DOI: 10.1057/s41262-017-0061-5

Wirtz, J., Holmqvist, J., & Fritze, M. P. (2020). Luxury services. In Journal of service management (Vol. 31, Issue 4, p. 665). Emerald Publishing Limited. https://doi.org/DOI: 10.1108/JOSM-11-2019-0342

Wittstruck, D., & Teuteberg, F. (2012). Integrating the concept of sustainability into the partner selection process: a fuzzy-AHP-TOPSIS approach. In International Journal of Logistics Systems and Management (Vol. 12, Issue 2, p. 195). Inderscience Publishers. https://doi.org/DOI: 10.1504/IJLSM.2012.047221

WTO & UdSi (Ed.). (2025). *Fashion and Cultural Tourism – Connecting Creators, Businesses and Destinations*. UN Tourism., DOI: 10.18111/9789284426287

Wu, Y. (2022, April). Research on the factors influencing the purchasing behavior of young luxury consumers: a case. In 2022 7th International Conference on Social Sciences and Economic Development (ICSSED 2022) (pp. 1338-1343). Atlantis Press. DOI: 10.2991/aebmr.k.220405.223

Wu, D., Ren, Z., Lei, H., & Zhao, S. (2025). Life cycle assessment of a luxury fashion brand's HPP event booth in the context of circular economy. *Scientific Reports*, *15*(1), 5153. DOI: 10.1038/s41598-024-84528-4 PMID: 39934184

Wu, S., & Kim, S. Y. (2023). The Co-creation of Luxury Brand Hedonic Experiences in Social Media: Luxury Brand Strategies from a Design Perspective. *Archives of Design Research*, *36*(3), 27–47. DOI: 10.15187/adr.2023.08.36.3.27

Xiang, D., & Cho, J. (2024). Impact of Yayoi Kusama's art on Commercial Design: A case study of brand Collaborations. *Asia-pacific Journal of Convergent Research Interchange*, *10*(9), 469–485. DOI: 10.47116/apjcri.2024.09.38

Xie, P., Shi, R., & Xu, D. (2023). Retailer service strategy on livestreaming platforms considering free riding behavior. In Annals of Operations Research. Springer Science+Business Media. https://doi.org/DOI: 10.1007/s10479-023-05201-z

Xie, Q., & Lou, C. (2020). Curating luxe experiences online? Explicating the mechanisms of luxury content marketing in cultivating brand loyalty. *Journal of Interactive Advertising*, *20*(3), 209–224. DOI: 10.1080/15252019.2020.1811177

Xue, J. (2024). The influence of brand culture and effects on consumer behavior. *SHS Web of Conferences, 199*, 03010. https://doi.org/DOI: 10.1051/shsconf/202419903010

Xue, M. T., & Chawdhary, R. (2023). Co-branding as a masstige strategy for luxury brands: Desirable or not? *Journal of Business Research, 158*, 113704. DOI: 10.1016/j.jbusres.2023.113704

Xu, L., & Mehta, R. (2022). Technology devalues luxury? Exploring consumer responses to AI-designed luxury products. *Journal of the Academy of Marketing Science, 50*(6), 1135–1152. DOI: 10.1007/s11747-022-00854-x

Yadav, T. (2024). AI in Neuromarketing: Understanding Consumer Emotions and Behavior through Machine Learning. *International Journal for Multidisciplinary Research, 6*(5), 28154. Advance online publication. DOI: 10.36948/ijfmr.2024.v06i05.28154

Yang, J., & Mundel, J. (2021). Effects of brand feedback to negative eWOM on brand love/hate: An expectancy violation approach. *Journal of Product and Brand Management, 31*(2), 279–292. DOI: 10.1108/JPBM-05-2020-2900

Yassin, C. A., & Soares, A. M. (2023). Understanding impulse buying: A two-faceted tale. *Journal of Consumer Marketing, 40*(7), 884–896. DOI: 10.1108/JCM-09-2020-4116

Yeo, P. (2021). Exhibiting Transformative Fashion: Digital Interventions to Enhance Display and Interpretation. *Costume, 55*(1), 97–120. DOI: 10.3366/cost.2021.0184

Yin, R. K. (2018). *Case study research and applications: Design and methods* (6th ed.). SAGE Publications.

Yoganathan, V., Osburg, V., & Bartikowski, B. (2021). Building better employer brands through employee social media competence and online social capital. *Psychology and Marketing, 38*(3), 524–536. DOI: 10.1002/mar.21451

Youn, S., Hwang, J., & Ju, N. (2025). Blockchain transparency in reducing second-hand luxury shopping uncertainty and the role of price consciousness. *Fashion and Textiles, 12*(1), 10. DOI: 10.1186/s40691-025-00420-3

Younus, M. (2025). The Economics of A Zero-Waste Fashion Industry: Strategies To Reduce Wastage, Minimize Clothing Costs, And Maximize & Sustainability. *Strategic Data Management and Innovation, 2*(01), 116–137. DOI: 10.71292/sdmi.v2i01.15

Your Majesty Co. (2024). *State of digital in fashion 2024*. https://yourmajesty.co/article/state-of-digital-in-fashion-2024

Yu, J., & Ko, E. (2021). UGC attributes and effects: implication for luxury brand advertising. In International Journal of Advertising (Vol. 40, Issue 6, p. 945). Taylor & Francis. https://doi.org/DOI: 10.1080/02650487.2021.1898773

Yuan, Y., & Xiao, T. (2022). Retailer's decoy strategy versus consumers' reference price effect in a retailer-Stackelberg supply chain. *Journal of Retailing and Consumer Services*, *68*, 103081. DOI: 10.1016/j.jretconser.2022.103081

Yu, B. (2021). How consumer opinions are affected by marketers: An empirical examination by deep learning approach. *Journal of Research in Interactive Marketing*, *16*(4), 601–614. DOI: 10.1108/JRIM-04-2021-0106

Yu, H., Lee, D., Cottingham, M., & Maeng, L. (2019). Consumer perceptions to luxury co-branding partnership in sport wearable market. *INTERNATIONAL JOURNAL OF HUMAN MOVEMENT SCIENCE*, *13*(03), 17–29. DOI: 10.23949/ijhms.2019.12.13.3.2

Yu, J., & Wang, J. (2024). Analysis of the Impact of luxury marketing methods on customer Consumption Psychology. *Advances in Economics Management and Political Sciences*, *81*(1), 76–83. DOI: 10.54254/2754-1169/81/20241325

Yu, Y., Rothenberg, L., & Moore, M. (2021). Exploring young consumers' decision-making for luxury co-branding combinations. *International Journal of Retail & Distribution Management*, *49*(3), 341–358. DOI: 10.1108/IJRDM-12-2019-0399

Zadeh, L. A. (1965). Fuzzy sets. *Information and Control*, *8*(3), 338–353. DOI: 10.1016/S0019-9958(65)90241-X

Zakiyah, N. T. A. (2023). The impact of AI-Driven approaches on product branding in ethical business practices. *SAMMAJIVA Jurnal Penelitian Bisnis Dan Manajemen*, *1*(2), 299–306. DOI: 10.47861/sammajiva.v1i2.949

Zancanella, G., Mazzù, M. F., & Baccelloni, A. (2024). Co-branding masstige within luxury brands: An empirical investigation based on grounded theory. In Atti Convegno.

Zein, S. E., Consolacion-Segura, C., & Huertas-Garcia, R. (2019). The role of sustainability in brand equity value in the financial sector. *Sustainability (Basel)*, *12*(1), 254. DOI: 10.3390/su12010254

Zhang, L. & Erturk, E. (2022). Potential lessons from Chinese businesses and platforms for online networking and marketing: An exploratory study. Social Sciences & Humanities Open. https://doi.org/ sciencedirect.comDOI: 10.1016/j.ssaho.2022.100274

Zhang, Q., Chen, J., & Lin, J. (2021). Market targeting with social influences and risk aversion in a co-branding alliance. *European Journal of Operational Research, 297*(1), 301–318. DOI: 10.1016/j.ejor.2021.05.022

Zhang, Y. (2024). Attention-based multilayer linear diffusion model. *Academic Journal of Computing & Information Science, 7*(4). Advance online publication. DOI: 10.25236/AJCIS.2024.070408

Zhang, Y. (2025). Cross-border co-branding influencing consumer purchases: Case of Luckin Coffee's Moutai Flavored Latte. In Chew, F. P. (Ed.), *Global Dialogue on Media Dynamics, Trends and Perspectives on Public Relations and Communication* (p. 6). CRC Press., DOI: 10.1201/9781003564966

Zhang, Y., Guo, X., & Qu, Y. (2025). Brand extension or co-branding: The consumer's attribution of responsibility to the crossover strategies of heritage brands. *Journal of Retailing and Consumer Services, 82*, 104133. DOI: 10.1016/j.jretconser.2024.104133

Zhao, Y. (2024). The study of the impact of the crossover co-branding marketing strategy in luxury firms. *Journal of Economics Business and Management*, 383–391. https://doi.org/DOI: 10.18178/joebm.2024.12.4.826

Zhou, Y., & Zhao, J. (2024). Leveraging computational systems for lifecycle management and enhancement of circular economy in fashion: A study on tracking, recycling, and reuse technologies. *Applied and Computational Engineering, 87*(1), 148–153. DOI: 10.54254/2755-2721/87/20241531

Zhu, D. (2022). The strategic and financial analysis of Hermès. *BCP Business & Management, 26*, 733–740. DOI: 10.54691/bcpbm.v26i.2033

Zhu, L. (2024). Marketing Strategy Analysis based on Consumer Psychology-Taking Tiffany as an example. *Advances in Economics Management and Political Sciences, 69*(1), 271–276. DOI: 10.54254/2754-1169/69/20231448

Zhu, Y., & Kanjanamekanant, K. (2020). No trespassing: Exploring privacy boundaries in personalized advertisement and its effects on ad attitude and purchase intentions on social media. *Information & Management, 58*(2), 103314. DOI: 10.1016/j.im.2020.103314

Zhu, Z. (2023). The impact of digital transformation on the fashion industry in the Post-Pandemic era. *BCP Business & Management, 44*, 696–705. DOI: 10.54691/bcpbm.v44i.4920

Zniva, R., Weitzl, W., & Lindmoser, C. (2023). Be constantly different! How to manage influencer authenticity. In Electronic Commerce Research (Vol. 23, Issue 3, p. 1485). Springer Science+Business Media. https://doi.org/DOI: 10.1007/s10660-022-09653-6

Zuo, Y. (2024). Exploring the Synergy: AI Enhancing Blockchain, Blockchain Empowering AI, and their Convergence across IoT Applications and Beyond. *IEEE Internet of Things Journal, 1*. Advance online publication. DOI: 10.1109/JIOT.2024.3507746

About the Contributors

Theodore Tarnanidis is a marketing scholar, adjunct professor at the International Hellenic University and researcher in Applications of D.Sc. and MCDA. Theodore has six years experience as a marketing and decision making practitioner. He made his post-doc research in the area of sustainable entrepreneurship from the University of Macedonia. He obtained a Ph.D. from the University of London Met., UK. He received his M.B.A from Liverpool University, UK and is a graduate from the University of Macedonia (Business Administration) and Alexander Technological Educational Institute (Marketing). His research focuses on International Marketing, Multi-cultural Marketing, Marketing Management, Consumer Science by Means of Rank-Coded Data, Preference Measurement Techniques, Quantitative Methods & Structural Equation Modelling. His work has been published in various internationally renowned scientific conferences (Academy of Marketing, European Marketing Academy, PROMETHEE Days, Hellenic Operational Research) and in journals (Journal of Business Ethics, World Review of Entrepreneurship, Management and Sustainable Development, Journal of Retailing and Consumer Services, Current Issues in Tourism, Marketing Science & Inspirations, European Journal of Educational Management, Management Science Letters).

Bruno Sousa is an Professor Coordenador of Marketing at Polytechnic University of Cávado and Ave (IPCA), Portugal and holds a PhD in Marketing and Strategy in Universidade do Minho, Portugal. Bruno Sousa is also Head of Masters Program - Tourism Management and Tourism Marketing (IPCA); CiTUR – Center for Tourism Research, Development and Innovation and UNIAG research member. Bruno Sousa teaching fields include marketing, strategy and tourism. Bruno Sousa research areas include brand management and consumer behavior, strategy implementation, niche marketing, niche tourism and brand attachment. He has published in the Journal of Enterprising Communities, Tourism

Management Perspectives, Current Issues in Tourism, Journal of Organizational Change Management, World Review of Entrepreneurship, Management and Sust. Development, among others. Email: bsousa@ipca.pt,

Maro Vlachopoulou is a Professor at the Department of Applied Informatics, Dean of the School of Information Sciences, University of Macedonia, Greece. Degree in Business Administration, Aristotle University of Thessaloniki, Greece (1978). Degree in Law, Aristotle University of Thessaloniki, Greece (1989). Postgraduate studies in Marketing, University of Mannheim, Germany (1981), and MBS in Business Administration, Aristotle University of Thessaloniki (1987). PhD in Marketing Information Systems, University of Macedonia, Department of Applied Informatics, Greece (1992). Visiting research professor at the University of Sunderland, UK (1995). Her professional expertise, research and teaching interests include: Digital Marketing, Marketing Information Systems, E-Business/ e-marketing models, emarketplaces, E-Business plan, new technologies and marketing, mobile Business / marketing, SEO & SEM, social media / networks and marketing, ERP (Enterprise Resource Planning), CRM (Customer Relationship Management) systems, Supply Chain Management (SCM) Systems, Knowledge management systems, e-supply chain management / logistics, with particular emphasis on specific applications (agrifood, e-trust, e-health, government, tourism, environment). She has published over 100 papers in scientific journals and organized numerous of conferences. Email: mavla@uom.edu.gr

Evridiki Papachristou is an Assistant Professor at International Hellenic University, School of Design Sciences, Faculty of Creative Design andClothing. She is also a research member of the CAD/CAM Lab (Technical University of Crete-School of Production Engineering & Management). Recent research projects involve a Conceptual Fashion Product Design Assisted by Artificial Intelligence - AI-CFPD, andi-mannequin a Digital Platform for the Design and Rapid Garment Prototyping - both co-financed by Greece and the EU. After many years spent researching and practicing the effective integration of 3D virtual prototype in the apparel industry, she is currently conducting applied research on the intersection of fashion and data mining and knowledge discovery, to further the technology and its adoption. Evridiki is also interested in the role of Industry 4.0 in creating institutional, social and cultural change towards more sustainable and circular practices. She is the scientific responsible for an Erasmus+ Capacity Building program, 3DGarT (3D Garment Training) and co-ordinator of Erasmus+ Capacity Building program 3D4U (3D for Ukraine). Her work has been published in various scientific journals (International Journal of Advanced Manufacturing Technologies, Journal of Textile Science & Engineering, Machines, Journal of Sustainable Engineering, Technologies,

Journal of Textile Engineering & Fashion Technology) and in numerous international conferences (AUTEX, FAIM, PI APPAREL, SETN, 3DBodyTech, DFI, ICMI, DCAC). She has also appeared as a 3D expert in WhichPLM.com, a fashion technology editorial. Email:evridikipapa@ihu.gr

Vijaya Kittu Manda has nearly 13+ years of experience in capital markets, financial planning, and investing. He is a Researcher at PBMEIT. He is an Advocate, a technocrat, an academician, a book writer, and a stock market enthusiast. He has 11 University Postgraduate Degrees in various disciplines. He is a Ph.D. in Management (Finance). His thesis was on Mutual Funds and their Market Competition. His thesis won the prestigious NSE-IEA Best Thesis Award in 2023. He is currently pursuing his second Ph.D. in Computer Science with focus on Blockchain. He has over 110+ Google Scholar publications and 785+ magazine article publications so far. He is the Chief Editor for a Management Book Series and the Editor of several Scopus-indexed books. He is a Peer Reviewer and is a Certified Peer Review Supervisor. He frequently reviews for leading publishers including IEEE, Springer, Emerald, Sage, amongst others. He sits as a Session Chair and an Advisor for various academic and industrial conferences. He is a guest speaker for colleges and universities. He serves as the host for a research talk show called PBMEIT Expert Insights.

***

Andi Subhan Amir is a faculty member at the Department of Communication Science, Faculty of Social and Political Sciences, Hasanuddin University. He earned his Bachelor's degree in Communication Science from Universitas Hasanuddin, followed by a Master's degree in Mass Communication from the university's Graduate School. He completed his doctoral studies at the Faculty of Communication Science, Universitas Padjadjaran.

Noelia Araújo Vila holds a Ph.D. in Tourism Management and Planning (2013) and a Bache-lor's degree in Business Administration and Management (2006). She also holds a Master's degree in SME Management, earning the award for the best master's thesis. She is a Full Professor in the Department of Business Organization and Marketing, Universi-ty of Vigo. She also teaches in the Master's Program in Tourism Management and Planning for Inland and Health Tourism. She has participated in various research projects related to the tourism and business sec-tors and is a member of the Emitur (Marketing and Tourism) research group at the University of Vigo. She is part of the Management Board of REDINTUR (Interuniversity Network of Tour-ism Postgraduate Programs) and the Executive Board of AECIT (Spanish Association of Sci-entific Tourism Experts).

Suaedi Fachruddin suaedif@gmail.com 0812 833 8583 Universitas Cokroaminoto Palopo Jl. Latamacelling, Tompotika, Kec. Wara, Kota Palopo, Sulawesi Selatan 91911 Dr. Suaedi, S.Pd., M.Si., born on December 31, 1969, in Suli, is a distinguished academic and educator currently serving as the Director of Politeknik Dewantara since 2023. He possesses an extensive educational background, including a Doctorate in Natural Resource and Environmental Management from Institut Pertanian Bogor (2007), a Master's degree in Resource Economics from Universitas Hasanuddin (2000), and a Bachelor's in Mathematics Education from IKIP Ujung Pandang (1993). With a career marked by significant leadership roles, Dr. Suaedi has previously held positions such as Rector of Universitas Cokroaminoto Palopo and Director of Akademi Teknologi Industri Dewantara. He has actively contributed to various educational associations, including serving as the Chairman of the Indonesian Lecturer Association for Tana Luwu and the Association of Private Higher Education Institutions in Sulawesi, Maluku, and Papua. Dr. Suaedi's research interests focus on sustainable development and education, leading to numerous publications in reputable journals. His commitment to enhancing education quality and promoting sustainable practices underscores his dedication to both academic excellence and societal advancement. His contributions to local governance and policy formulation further reflect his integral role in fostering educational reform and community engagement.

Ahmad Salman Farid is a Lecturer at the Islamic Broadcasting Communication Study Program in STAIN Mandailing Natal in North Sumatra, Indonesia. He is a prolefic researcher with interests on advertising marketing, journalism, islamic communication, environmental and sustainability issues, and media management. He sits as Session Chair at various International Conference. He is also an Editor for various diversified international journals.

Venkatesh Ganapathy a seasoned academician with over two decades experience in the corporate world. Mentored thousands of students pursuing MBA in soft skills and pre-placement. Teaching courses in marketing, strategy and supply chain management.

Jomon Jose M is a passionate academician, researcher, and thought leader with a deep interest in the intersections of consumer behavior, sustainability, marketing innovation, and human resource development. With extensive experience in teaching and academic research, Jomon has contributed significantly to management education by developing and delivering insightful lectures on entrepreneurship, marketing, human resource management, project management, and financial systems.

Jyothis Rachel Mathews is an accomplished academician serving as an Assistant Professor with expertise in the fields of business management and commerce. Known for her dedication to student development and academic excellence, she brings a blend of theoretical knowledge and practical experience into the classroom, fostering an engaging learning environment.

Harman Preet is a Assistant Professor at Chandigarh Group of Colleges, Jhanjeri, Mohali, Punjab, specializing in Marketing & Finance. With a background in Academics & Research, she has contributed to research in Augmented Reality & Digitization in Transforming life. Her current work focuses on Co Branding Strategies for Smart Luxury Products. Harmanpreet has published in peer-reviewed journals and presented at international conferences, aiming to bridge the gap between academic research and real-world applications.

Rismawati is a distinguished academic and consultant with a robust focus on sustainability reporting. She currently holds the position of lecturer and Head of the Quality Assurance Institute at Universitas Muhammadiyah Palopo, where she has been pivotal in ensuring the quality of university processes and outputs since February 2020. Additionally, she serves as an Assurer for Sustainability Reporting at the National Center for Sustainability Reporting in Jakarta, specializing in the evaluation of sustainability reports for mining companies across Indonesia. Her academic qualifications are extensive, comprising a Bachelor's degree in Economics and Accounting, a Master's degree in Accounting Science, and a Doctorate in Accounting from Universitas Brawijaya, Malang. Rismawati is also a certified specialist in sustainability reporting, risk management, and tax planning, holding several prestigious certifications including Certified Sustainability Reporting Specialist (CSRS), Certified Risk Management Practitioner (CRMP), Certified Management Accountant (CMA) and Certified Enterprise Risk Analysis (CERA). In her professional capacity, she has provided consultancy services in financial management and tax compliance for companies such as PT. Reski Barito Timur and PT. Helay Raya. Furthermore, she has authored multiple publications on Social Accounting, Sustainability Accounting, Management Accounting and CSR, contributing significantly to the academic community. Her interests in reading, writing, and travel reflect her commitment to lifelong learning and professional growth in her field.

Lucía Rubio Escuderos is a sociologist, researcher, and lecturer at the University of Valencia. She holds a Master's Degree in Tourism Management and Planning and a PhD in Tourism. Her academic career includes international and national research stays, notably at the University of Derby (United Kingdom, 2018) and the University of Vigo, where she conducted research as a Margarita Salas postdoctoral fellow (2023–2024). Her research focuses on social sustainability and social tourism, with a particular emphasis on accessible tourism for people with disabilities and older adults, approached from a sociological perspective.

Sushila Sahani, is an assistant professor in the School of Fashion Design and Technology at Lovely Professional University (India). She holds a doctorate in Apparel and Textile Science. Her academic and professional journey integrates expertise in textiles, ranging from material properties to sustainable innovations. Her research interests extend to fashion design, intellectual property rights, and ethical consumerism. Her research and educational work explore the connections between textile knowledge and design thinking, highlighting the economic, cultural, and sustainable impact of the fashion and textile industry.

Estela Vilhena graduated in 1997 in Applied Mathematics and Computer Science from the University of Aveiro. In 2003 acquired the degree of Master of Applied Statistics and Modelling from the Faculty of Engineering of Porto University and in 2014 acquired the title of PhD in Biomedical Sciences from the Institute of Biomedical Sciences Abel Salazar, University of Porto. Between 2014 and 2024 was a member of the research group Biostatistics and Measurements in Health Research Unit in Epidemiology, at ISPUP. Since 2000, she has been at the Polytechnic Institute of Cávado and Ave - IPCA, Portugal, where she is an Associate Professor at the Department of Applied Sciences and Industrial Engineering of the School of Technology. She is an Integrated Member / Researcher at 2Ai - Applied Artificial Intelligence Laboratory of the School of Technology (since 2018) and Member of the Intelligent Systems Associate Laboratory - LASI (since 2021). She is the director of the Master in Integrated Systems for Environmental, Quality, and Safety Management (since 2017). Her investigation is focused on developing, implementing, and disseminating mathematical and statistical methods of analysis in public health, specifically, statistical modelling and measuring health. Structural equation models are the focus of her theoretical models. The research interest lines are also on Statistics Applied to Human-Machine Interaction: cognitive science and applications; artificial intelligence; and cognitive psychology.

Anuradha Yadav is an Assistant Professor at REVA University, Bengaluru, India. She is earlier with the Department of Management Studies at Dayananda Sagar College of Engineering (DSCE). She is an MBA, MA, PGDJM and a Ph.D. in Marketing. Her teaching and research focus is on Product and Brand Management, Strategic Management, and Marketing Management. She guided several UG and PG students on Marketing Management topics. She is a prolific author of several research papers and book chapters. She frequently presents her research at Conferences, organizes conferences, and is an Associate Editor. She also acts as a Conference Chairperson and Resource Person at various Research Conferences.

Index

A

AI Personalization 275, 282, 298

B

Behavioral Economics 105, 117, 121, 137
brand attachment 29, 71, 295, 315, 316, 317, 320, 321, 322, 323, 325, 327, 330, 331, 332
Brand collaboration 71, 162, 232, 233
Brand Differentiation 74, 327
Brand Dilution 5, 7, 9, 11, 13, 16, 17, 19, 35, 38, 50, 53, 55, 56, 81, 82, 137, 228, 230, 279
Brand Equity 1, 4, 7, 8, 11, 19, 20, 23, 24, 31, 36, 38, 44, 45, 46, 47, 50, 53, 55, 56, 59, 62, 75, 79, 80, 83, 90, 91, 92, 96, 102, 110, 128, 132, 133, 146, 162, 165, 195, 200, 204, 210, 248, 260, 278, 286, 291, 292, 293, 296, 298, 300, 301, 305, 311, 313, 321, 329
Brand Strategies 151, 164, 259, 282

C

Circular Economy 4, 85, 86, 88, 92, 93, 96, 165, 170, 172, 180, 181, 182, 186, 187, 188, 199, 201, 248, 265, 267, 276, 283, 288
Co-Branding 1, 3, 4, 5, 7, 8, 9, 10, 11, 12, 13, 14, 15, 16, 17, 18, 19, 20, 21, 22, 23, 24, 25, 26, 27, 28, 31, 35, 36, 37, 38, 39, 40, 41, 42, 43, 44, 45, 46, 47, 48, 49, 50, 51, 52, 53, 54, 55, 56, 57, 58, 59, 60, 61, 62, 63, 65, 66, 67, 68, 69, 75, 78, 79, 80, 81, 82, 83, 84, 90, 94, 96, 99, 100, 102, 103, 106, 107, 108, 109, 110, 111, 112, 113, 114, 115, 116, 117, 119, 120, 121, 122, 123, 124, 125, 126, 127, 129, 130, 131, 132, 133, 134, 135, 136, 137, 138, 139, 143, 144, 147, 148, 157, 161, 162, 163, 165, 167, 168, 169, 175, 176, 177, 178, 179, 180, 181, 182, 183, 184, 185, 187, 189, 190, 191, 192, 193, 194, 195, 198, 201, 203, 204, 206, 208, 209, 210, 211, 212, 214, 215, 220, 221, 222, 224, 225, 226, 227, 228, 231, 232, 233, 234, 235, 236, 237, 238, 239, 240, 241, 242, 243, 244, 246, 247, 248, 250, 251, 259, 260, 263, 271, 275, 276, 277, 278, 280, 282, 283, 288, 291, 292, 293, 294, 296, 297, 298, 299, 300, 301, 302, 303, 304, 305, 306, 307, 308, 309, 310, 311, 312, 313, 316, 319
Co-Branding Strategies 5, 10, 11, 18, 23, 28, 36, 47, 49, 53, 56, 59, 60, 62, 65, 66, 67, 80, 115, 125, 126, 127, 168, 177, 178, 194, 203, 231, 233, 235, 242, 244, 247, 291, 302, 307, 316
Co-Branding Strategy 19, 21, 24, 42, 43, 44, 47, 48, 58, 59, 62, 81, 84, 206, 208, 235, 247
Co-creation 48, 65, 66, 67, 68, 69, 70, 71, 72, 73, 74, 79, 90, 92, 94, 95, 96, 136, 140, 142, 143, 148, 155, 159, 160, 163, 164, 308
Co-marketing 65, 66, 67, 68, 69, 74, 75, 76, 77, 78, 79, 90, 96
consumer behavior 10, 20, 27, 36, 47, 56, 104, 106, 117, 121, 146, 150, 151, 155, 158, 164, 205, 206, 255, 257, 296, 297, 298, 303, 308, 315, 316, 317, 318, 319, 320, 329, 331
Consumer-Brand Relationships 160, 315, 316, 317, 318
Consumer engagement 1, 4, 28, 31, 36, 47, 65, 67, 69, 71, 73, 76, 77, 85, 90, 96, 99, 106, 117, 123, 140, 142, 145, 146, 149, 150, 155, 158, 162, 212, 230, 248, 271, 277, 278, 279, 282, 283, 291, 298, 307, 311, 312, 318, 328
Consumer Perception 1, 8, 24, 36, 43, 44, 56, 82, 83, 96, 169, 204, 205, 208, 211, 212, 224, 225, 227, 228, 229, 231, 234, 241, 248, 260, 263, 268, 278, 291, 310, 311

Consumer Psychology 103, 112, 136, 137, 153, 158, 159, 162, 166
Corporate Responsibility 167, 174, 189
Crisis management 66, 67, 68, 86

D

Digital Fashion 19, 27, 28, 58, 110, 149, 259, 286, 291, 292, 294, 295, 297, 305, 307, 308, 309, 311
Digital Transformation 3, 4, 25, 32, 33, 71, 88, 108, 166, 180, 198, 253, 286

E

Ethical Sourcing 3, 4, 88, 126, 212, 261, 264, 265, 278

G

Greenwashing 167, 169, 173, 174, 175, 176, 177, 178, 181, 188, 191, 193, 194, 197, 201, 276

H

hospitality 29, 33, 36, 39, 52, 57, 70, 83, 85, 88, 90, 94, 143, 145, 161, 162

I

Innovation 2, 3, 4, 5, 6, 7, 9, 10, 11, 12, 13, 14, 15, 16, 18, 19, 20, 21, 25, 26, 27, 28, 35, 37, 38, 39, 40, 41, 44, 45, 46, 49, 50, 51, 52, 53, 54, 55, 56, 58, 59, 62, 63, 66, 69, 70, 71, 73, 75, 76, 78, 79, 80, 82, 84, 85, 86, 89, 90, 95, 96, 99, 100, 102, 103, 104, 105, 106, 107, 108, 109, 110, 111, 112, 113, 114, 115, 116, 120, 121, 125, 126, 127, 129, 130, 131, 133, 134, 135, 136, 139, 140, 141, 142, 143, 145, 168, 169, 176, 179, 181, 187, 188, 189, 193, 194, 204, 208, 209, 210, 211, 212, 213, 223, 224, 225, 227, 229, 232, 237, 239, 242, 244, 245, 246, 248, 254, 259, 260, 261, 263, 264, 267, 269, 271, 273, 274, 275, 276, 278, 280, 282, 283, 284, 285, 289, 291, 293, 294, 295, 296, 298, 299, 300, 301, 302, 304, 306, 308, 311, 315, 316, 317, 319, 321, 322, 325, 328, 330, 331

L

Luxury 1, 2, 3, 4, 5, 6, 7, 8, 9, 10, 11, 12, 13, 14, 15, 16, 17, 18, 19, 20, 21, 22, 23, 24, 25, 26, 27, 28, 29, 30, 31, 32, 33, 34, 35, 36, 37, 38, 39, 40, 41, 42, 43, 44, 45, 46, 47, 48, 49, 50, 51, 52, 53, 54, 55, 56, 57, 58, 59, 60, 61, 62, 63, 65, 66, 67, 68, 69, 70, 71, 72, 73, 74, 75, 76, 77, 78, 79, 80, 81, 82, 83, 84, 85, 86, 87, 88, 89, 90, 92, 93, 94, 95, 96, 97, 99, 100, 101, 102, 103, 104, 105, 106, 107, 108, 109, 110, 111, 112, 113, 114, 115, 116, 117, 118, 119, 120, 121, 122, 123, 124, 125, 126, 127, 128, 129, 130, 131, 132, 133, 134, 135, 136, 137, 138, 139, 140, 141, 142, 143, 144, 145, 146, 147, 148, 149, 150, 151, 152, 153, 154, 155, 156, 157, 158, 159, 160, 161, 162, 163, 164, 165, 167, 168, 169, 170, 171, 172, 173, 174, 175, 176, 177, 178, 179, 180, 181, 182, 183, 185, 186, 187, 188, 189, 190, 191, 192, 193, 194, 195, 196, 197, 198, 199, 200, 201, 203, 204, 205, 206, 208, 209, 210, 211, 212, 213, 214, 215, 221, 222, 223, 224, 225, 226, 227, 228, 229, 230, 231, 232, 233, 234, 235, 236, 237, 239, 240, 241, 242, 244, 245, 246, 247, 248, 249, 252, 253, 254, 255, 256, 257, 259, 260, 261, 262, 263, 264, 265, 266, 267, 268, 269, 270, 271, 272, 273, 274, 275, 276, 277, 278, 279, 280, 281, 282, 283, 284, 285, 286, 287, 288, 291, 292, 294, 295, 296, 298, 300, 301, 302, 303, 304, 305, 306, 307, 308, 309, 310, 311, 313, 315, 316, 317, 318, 319, 320,

321, 322, 323, 324, 325, 326, 327, 328, 329, 330, 331, 332
Luxury-tech partnerships 249

M

Multi-Criteria Decision-Making 203, 208

N

Neuromarketing 116, 117, 119, 120, 137, 139, 149, 150, 154, 158, 162, 164
Neutrosophic Sets 203, 206, 208, 217, 218, 219, 220, 221, 222, 228, 231, 232, 241, 244, 254

O

Operational Complexities 54, 56

R

Reputational Risk 53, 201

S

SDGs 88, 157, 167, 169, 170, 179, 182, 185, 188, 191, 201
Smart luxury 1, 2, 3, 10, 11, 12, 13, 14, 16, 18, 19, 20, 21, 23, 24, 25, 26, 27, 28, 35, 36, 39, 40, 41, 43, 50, 51, 52, 55, 56, 63, 65, 66, 67, 68, 69, 75, 77, 78, 79, 81, 82, 84, 90, 91, 96, 97, 100, 101, 104, 106, 108, 110, 112, 113, 114, 115, 116, 117, 118, 121, 122, 124, 127, 128, 129, 130, 134, 135, 136, 137, 203, 204, 206, 208, 209, 210, 211, 212, 213, 214, 221, 222, 224, 226, 227, 228, 230, 234, 237, 239, 241, 242, 244, 248, 259, 260, 262, 263, 264, 271, 272, 275, 278, 279, 282, 283, 291, 294, 295, 303, 307, 316
Smart Luxury Products 1, 2, 3, 18, 19, 35, 36, 39, 40, 41, 50, 51, 52, 55, 56, 63, 100, 115, 136, 203, 208, 209, 210, 211, 212, 213, 214, 227, 228, 239, 241, 259, 263, 291, 294, 295, 307
Sustainable Luxury 15, 20, 85, 140, 152, 168, 171, 198, 199, 232
Symbolic Value 132, 167, 169, 173, 174, 175, 176, 177, 192, 194, 195, 320

W

Weighted Aggregation Operators 204, 208, 236, 238, 248

www.ingramcontent.com/pod-product-compliance
Lightning Source LLC
LaVergne TN
LVHW080607300926
849162LV00005B/1074
9798337330105